# Microbial Communities in Cultural Heritage and Their Control

# Microbial Communities in Cultural Heritage and Their Control

Editors

**Filomena De Leo**
**Valme Jurado**

MDPI • Basel • Beijing • Wuhan • Barcelona • Belgrade • Manchester • Tokyo • Cluj • Tianjin

*Editors*
Filomena De Leo
University of Messina
Italy

Valme Jurado
Instituto de Recursos Naturales y Agrobiologia
Spain

*Editorial Office*
MDPI
St. Alban-Anlage 66
4052 Basel, Switzerland

This is a reprint of articles from the Special Issue published online in the open access journal *Applied Sciences* (ISSN 2076-3417) (available at: https://www.mdpi.com/journal/applsci/special_issues/microbial_communities_cultural_heritage).

For citation purposes, cite each article independently as indicated on the article page online and as indicated below:

LastName, A.A.; LastName, B.B.; LastName, C.C. Article Title. *Journal Name* **Year**, *Volume Number*, Page Range.

**ISBN 978-3-0365-3050-5 (Hbk)**
**ISBN 978-3-0365-3051-2 (PDF)**

Cover image courtesy of Valme Jurado

# Contents


**About the Editors** . . . . . . . . . . . . . . . . . . . . . . . . . . . . . . . . . . . . . vii

**Filomena De Leo and Valme Jurado**
Editorial for the Special Issue "Microbial Communities in Cultural Heritage and Their Control"
Reprinted from: *Appl. Sci.* **2021**, *11*, 11411, doi:10.3390/app112311411 . . . . . . . . . . . . . . . . 1

**Zuzana Kisová, Matej Planý, Jelena Pavlović, Mária Bučková, Andrea Puškárová, Lucia Kraková, Magdaléna Kapustová, Domenico Pangallo and Katarína Šoltys**
Biodeteriogens Characterization and Molecular Analyses of Diverse Funeral Accessories from XVII Century
Reprinted from: *Appl. Sci.* **2020**, *10*, 5451, doi:10.3390/app10165451 . . . . . . . . . . . . . . . . . 5

**Yin Jia, Liuyu Yin, Fengyu Zhang, Mei Wang, Mingliang Sun, Cuiting Hu, Zijun Liu, Yue Chen, Jie Liu and Jiao Pan**
Fungal Community Analysis and Biodeterioration of Waterlogged Wooden Lacquerware from the Nanhai No. 1 Shipwreck
Reprinted from: *Appl. Sci.* **2020**, *10*, 3797, doi:10.3390/app10113797 . . . . . . . . . . . . . . . . . 25

**Fernando Bolívar-Galiano, Clara Abad-Ruiz, Pedro Sánchez-Castillo, Maurizio Toscano and Julio Romero-Noguera**
Frequent Microalgae in the Fountains of the Alhambra and Generalife: Identification and Creation of a Culture Collection
Reprinted from: *Appl. Sci.* **2020**, *10*, 6603, doi:10.3390/app10186603 . . . . . . . . . . . . . . . . . 39

**Giulia Caneva, Daniela Isola, Hyun Ju Lee and Yong Jae Chung**
Biological Risk for Hypogea: Shared Data from Etruscan Tombs in Italy and Ancient Tombs of the Baekje Dynasty in Republic of Korea
Reprinted from: *Appl. Sci.* **2020**, *10*, 6104, doi:10.3390/app10176104 . . . . . . . . . . . . . . . . . 57

**Silvia Sparacello, Giuseppe Gallo, Teresa Faddetta, Bartolomeo Megna, Giovanna Nicotra, Beatrice Bruno, Belinda Giambra and Franco Palla**
*Thymus vulgaris* Essential Oil and Hydro-Alcoholic Solutions to Counteract Wooden Artwork Microbial Colonization
Reprinted from: *Appl. Sci.* **2021**, *11*, 8704, doi:10.3390/app11188704 . . . . . . . . . . . . . . . . . 83

**Lucrezia Gatti, Federica Troiano, Violetta Vacchini, Francesca Cappitelli and Annalisa Balloi**
An In Vitro Evaluation of the Biocidal Effect of Oregano and Cloves' Volatile Compounds against Microorganisms Colonizing an Oil Painting—A Pioneer Study
Reprinted from: *Appl. Sci.* **2021**, *11*, 78, doi:10.3390/app11010078 . . . . . . . . . . . . . . . . . . 97

**Lorenza Rugnini, Giada Migliore, Flavia Tasso, Neil Thomas William Ellwood, Anna Rosa Sprocati and Laura Bruno**
Biocidal Activity of Phyto-Derivative Products Used on Phototrophic Biofilms Growing on Stone Surfaces of the Domus Aurea in Rome (Italy)
Reprinted from: *Appl. Sci.* **2020**, *10*, 6584, doi:10.3390/app10186584 . . . . . . . . . . . . . . . . . 109

**Pilar Bosch-Roig, Lourdes Pérez-Castro, Ángeles Fernández-Santiago and Ignacio Bosch**
High Dimension Granite Pavement Bio-Desalination Practical Implementation
Reprinted from: *Appl. Sci.* **2021**, *11*, 6458, doi:10.3390/app11146458 . . . . . . . . . . . . . . . . . 125

**Julio Romero-Noguera, Rafael Bailón-Moreno and Fernando Bolívar-Galiano**
Varnishes with Biocidal Activity: A New Approach to Protecting Artworks
Reprinted from: *Appl. Sci.* **2020**, *10*, 7319, doi:10.3390/app10207319 . . . . . . . . . . . . . . . . . . **143**

**Sandra Lo Schiavo, Filomena De Leo and Clara Urzì**
Present and Future Perspectives for Biocides and Antifouling Products for Stone-Built Cultural Heritage: Ionic Liquids as a Challenging Alternative
Reprinted from: *Appl. Sci.* **2020**, *10*, 6568, doi:10.3390/app10186568 . . . . . . . . . . . . . . . . . . **159**

**Valme Jurado, Yolanda del Rosal, Jose Luis Gonzalez-Pimentel, Bernardo Hermosin and Cesareo Saiz-Jimenez**
Biological Control of Phototrophic Biofilms in a Show Cave: The Case of Nerja Cave
Reprinted from: *Appl. Sci.* **2020**, *10*, 3448, doi:10.3390/app10103448 . . . . . . . . . . . . . . . . . . **177**

**Mikolaj Dziurzynski, Karol Ciuchcinski, Magdalena Dyda, Anna Szych, Paulina Drabik, Agnieszka Laudy and Lukasz Dziewit**
Assessment of Bacterial Contamination of Air at the Museum of King John III's Palace at Wilanow (Warsaw, Poland): Selection of an Optimal Growth Medium for Analyzing Airborne Bacteria Diversity
Reprinted from: *Appl. Sci.* **2020**, *10*, 7128, doi:10.3390/app10207128 . . . . . . . . . . . . . . . . . . **191**

**Xinghua Ding, Wensheng Lan and Ji-Dong Gu**
A Review on Sampling Techniques and Analytical Methods for Microbiota of Cultural Properties and Historical Architecture
Reprinted from: *Appl. Sci.* **2020**, *10*, 8099, doi:10.3390/app10228099 . . . . . . . . . . . . . . . . . . **207**

# About the Editors

**Filomena De Leo**, Professor of Microbiology at Messina University. Website: https://www.unime.it/it/persona/filomena-de-leo/curriculum. Filomena De Leo graduated with a degree in Biological Sciences, specialized in "Applied Microbiology" and she earned a post graduate degree (PhD) in Microbial Biotechnologies. She is an Associate Professor of Microbiology at the Department of Chemical, Biological, Pharmaceutical and Environmental Sciences of the University of Messina, Italy, a professor of "General Microbiology", degree course CdL in "Biotechnology", and a professor of "Nutrimicrobiomic" degree course CdLM in "Biology of Health, Applied Technologies and Nutrition". She is a member of the Council of PhD in "Applied Biology and Experimental Medicine" of the University of Messina, Italy. She is a member of the Reference Scientific Community of the "Stazione Zoologica Anton Dohrn", Naples, Italy (http://www.szn.it/index.php/it/chi-siamo/comunita-scientifica-di-riferimento). She is a member of the University Research Center for the study of extreme environments and extremophiles (CUR AEE), and she is a member of the National Consortium for Marine Sciences (CoNISMa). Her main interest is the study of the biodiversity, ecology and genetic diversity of bacterial and fungal populations involved in the biodeterioration of cultural heritage (especially stone monuments) and more recently, she has been interested in the diversity of halophilic microorganisms in deep sea environments. One special area of interest is the taxonomy of black fungi, which are among the most harmful microorganisms that cause the biodeterioration of stone monuments. Due to this, she collaborates with Prof. Sybren de Hoog (expert on taxonomy of black fungi) of the Westerjeink Institute (ex Centraalbureau voor Schimmelcultures -CBS) of Utrecht, the Netherlands, where she has been a guest several times. She described two new fungal species and four new bacterial species isolated from monuments. She has participated in numerous national and international research projects both as Scientific Responsible and as a member of the Research Unit. All of these studies are part of the Environmental Microbiology group, and some topics include biotechnological applications. Prof Filomena De Leo has published 49 articles in journals, 16 monographs (or book chapters), 19 Proceedings of International Congress and 54 posters and oral communications. These publications are visible on Researchgate: https://www.researchgate.net/profile/Filomena_De_Leo

**Valme Jurado** is a research scientist at the Institute for Natural Resources and Agrobiology of Seville (IRNAS-CSIC), Spain. Dr. Jurado is part of the Environmental Microbiology and Cultural Heritage group, where she is developing a research topic focused on the study of microbial biodiversity and the implications of microorganisms in cultural heritage. She is a member of the Network of Experts of the Campus of International Excellence in Heritage Project. Her areas of expertise are environmental microbiology and cultural heritage. She is involved in biodiversity studies, the description of new bacterial and fungal species of elements declared of cultural interest (BIC) and the monitoring and control of microbial invasions in caves in Spain and France that have cave paintings. She has over 130 publications, of which 88 are scientific articles, 26 book chapters, 4 complete books as author or co-author and 14 scientific–technical reports. She has directed 5 doctoral theses. She is an invited professor at the University of Seville, where she received her Master´s Degree. She has taught specialization courses in Microbiology and Cultural Heritage. She received a positive evaluation from the National Agency for Quality Assessment and Accreditation for Assistant Professor Doctor, Contracted Professor Doctor and Professor Private University. She has participated or participates in 13 research projects and 15 research contracts with private companies and public administrations, of which she has directed 9 for the study of the microbiology of cultural heritage.

applied sciences

MDPI

*Editorial*

# Editorial for the Special Issue "Microbial Communities in Cultural Heritage and Their Control"

**Filomena De Leo [1,*] and Valme Jurado [2,*]**

1 Department of Chemical, Biological, Pharmaceutical and Environmental Sciences, University of Messina, 98122 Messina, Italy
2 Instituto de Recursos Naturales y Agrobiologia, IRNAS-CSIC, 41012 Sevilla, Spain
* Correspondence: fdeleo@unime.it (F.D.L.); vjurado@irnase.csic.es (V.J.); Tel.: +39-0906765201 (F.D.L.); +34-954624711 (V.J.)

**Abstract:** This editorial focuses on the studies published within the present Special Issue presenting advances in the field of biodeterioration of cultural heritage caused by microbial communities with a particular focus on new methods for their elimination and control.

**Keywords:** cultural heritage; biodeterioration; microorganisms; biofilms; control

**Citation:** De Leo, F.; Jurado, V. Editorial for the Special Issue "Microbial Communities in Cultural Heritage and Their Control". *Appl. Sci.* **2021**, *11*, 11411. https://doi.org/10.3390/app112311411

Received: 26 November 2021
Accepted: 30 November 2021
Published: 2 December 2021



## 1. Introduction

Cultural heritage plays a key role in understanding the history of humankind; therefore, adopting appropriate strategies for its conservation is essential. Microorganisms, such as bacteria, fungi and microalgae, usually organized on the surface in microbial communities as "biofilms", can cause serious problems in the conservation of cultural heritage, making the adoption of prevention and conservation strategies a critical issue. The colonization process and the subsequent biodeterioration phenomena strictly depend on the characteristics of the microorganisms, the type of material and the environmental conditions. So, to develop new strategies to eliminate and/or mitigate the presence of biodeteriogens on monuments, it is often necessary to perform interdisciplinary studies aiming to know the state of conservation of the artifact and implement targeted maintenance strategies.

Controlling microorganisms in cultural heritage is a major challenge for microbiologists. The search for appropriate methods to stop progressive microbial attacks without provoking secondary, and perhaps more dangerous, biological succession and biodeterioration is a must.

The aim of this Special Issue was to bring together recent studies presenting cutting edge advances in the field of biodeterioration of cultural heritage caused by microbial communities and their control.

## 2. Summary of the Special Issue Contents

The Special Issue can be subdivided into three groups according to the topics covered by the articles. Four articles focused on study cases aiming to identify the microorganisms involved in the biodeterioration processes. Seven articles focused on "Green methods" of treatments, of which three investigated the biocidal activity of natural products derived from plants, one each on the biocidal activity of varnishes combined with chemical products, the application of a biocleaning procedure on granite, the potentiality of natural biological control of phototrophic biofilm in cave, and a review on the most innovative antimicrobials among which nano- and bio-technologies play a main role. Finally, two papers focused on the methods for studying biodeterioration.

### 2.1. Study Cases

A proper approach that aims to eliminate or mitigate the biodeteriogenic effect that microorganisms have on an artifact should take into account the chemical and physical characteristics of the material, the environmental condition, the description of the alterations and the identification of the main biodeteriogenic microorganisms involved. To identify the complex microbiota composed of dangerous deteriorative microorganisms, it is necessary to apply a multiphasic approach that includes different techniques such as microscopy, cultural dependence and cultural independent analyses as Kizová et al. [1] performed in their contribution on funeral texting items from the 17th century. In particular, cultural methods and the high-throughput sequencing method followed by biodegradative tests and biocide susceptibility of the isolates are useful for a targeted intervention [2].

Another important issue is the creation of a database with the main biodeteriogens and their maintenance in a culture collection. This should be useful not only to obtain a deepened knowledge of the ecophysiological and molecular characteristics of the microorganisms involved in the biodeterioration phenomena, but also to plan trials in situ and in laboratory conditions to develop new strategies for their control. In this regard, Bolívar-Galiano et al. [3] isolated 10 genera of green algae and 13 genera of cyanobacteria from the fountains of Alhambra and Generalife in Spain, thereby constituting the foundation for the creation of a reference collection of living algae.

In the perspective to carry out an intervention of restoration and conservation of monuments or artifact with evident alterations due to the growth of microorganisms, it is always advisable to carry out microbiological analyses before and after treatments and plan subsequent monitoring over time. Caneva et al. [4] in their contribution compared the data coming from the microbiological analyses of Etruscan tombs in Italy and ancient tombs in the Republic of Korea, as well as the conservative treatments carried out at these sites, with a special focus on preventive intervention. Although the hypogean environments are characterized by constant microclimatic conditions with relatively high humidity and constant temperature, the collected data confirmed the complexity of the microbial communities in hypogea as the result of variable environmental values and edaphic conditions. Events that have occurred over time regarding the tombs such as opening, flooding, upper vegetation, visitors, treatment, etc., also played an important role in the biodeterioration processes. In general, they observed the prevalence of autotrophic microorganisms, only if the lighting was sufficient, whereas fungi and bacteria were widespread.

### 2.2. Green Conservation of Cultural Heritage

In the field of cultural heritage, a very important task is the use of methods of control and treatments that are efficient but not harmful for the material, operator and environment. For this reason, in the last decade scientists have been looking for "eco-friendly" control methods as an alternative to traditional ones such as: aromatic compounds for wood treatments [5]; oregano and clove essential oil as a cleaning method for canvas painting [6]; plant derivatives such as a liquorice leaf extract and lavender essential oil as biocides against phototrophic biofilm growing on stones [7]; water-based gel delivery systems and heating systems for high dimension granite pavement bio-desalination [8], are some examples reported in this Special Issue that seem to be promising for possible future applications.

In other cases, the association of a traditional chemical product such as benzalkonium chloride, o phenylphenol, and tributyltin naphthenate with natural varnishes should be useful to protect paintings and polychrome sculptures from environmental fungi and bacteria without altering the original materials or the visual appearance of the artworks [9].

Another research field concerns the use of new synthetic molecules to counteract the growth of microorganisms and biofilm formation on stone monuments, such as Ionic Liquids technologies, which can contribute to the production of new formulations of antifouling and antimicrobial surface coatings. These technologies must be defined as eco-friendly [10].

In addition, to biologically derived biocides and chemical methods, natural biological control has been proposed by Jurado et al. [11]. The study of phototrophic biofilms on speleothems revealed the presence of different predators within the complex microbial communities that act naturally controlling the development of photosynthetic-based biofilms in caves.

### 2.3. Method of Studying

The different kinds of analyses of cultural heritage usually are not standardized methods; however, sampling, microscopy, cultural and molecular analysis procedures are consolidated enough over time to be now used as "Standard" methods by most scientists. Regardless, there are guidelines known as UNINormal techniques, the catalogue for which is accessible on the website [12]. UNINormal collects also over ten years of activities of "The Italian Normal Committee" on cultural heritage and contains information about the descriptions of alterations, methods of cultural analyses and biocide test procedures.

New methods of analyses such as high-throughput next generation sequencing (NGS) have opened new frontiers for the study of the state of conservation and treatments of monuments and cultural heritage items, bringing up new issues regarding their use and standardization. In this Special Issue, Dziurinsky et al. [13] used metabarcoding analyses to select the best culture media combinations to assess the bacterial contamination of air in the Museum of King Jhon III's Palace in Wilanow (Warsaw, Poland). Ding et al. [14] reviewed the analytical method in use to study the microbiota in cultural properties and historical architecture,, from sampling to NGS, which, in combination with cultural and biochemical functional analyses, can provide useful information for further verification of the biochemical and ecophysiological roles of microorganisms in deterioration.

## 3. Conclusions

Due to the complexity of the topic, the approach to the study of microbial communities responsible for the biodeterioration of cultural heritage cannot be improvised. Descriptions of alterations, methods of study and sampling, implementations of modern and advanced methodologies coming from multidisciplinary approaches are a necessity to study and maintain the cultural heritage in a good state of conservation. Certainly, the eradication and mitigation of bioteriogenic microorganisms represents the main challenge for microbiologists. Although there is no valid solution for every situation, the acquisition of new eco-sustainable findings and the exchange of experiences in the field can definitely represent a step forward.

**Author Contributions:** Conceptualization, F.D.L. and V.J.; writing—original draft preparation, F.D.L. and V.J.; writing—review and editing, F.D.L. and V.J.; visualization, F.D.L. and V.J.; supervision, F.D.L. and V.J. All authors have read and agreed to the published version of the manuscript.

**Funding:** This research received no external funding.

**Institutional Review Board Statement:** Not applicable.

**Informed Consent Statement:** Not applicable.

**Data Availability Statement:** Not applicable.

**Acknowledgments:** This publication was only possible with the valuable contributions from the authors, reviewers, and the editorial team of Applied Sciences. In particular, we would like to thank Frederic Yuan, Assistant Editor, for his tireless and efficient support.

**Conflicts of Interest:** The authors declare to be Guest Editors of the present Special Issue and there are no other conflict of interest.

## References


1. Kizová, Z.; Planý, M.; Pavlović, J.; Bučková, M.; Puškárová, A.; Kraková, L.; Kapustová, M.; Pangallo, D.; Šoltys, K. Biodeteriogens Characterization and Molecular Analyses of Diverse Funeral Accessories from XVII Century. *Appl. Sci.* **2020**, *11*, 5451. [CrossRef]
2. Jia, Y.; Yin, L.; Zhang, F.; Wang, M.; Sun, M.; Hu, C.; Liu, Z.; Chen, Y.; Liu, J.; Pan, J. Fungal Community Analysis and Biodeterioration of Waterlogged Wooden Lacquerware from the Nanhai No. 1 Shipwreck. *Appl. Sci.* **2020**, *10*, 3797. [CrossRef]
3. Bolívar-Galiano, F.; Abad-Ruiz, C.; Sánchez-Castillo, P.; Toscano, M.; Romero-Noguera, J. Frequent Microalgae in the Fountains of the Alhambra and Generalife: Identification and Creation of a Culture Collection. *Appl. Sci.* **2020**, *10*, 6603. [CrossRef]
4. Caneva, G.; Isola, D.; Lee, H.J.; Chung, Y.J. Biological Risk for Hypogea: Shared Data from Etruscan Tombs in Italy and Ancient Tombs of the Baekje Dynasty in Republic of Korea. *Appl. Sci.* **2020**, *10*, 6104. [CrossRef]
5. Sparacello, S.; Gallo, G.; Faddetta, T.; Megna, B.; Nicotra, G.; Bruno, B.; Giambra, B.; Palla, F. Thymus vulgaris Essential Oil and Hydro-Alcoholic Solutions to Counteract Wooden Artwork Microbial Colonization. *Appl. Sci.* **2021**, *11*, 8704. [CrossRef]
6. Gatti, L.; Troiano, F.; Vacchini, V.; Cappitelli, F.; Balloi, A. An In Vitro Evaluation of the Biocidal Effect of Oregano and Cloves' Volatile Compounds against Microorganisms Colonizing an Oil Painting—A Pioneer Study. *Appl. Sci.* **2021**, *11*, 78. [CrossRef]
7. Rugnini, L.; Migliore, G.; Tasso, F.; Ellwood, N.T.W.; Sprocati, A.R.; Bruno, L. Biocidal Activity of Phyto-Derivative Products Used on Phototrophic Biofilms Growing on Stone Surfaces of the Domus Aurea in Rome (Italy). *Appl. Sci.* **2020**, *10*, 6584. [CrossRef]
8. Bosch-Roig, P.; Pérez-Castro, L.; Fernández-Santiago, A.; Bosch, I. High Dimension Granite Pavement Bio-Desalination Practical Implementation. *Appl. Sci.* **2021**, *11*, 6458. [CrossRef]
9. Romero-Noguera, J.; Bailón-Moreno, R.; Bolívar-Galiano, F. Varnishes with Biocidal Activity: A New Approach to Protecting Artworks. *Appl. Sci.* **2020**, *10*, 7319. [CrossRef]
10. Lo Schiavo, S.; De Leo, F.; Urzì, C. Present and Future Perspectives for Biocides and Antifouling Products for Stone-Built Cultural Heritage: Ionic Liquids as a Challenging Alternative. *Appl. Sci.* **2020**, *10*, 6568. [CrossRef]
11. Jurado, V.; del Rosal, Y.; Gonzalez-Pimentel, J.L.; Hermosin, B.; Saiz-Jimenez, C. Biological Control of Phototrophic Biofilms in a Show Cave: The Case of Nerja Cave. *Appl. Sci.* **2020**, *10*, 3448. [CrossRef]
12. Uni. Un Mondo Fatto Bene. Available online: http://store.uni.com/catalogo/index.php/home/ (accessed on 1 December 2021).
13. Dziurzynski, M.; Ciuchcinski, K.; Dyda, M.; Szych, A.; Drabik, P.; Laudy, A.; Dziewit, L. Assessment of Bacterial Contamination of Air at the Museum of King John III's Palace at Wilanow (Warsaw, Poland): Selection of an Optimal Growth Medium for Analyzing Airborne Bacteria Diversity. *Appl. Sci.* **2020**, *10*, 7128. [CrossRef]
14. Ding, X.; Lan, W.; Gu, J.-D. A Review on Sampling Techniques and Analytical Methods for Microbiota of Cultural Properties and Historical Architecture. *Appl. Sci.* **2020**, *10*, 8099. [CrossRef]

*applied sciences*

MDPI

*Article*

# Biodeteriogens Characterization and Molecular Analyses of Diverse Funeral Accessories from XVII Century

**Zuzana Kisová [1], Matej Planý [1], Jelena Pavlović [1], Mária Bučková [1], Andrea Puškárová [1], Lucia Kraková [1], Magdaléna Kapustová [1], Domenico Pangallo [1,2,*] and Katarína Šoltys [3,4]**

1 Institute of Molecular Biology, Slovak Academy of Sciences, Dúbravská Cesta 21, 84551 Bratislava, Slovakia; zuzana.kisova@savba.sk (Z.K.); matej.plany@savba.sk (M.P.); jelena.pavlovic@savba.sk (J.P.); maria.buckova@savba.sk (M.B.); andrea.puskarova@savba.sk (A.P.); lucia.krakova@savba.sk (L.K.); magdalena.kapustova@savba.sk (M.K.)

2 Caravella, s.r.o., Tupolevova 2, 85101 Bratislava, Slovakia

3 Department of Microbiology and Virology, Faculty of Natural Sciences, Comenius University in Bratislava, Ilkovičova 3278/6, 841 04 Bratislava, Slovakia; katarina.soltys@uniba.sk

4 Faculty of Natural Sciences, Science Park, Comenius University in Bratislava, Ilkovičova 8, 84104 Bratislava, Slovakia

* Correspondence: domenico.pangallo@savba.sk

Received: 1 July 2020; Accepted: 3 August 2020; Published: 6 August 2020

**Abstract:** A historical crypt offers us a particular view of the conditions of some buried materials (in this case textiles) and the various biogenic phenomena to which they were subjected over the centuries. In addition, significant knowledge can come by studying the DNA of buried objects which allows the recognition of materials, but also to reveal some practice of the funeral ceremony. In this study, the deteriorating microbial communities colonizing various funeral textile items were identified and characterized using microscopic observation, cultivation, polymerase chain reaction (PCR) and sequencing, hydrolytic tests; and culture-independent analysis (high-throughput sequencing, MinION platform). Different PCR assays and consequent sequencing of amplicons were employed to recognize the animal origin of bodice reinforcements and the type of plant used to embellish the young girl. The analysis of ancient DNA (aDNA from animal and plant) was also completed by the application of high-throughput sequencing through Illumina platform. The combination of all these techniques permitted the identification of a complex microbiota composed by dangerous degradative microorganisms able to hydrolyze various organic substrates such as fibroin, keratin, and cellulose. Bacteria responsible for metal corrosion and bio-mineralization, and entomopathogenic and phytopathogenic fungi. The analysis of aDNA identified the animal component used in bodice manufacturing, the plant utilized as ornament and probably the season of this fatal event.

**Keywords:** microbial community; biodeterioration; high-throughput sequencing; MinION approach; SEM; aDNA; animal; plant

## 1. Introduction

At the beginning of a restoration and conservation trial it is necessary, when it is possible, to analyze the target items in order to know well their characteristics and their degree of conservation. Among the various possible investigations, the microorganisms' identification and their degradation abilities are important to perform using culture-dependent and –independent approaches [1,2]. In fact, the responsibility of microbial communities on the deterioration of cultural heritage items is largely accepted by the scientific community and operators [3–5]. Regarding the microbial action on fabric produced by natural fibers there is a widespread opinion that filamentous fungi are responsible for the biodeterioration of cellulosic textiles (cotton, linen, hemp, and jute), while bacteria primarily degrade animal fibers (silk and wool) [4,6]. In the past, some attempts were tried to investigate silk biodegradation in laboratory conditions [7] using bacterial strains. A deep analysis of bacterial community present on museum silk velvet was also attempted [8]. Only recently a study described the role of mold in silk deterioration, although the studied items were manuscript pages where the silking technique was applied [9]. The silk objects investigated in this manuscript are buried clothes, therefore they were subjected to particular conditions, different to those described by previous works. Moreover, to our knowledge a complete microbial investigation (including culture-dependent and culture–independent approaches) was never performed for these kinds of objects coming from this specific environment.

Various molecular biology techniques can help also to well recognize some materials, which other methods are not able to properly identify. Many cultural heritage items made by organic materials, of animal and plant origin, can be classified by the DNA analysis [10,11]. Another aim of our study was the DNA recognition of some no-well identified items.

The basilica of St. Giles is the most important monument on the principal square in Bardejov (Slovakia), which is inscribed in the UNESCO World Cultural and Natural Heritage. At the beginning of the second half of the XVII century a young girl was buried under the presbytery of this basilica. In addition to the bone remains of the girl, the archaeological study also brought to light the funeral clothes and accessories worn by the girl. Several of these girl's items, mainly diverse textile objects, should be restored in order to set up a permanent exhibition of the finds discovered during the various archaeological studies.

We have focused our attention on those cloth objects which on their surfaces seemed to have biodeterioration phenomena and also on establishing the origin of the materials used to manufacture the stiffening material of bodice and the decorative plant relic on the girl's head.

The microbiological survey was performed combining microscopic observation, culture-dependent, hydrolytic plate assays, and culture-independent approaches (MinION platform). The DNA analysis of bodice sticks and plant decoration was attempted using specific polymerase chain reaction (PCR) assays and high-throughput sequencing.

## 2. Materials and Methods

### 2.1. Sampling Strategy, Microbiological Cultivation, and Microscopic Observation

Several samples were taken from the three textile objects better conserved: the bodice, the cap, and the head-band. We have tried to pick up samples from parts that showed some biodeterioration phenomena under the constant supervision of restorers. The sampling campaign was carried out 4 days after the archaeological excavation. The objects were conserved in clean plastic boxes inside the sacristy of the church.

The microbiological sampling was performed using nitrocellulose membranes (5 samples; a–f) [12]. Three samples were obtained from the silk bodice; two from the front of the bodice: (a) from the ribbon; (b) from the body (Figure 1A). The third from a portion of tissue covering the stiff reinforcement (c), the sampling was performed on the backside of the bodice (Figure 1B). The other three samples were taken from the items which adorned the head of the girl: (d) from the silk bobbin lace; (e) from the linen coronet; (f) a little fragment of the decorative plant sprig (Figure 1C). A small piece of the stick used as stiff material (g) of the bodice was collected also for molecular analysis (Figure S1).

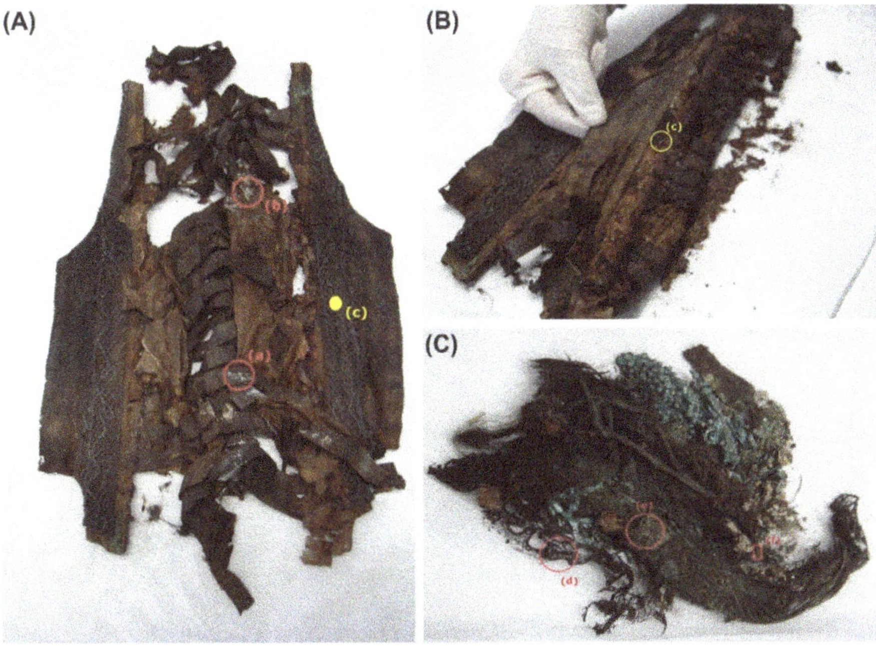


**Figure 1.** (**A**) Front side of the bodice with the samples a and b. (**B**) Back side of the bodice with the sample c. (**C**) Decorative accessories of the head, samples d, e, and f.

The nitrocellulose membranes were cut in several small pieces, suspended in 2 mL of physiological solution and shaken overnight. This suspension was used for preparing the decimal dilution. Two-hundred microliters of each dilution were spread in diverse agar media. The bacteria were isolated on plates of Actinomycete Isolation Agar (AIA; Himedia, Mumbai, India); Reasoner's 2A (R2A; Himedia) and LB10 agar (peptone 1 g l−1, yeast extract 0.5 g l−1, NaCl 1 g l−1, agar 15 g l−1; [13]). The agar media suitable for the cultivation of fungi were Malt Extract Agar (MEA; Himedia) and Dichloran Rose Bengal Chloramphenicol (DRBC; Himedia). The plates were incubated at room temperature (24–26 °C) and checked for the microorganisms' growth for a minimum of two weeks. In the isolation plates a limited number of microorganisms grew. Therefore, all these microorganisms were purified performing several inoculation steps on R2A and MEA plates. The purified bacteria and fungi were maintained on plates of R2A and MEA, respectively.

Portions of the 25 mm nitrocellulose membranes, used for the microbiological sampling, were examined by scanning electron microscope (SEM; Jeol JSM 6610, Tokyo, Japan). Prior to SEM observation, the samples were sputtered with gold ions [12].

### 2.2. DNA Identification of Isolated Microorganisms

The DNA from bacteria was extracted using the kit Instagene (Biorad, Hercules, CA, USA) following the instruction of the manufacturer. The DNeasy Plant Mini kit (Qiagen, Hilden, Germany) was utilized to extract the fungal DNA.

The 16S rRNA gene was amplified, using the primers 27f (5′–AGA GTT TGA TCC TGG CTC AG-3′) and 685r (5′-TCT ACG CAT TTC ACC GCT AC-3′ [14]), for identifying the isolated bacteria by sequencing. The fungal isolates were identified by the amplification and consequent sequencing of the internal transcribed spacer (ITS) fragment using the primers ITS1 (5′-TCC GTA GGT GAA CCT GCG G-3′) and ITS4 (5′-TCC TCC GCT TAT TGA TAT GC-3′ [15]).

Twenty-five microliters of PCR mixture contained 50 pmol of each primer, 200 µmol l−1 of dNTP (Life Technologies, Gaithersburg, MD, USA), 1.5 U HotStar Taq plus DNA polymerase (Qiagen), 1× PCR buffer and 3 µL of the extracted bacterial or fungal DNA. The PCR program consisted of an initial denaturation at 94 °C for 5 min, followed by 30 cycles (denaturation at 94 °C for 30 s, annealing at 54 °C for 45 s, extension at 72 °C for 1 min) and a final polymerization step at 72 °C for 10 min.

PCR products from both fungal and bacterial isolates were purified using ExoSAP-IT (Affymetrix, Cleveland, OH, USA) and sequenced at a commercial facility (Eurofins Genomics, Ebersberg, Germany). The obtained sequences were directly compared with those in GenBank using BLAST program (http://blast.ncbi.nlm.nih.gov/Blast.cgi) and were subsequently deposited in GenBank under the accession numbers MT256363-MT256375 (bacterial isolates) and MT256388-MT256394 (fungal isolates).

### 2.3. *Hydrolytic Extracellular Enzyme Assays*

A loop for bacterial cells and a needle for fungal inoculation were used to transfer freshly grown cultures from the medium of maintenance to the center of Petri dishes with specific agar media, in order to assay their hydrolytic abilities. Congo red agar [13] (0.5 g l−1 $KH_2PO_4$, 0.25 g l−1 $MgSO_4$, 2 g l−1 cellulose, 0.2 g l−1 Congo red, 2 g l−1 gelatin, 15 g l−1 agar; pH 6.8–7.2), Gelatin agar [13] (0.5 g l−1 $KH_2PO_4$, 0.25 g l−1 $MgSO_4$, 4 g l−1 gelatin, 15 g l−1 agar), R2A supplemented with gelatin (18.12 g l−1 R2A agar and 4 g l−1 gelatin) and Fibroin agar [1] (1 mL of fibroin and 1.5 g of agar were mixed in 100 mL of distilled water) were used to assess the cellulolytic, proteolytic (gelatin-based agars) and silk deterioration properties of isolates, respectively. The keratinolytic activity was assayed through a keratin medium [2] (0.5 g l−1 $KH_2PO_4$, 0.25 g l−1 $MgSO_4$, 4 g l−1 keratin, 15 g l−1 agar). All chemicals except hydrolysed keratin (Vipo a.s., Partizánske, Slovakia) were bought from Sigma-Aldrich. A positive reaction was manifested by a clear zone around the investigated colony.

The hydrolytic assays were performed in triplicate. All plates were incubated at room temperature (24–26 °C) for a maximum of 10 days. The positive hydrolytic reaction was represented by a clear zone around the microbial colony. The size (mm) of each hydrolytic area was measured from the edge of the colony to the edge of the zone. The microbial colony was not included in the measurement.

### 2.4. *Total Microbial Community DNA Extraction and PCR Amplifications*

The rest of the suspension (described on paragraph "Sampling strategy, microbiological cultivation and microscopic observation"), containing very small membrane pieces, was centrifuged and from the obtained pellets the DNA was extracted using DNeasy PowerSoil extraction kit (Qiagen) following the protocol provided by the producer.

The bacterial 16S rRNA gene was amplified with the primers 27f (as above) and 1492r (5′-GGT ACC TTG TTA CGA CTT-3′ [14]), while the fungal community was analyzed by the amplification of the ITS fragment (ITS1 and ITS4 primers, as above) and the 28S rRNA gene by the primers NL1 (5′-GCATAT CAATAA GCG GAG GAA AAG-3′) and NL4 (5′-GGT CCG TGT TTC AAG ACG G-3′ [16]). The PCR products were purified using Agencourt AMPure XP beads (Beckman Coulter, Brea, CA, USA).

### 2.5. *Library Preparation for Microbial Communities' Analysis*

The library for nanopore sequencing by MinION platform was made from purified amplicons using Ligation Sequencing Kit 1D (SQK-LSK108) as described previously [17]. Briefly, the process comprised the following parts: (i) the ends of amplicons were prepared by Ultra II End-prep reaction buffer and Ultra II End-prep enzyme mix from the NEBNext Ultra II End Repair/da-Tailing Module (New England Biolabs, Ipswich, MA, USA), (ii) ligation step with Barcode Adapters was performed using previous End-prep. DNA, Barcode Adapter (PCR Barcoding kit 96, Oxford Nanopore Technologies, Oxford, UK) and Blunt/TA Ligase Master Mix (NEB) and (iii) Barcoding PCR was performed using one of PCR Barcode primer (BC1-BC96) from PCR Barcoding Expansion Pack 1–96 (EXP-PBC096, ONT), LongAmp Taq 2x master mix (NEB) and Nuclease-free water (Qiagen). For the Barcoding PCR, the following steps were used: initial denaturation step (95 °C, 3 min), followed by 19 cycles (95 °C for 15 s.; 62 °C for 15 s.; 65 °C for 2 min) and final extension step (65 °C for 10 min).

The intermediate purification steps by Agencourt AMPure XP magnetic beads (Beckman Coulter) and the DNA fluorometric quantification procedures with Qubit v2 Fluorometer (Invitrogen, Waltham, MA, USA) using Qubit dsDNA HS Assay Kit (Thermo Fisher Scientific, Waltham, MA, USA) were performed between each of the three abovementioned parts.

### 2.6. Pooled Library Preparation and Flow-Cell Processing

Each sample labeled by unique barcode was equimolar pooled to final 1 μg of DNA in 45 μL Nuclease free water (Qiagen). The procedure of second end- preparation using NEBNext Ultra II End Repair/da-Tailing Module (New England Biolabs) chemistry and control DNA CS (DCS; Ligation Sequencing kit 1D, Oxford Nanopore Technologies), and also the second adapter ligation procedure of thus end-prepared DNA using Adapter Mix 1D (AMX1D; Ligation Sequencing kit 1D, ONT) and Blunt/TA Ligase Master Mix (NEB) were done following the instructions in 1D PCR barcoding (96) genomic DNA (SQK-LSK108) protocol.

Also the final cleaning procedure using Agencourt AMPure XP beads (Beckman Coulter), Adapter Bead Binding buffer (ABB; Ligation Sequencing kit 1D, Oxford Nanopore Technologies) and Elution Buffer (ELB; Ligation Sequencing kit 1D, Oxford Nanopore Technologies) was performed following the recommendations of the manual supplied by the manufacturers (1D PCR barcoding (96) genomic DNA SQK-LSK108).

The prepared library was stored on ice during the priming of the SpotON flow cell (FLO-MIN 106D R9 Version, Oxford Nanopore Technologies). Seventy five microliters of sample composed from Running Buffer (RBF, Oxford Nanopore Technologies), Loading Beads (LLB; Oxford Nanopore Technologies), Nuclease-free water and DNA library was loaded to the flow cell via the SpotON sample port according manufacturer's recommendations.

### 2.7. Sequencing Performance and Evaluation

Sequencing run was performed for 48 h on MinION sequencer (Oxford Nanopore Technologies) using R9 flow cell, connected to the appropriate personal computer meeting the recommended criteria. The MinKNOW™ software was used to check the number of active pores and used with the default settings. After sequencing, the Fast5.tmp files were converted to Fast5 files via command prompt. Reads were then base called using Albacore (Oxford Nanopore Technologies) and split by barcodes with EPI2me Desktop Agent (Oxford Nanopore Technologies Metrichor). Taxonomic classification and quantitative analysis of reads derived from 28S rRNA amplicons were performed using "What's in my pot" tool (WIMP, Oxford Nanopore Technologies). In the case of bacterial taxa classification and quantification, the workflow for 16S rRNA gene encoded amplicon analysis in EPI2ME Agent was chosen. The minimal quality score in both cases was set to 7.

### 2.8. DNA Analysis of Plant Sprig and Stiff Material

The DNA from the samples f (decorative plant sprig) and g (stiff material of the bodice) was extracted by the kit ChargeSwitch Forensic DNA Purification Kit (Invitrogen) following the recommendations and instructions of manufacturer.

The obtained DNA of plant sprig was amplified using different combinations of primers oriented on the chloroplast trnL (UAA) intron (trnL_c-trnL_d, size of PCR product about 577 bp; trnL_c-trnL_h, product size about 187 bp; trnL_g-trnL_h, product size about 79 bp; trnL_g-trnL_d, product size about 474 bp [18]).

The DNA of stiff material was amplified with the primers and PCR protocols developed by [19] and oriented on the mitochondrial DNA D-loop region. We have tried the PCR assays comprising the pair of primers Pair 1, Pair 2, and Pair 3, but, also, we have combined the primers in this way: Pair1_Fw–Pair3_Rv; Pair2_Fw–Pair3_Rv.

The PCR products were separated on 1.5% agarose gel in TAE buffer. Gels were stained with ethidium bromide and visualized under UV light. The expected bands were excised from the gel and purified by QIAquick Gel Extraction kit (Qiagen). Then the purified bands were sequenced on both strands by a commercial facility (Eurofins Genomics). The resulting sequences were directly compared with those in GenBank using BLAST program (http://blast.ncbi.nlm.nih.gov/Blast.cgi).

### *2.9. Illumina Sequencing of Plant Sprig and Stiff Material DNA*

The DNA extracted from the samples f (decorative plant sprig) and g (stiff material of the bodice) was sequenced by the MiSeq system (Illumina, San Diego, CA, USA). Totally 20 ng of isolated double-stranded DNA was used for transposon-based sample fragmentation with Nextera XT DNA Library Prep Kit (Illumina, CA, USA). Samples were directly amplified and indexed using Nextera XT Index Kit (Illumina, CA, USA) according to the manufacturer's protocol. For PCR product purification 1,8x sample volume of Agencourt AMPure XP magnetic beads (Beckman Coulter, Brea, CA, USA) was used. The final DNA library quantity was determined with Qubit 2.0 Fluorometer (ThermoFisher Scientific, MA, USA) while for quality (fragment length) assessment Agilent 2100 Bioanalyzer was used (Agilent Technologies, Santa Clara, CA, USA). DNA libraries were diluted to 4 nM and pooled. Paired-end sequencing 2 × 300 was performed using Illumina MiSeq platform (Illumina, San Diego, CA, USA).

The data obtained from shot-gun sequencing of both samples were processed using CLC Genomics Workbench software v 9.5.2 (Qiagen, Germany). After trimming (quality limit 0.02, min.read length 50 bp) and quality control, de novo assembly of sequences, length fraction 0.9 (animal)/0.8 (plant), similarity fraction 1.0 (animal)/0.9 (plant)) was executed. Data obtained for stiff sample were mapped against GenBank WGS database using BLAST program with megablast settings mapping against taxon Equus. Contigs obtained for the plant sample were mapped against plant part of the GenBank database also using megablast settings.

## 3. Results

### *3.1. Microbial Isolates, Hydrolytic Properties, and SEM Observation*

The SEM analysis displayed the presence of fungi in various samples (Figure 2), in fact it is possible to note conidia, spores, and hyphae. The occurrence of bacteria in several samples is confirmed, indirectly, by the observation of biogenic crystals (Figure 2C,D).

Bacteria were isolated from the samples a (bodice, from the ribbon), b (bodice, from the body), d (silk bobbin lace cap), and e (linen head-band). The most isolated genera were *Moraxella*, *Kocuria*, and *Paracoccus*. The only bacterial isolates able to hydrolyze all the tested substrates was *Micrococcus yunnanensis* B2. Various bacteria degrade four substrates; three isolates (the two *Moraxella* isolates and *Paracoccus chinensis*) were negative to all assays. A slight cellulolytic activity was shown only by the two *Kocuria* isolates and *Micrococcus yunnanensis* B2. The most effective keratinolytic isolate was *Paenibacillus illinoisensis* D1 which displayed also an extensive protease ability together with *Micrococcus yunnanensis* B2. *Paenibacillus illinoisensis* D1 evidenced also the best fibroinase activity followed by *Sphingomonas* sp. D2. Usually the isolates positive to proteolytic assays can degrade also the silk fibroin (Table 1).

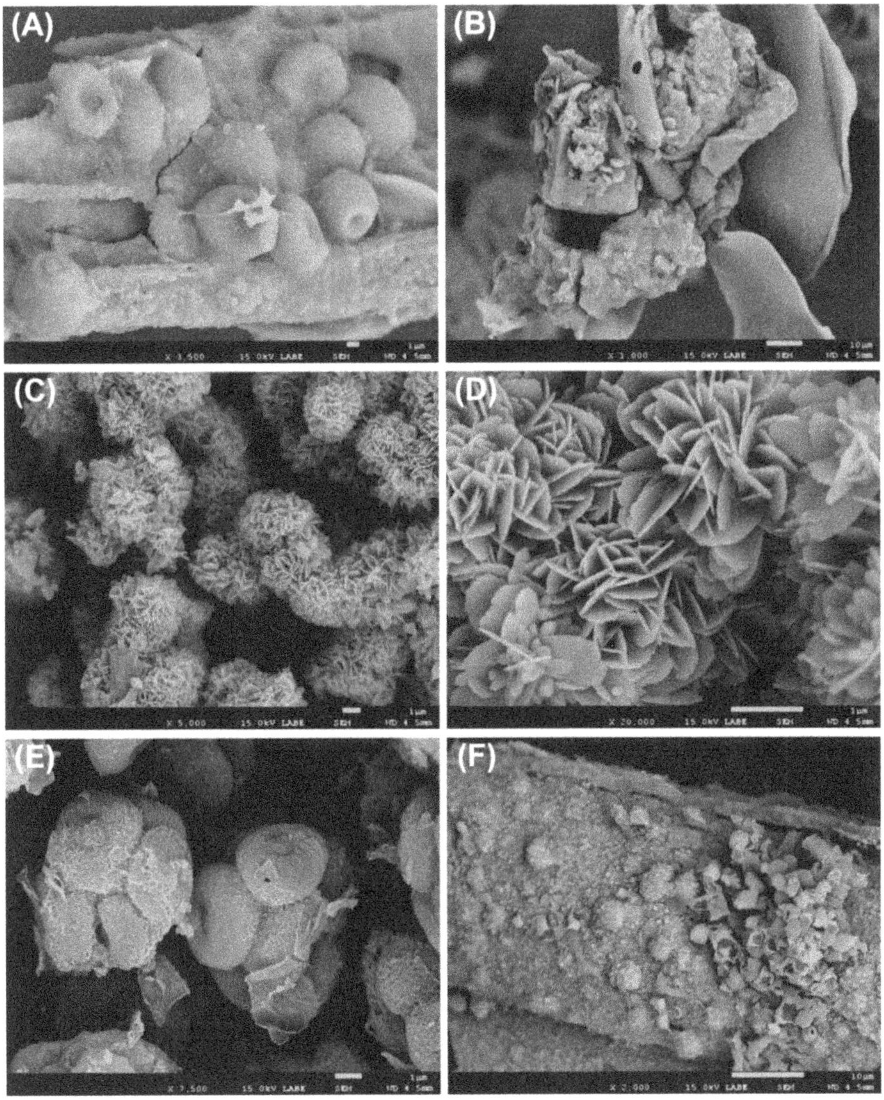

**Figure 2.** Scanning electron microscope (SEM) observation of nitrocellulose membranes. (**A**) Conidia, (**B**) fungi colonizing insect remains, (**C**,**D**) biogenic crystals and their morphological details, (**E**) ascospores, (**F**) presence of fungal conidia and spores.

From sample c (bodice, tissue covering the stiff reinforcement) only filamentous fungi were isolated belonging mainly to genera *Aspergillus* and *Penicillium*. The yeast *Sporidiobolus metaroseus* B2-F and *Penicillium commune* B1-F, recovered from sample b, were the only eukaryotic microorganisms able to produce positive results for all the hydrolytic assays. Three fungi over seven isolates displayed cellulolytic ability, on the contrary all the fungal members have significant fibroinase property.

**Table 1.** DNA identification and hydrolytic properties of bacterial and fungal isolates.

| Sample | Strain Code | Percent Identity | Keratinase Activity (mm) | Cellulase Activity (mm) | Proteinase [1] Activity (mm) | Proteinase [2] Activity (mm) | Fibroinase Activity (mm) |
|---|---|---|---|---|---|---|---|
| | | | **Bacteria** | | | | |
| a, bodice | A1 | MT071578 *Moraxella* sp. 100% | 0 | 0 | 0 | 0 | 0 |
| | A2 | CP024180 *Moraxella osloensis* 100% | 0 | 0 | 0 | 0 | 0 |
| | A3 | KY434635 *Kocuria* sp. 99.83% | 1 ± 0.3 | 1 ± 0.2 | 4 ± 0.6 | 2 ± 0.4 | 0 |
| | A4 | MT012185 *Rhodococcus* sp. 100% | 0 | 0 | 2 ± 0.3 | 0 | 4 ± 1 |
| b, bodice | B1 | EF072311 *Massilia* sp. 99.22% | 7 ± 0.8 | 0 | 7 ± 1.2 | 4 ± 1.2 | 10 ± 1 |
| | B2 | KY816357 *Micrococcus yunnanensis* 99.67% | 12 ± 0.5 | 2 ± 0.4 | 10 ± 1 | 15 ± 0.4 | 7 ± 0.6 |
| d, bobbin lace cap | D1 | FJ999728 *Paenibacillus illinoisensis* 100% | 14 ± 0.2 | 0 | 5 ± 1 | 15 ± 1.5 | 20 ± 1 |
| | D2 | KF815549 *Sphingomonas* sp. 100% | 5 ± 1 | 0 | 1 ± 0.6 | 4 ± 0.4 | 12 ± 1.3 |
| | D4 | MN396262 *Roseomonas mucosa* 99.61% | 2 ± 0.5 | 0 | 4 ± 0.8 | 1 ± 0.8 | 8 ± 0.6 |
| e, coronet | E1 | HM480273 *Staphylococcus* sp. 99.85% | 0 | 0 | 1 ± 0.4 | 0 | 0 |
| | E2 | KF424781 *Paracoccus chinensis* 99.14% | 0 | 0 | 0 | 0 | 0 |
| | E3 | MF804955 *Paracoccus* sp. 99.58% | 5 ± 0.7 | 0 | 3 ± 0.6 | 8 ± 0.8 | 2 ± 0.6 |
| | E4 | MN704104 *Kocuria* sp. 99.83% | 1 ± 0.6 | 1 ± 0.5 | 6 ± 1 | 0 | 9 ± 0.7 |
| | | | **Fungi** | | | | |
| b, bodice | B1-F | KY552625 *Penicillium commune* 99.37% | 4 ± 0.7 | 0 | 4 ± 1 | 4 ± 1.2 | 10 ± 1 |
| | B2-F | KY105470 *Sporidiobolus metaroseus* 100% | 9 ± 0.4 | 2 ± 0.8 | 5 ± 1 | 10 ± 1 | 20 ± 1.5 |
| c, bodice | C1-F | KP067214 *Aspergillus versicolor* 100% | 4 | 0 | 0 | 5 ± 0.6 | 10 ± 1 |
| | C2-F | MN905799 *Penicillium* sp. 100% | 3 | 2 | 2 ± 0.7 | 5 ± 0.5 | 4 ± 0.5 |
| | C3-F | MT102669 *Aspergillus* sp. 100% | 0 | 0 | 0 | 0 | 12 ± 0.9 |
| | C4-F | MH393399 *Penicillium* granulatum 98.94% | 7 | 0 | 8 ± 1 | 5 ± 0.3 | 12 ± 1 |
| | C5-F | MH424607 *Alternaria alternata* 99.64% | 5 ± 0.3 | 2 ± 0.4 | 0 | 5 ± 0.4 | 9 ± 0.6 |

Samples; Bodice: (a) from the ribbon; (b) from the body; (c) from a portion of tissue covering the stiff reinforcement; (d) silk bobbin lace cap; and (e) linen coronet. (1) R2A medium with gelatin; (2) Gelatin medium.

### 3.2. MinION Analysis of Bacterial 16S rRNA Gene

The results of all four samples (a, b, c, and d), examined by EPI2ME Desktop Agent software using "FASTQ 16S" tool are visualized, expressing the percentage of taxa abundance on genera degree, as bar plots in Figure 3. The percentages of reads assigned to given taxa were calculated from number of classified reads. The number of classified reads in order of samples a, b, c, and d were 46,671, 378,356, 38,042, and 64,707, respectively. Because of microbial diversity, the percentage of given taxa of each samples were counted together, and the value over 2% of this sum were used as threshold. This is the reason why the number of "other" taxa forms the biggest amount of all four analyzed samples. In the order of samples a, b, c, and d "other" taxa comprised 63.99%, 66.07%, 53.65%, and 64.40%, respectively.

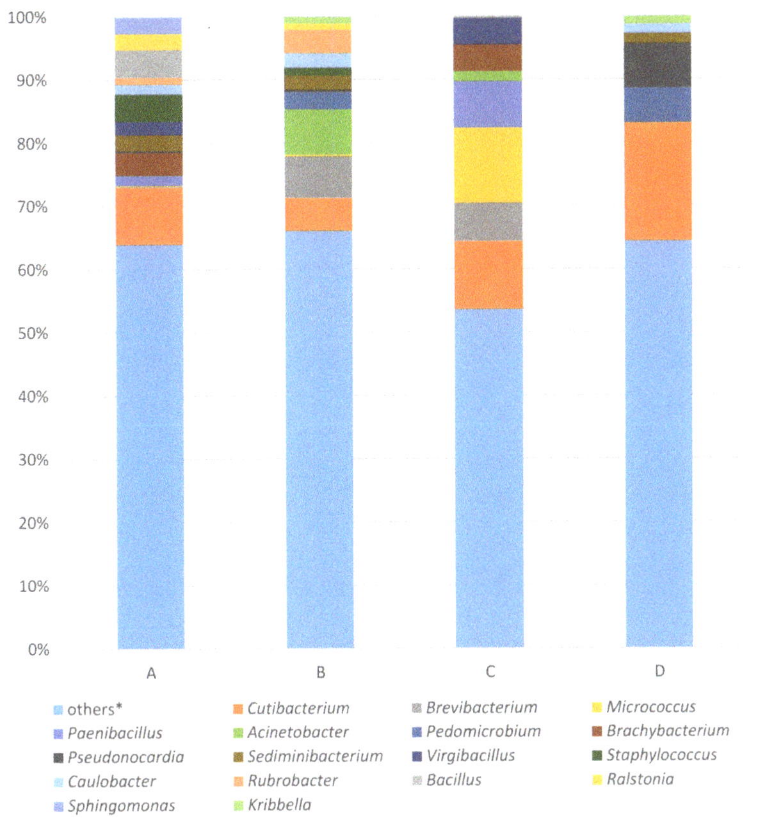


**Figure 3.** Results of bacterial 16S rRNA gene analysis of samples A (the ribbon of the bodice), B (body of the bodice), C (tissue covering the stiff reinforcement of the bodice), D (silk bobbin lace cap) using MinION. * Others: the percentages of reads assigned to taxa under (below) the value of 2% in the sum of all four samples.

The most abundant taxa identified in all four samples were *Cutibacterium* (a-9.03%; b-5.24%; c-10.74%; d-18.72%), *Brevibacterium* (0.09%; 6.57%; 6.00%; 0.00%), *Micrococcus* (0.17%; 0.31%; 11.95%; 0.00%), *Paenibacillus* (1.57%; 0.00%; 7.36%; 0.00%) and *Acinetobacter* (0.01%; 7.15%; 1.53%; 0.01%). The top 10 most abundant taxa enclosed genera *Pedomicrobium* (0.00%; 2.67%; 0.00%; 5.42%), *Brachybacterium* (3.73%; 0.00%; 4.19%; 0.00%), *Pseudonocardia* (0.20%; 0.37%; 0.01%; 7.06%), *Sediminibacterium* (2.53%; 2.27%; 0.00%; 1.63%), and *Virgibacillus* (2.12%; 0.00%; 4.25%; 0.00%).

### 3.3. MinION Analysis of Fungal 28S rRNA Gene

Fungal 28S rRNA gene analysis produced results only with samples a, b, and c. The results of this analysis are summarized in Figure 4. In the order of samples a, b, and c, the most occurring fungal taxa were *Malassezia* (5.59%; 4.08%; 51.77%), *Aspergillus* (15.46%; 20.08%; 0.25%), *Metarhizium* (2.93%; 8.41%; 5.22%), *Pestalotiopsis* (0.75%; 10.11%; 0.02%) and *Phycomyces* (6.75%; 0.73%; 0.10%) followed by *Serpula* (4.44%; 0.62%; 0.36%), *Colletotrichum* (1.69%; 1.02%; 0.52%), *Talaromyces* (1.30%; 1.68%; 0.12%), *Leptosphaeria* (0.64%; 1.50%; 0.83%), and *Agaricus* (1.37%; 0.11%; 1.41%), which completed the top 10 most abundant fungal taxa.

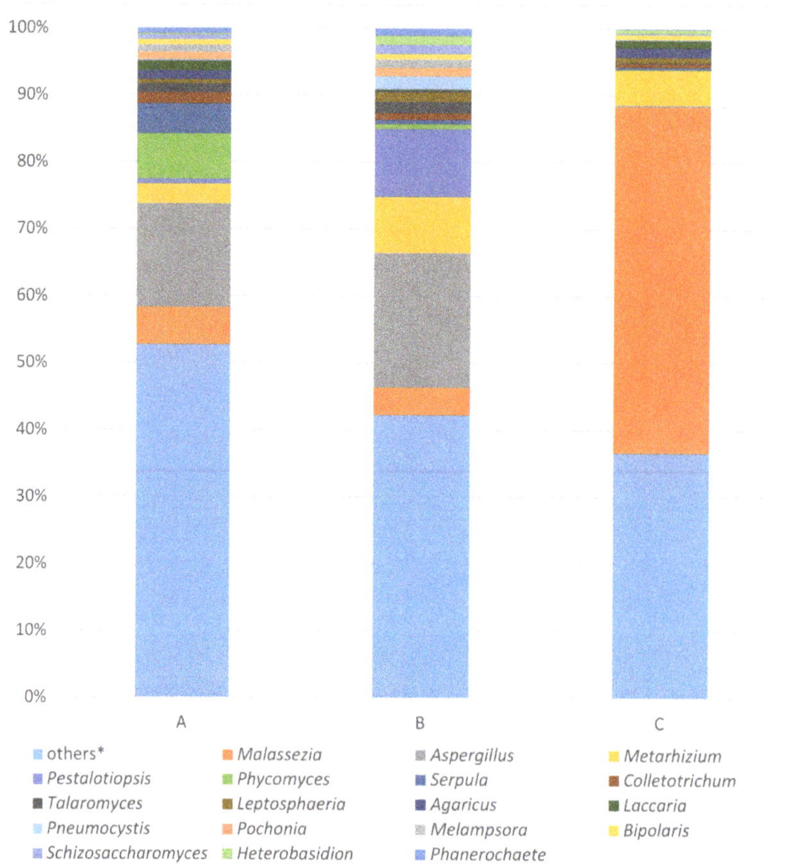


**Figure 4.** Results of fungal 28S rRNA analysis of samples A (the ribbon of the bodice), B (body of the bodice), C (tissue covering the stiff reinforcement of the bodice) using MinION. * Others: the percentages of reads assigned to taxa under (below) the value of 2% in the sum of all four samples.

### 3.4. MinION Analysis of Fungal ITS Fragment

The fungal ITS marker amplified with the samples c, d, and e. In this order of samples, *Metarhizium* (5.23%; 31.36%; 26.56%), *Colletotrichum* (1.75%; 7.05%; 5.23%), *Wallemia* (0.04%; 0.41%; 6.27%), *Baudoinia* (0.45%; 2.37%; 3.68%), followed by *Neofusicoccum* (4.56%; 0.21%; 0.86%), *Phycomyces* (0.20%; 0.18%; 4.94%), Botrytis (2.50%; 0.58%; 0.54%), *Torulaspora* (0.01%; 0.00%; 3.56%), *Parastagonospora* (0.46%; 2.51%;

0.56%), and *Aspergillus* (0.80%; 1.51%; 1.13%) were identified as the most abundant fungal by WIMP tool whose results are expressed as bar plots in Figure 5.

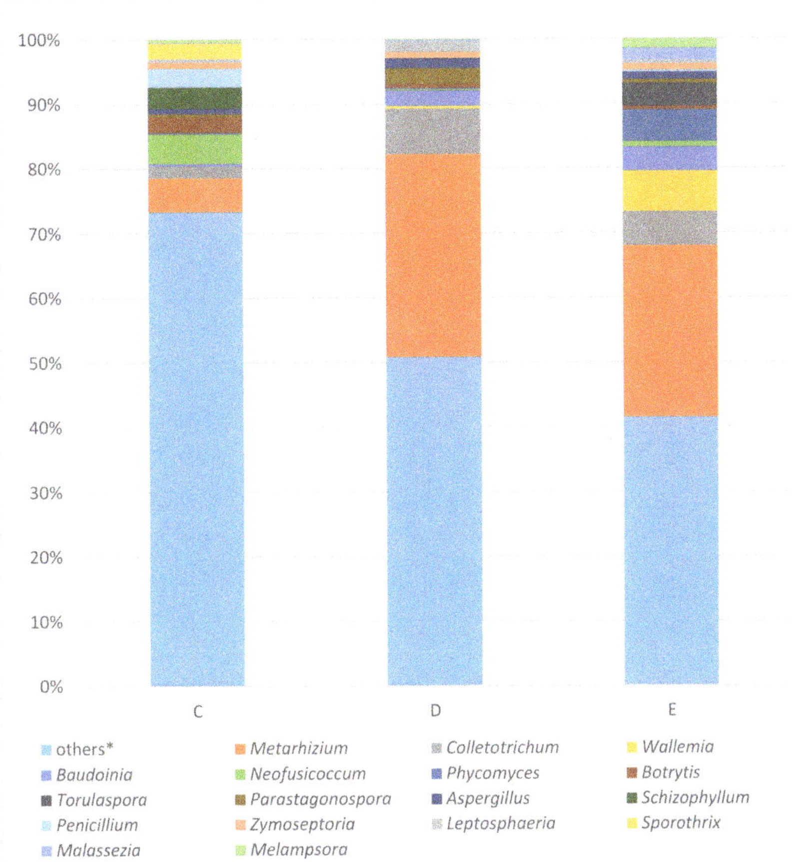


**Figure 5.** Results of fungal ITS analysis of samples C (tissue covering the stiff reinforcement of the bodice), D (silk bobbin lace cap), and E (linen coronet) using MinION. * Others: the percentages of reads assigned to taxa under (below) the value of 2% in the sum of all four samples.

### *3.5. DNA Analysis of Animal and Plant Samples*

In order to identify the animal origin of the bodice stiff material (sample g), we have tried different combination of primers, but unfortunately all PCR assays produced negative results. Using MiSeq platform (Illumina), with this sample, we obtained 1240 contigs, with a length range of 69–25, 511 bp, several of them matching *Equus caballus*.

Positive PCR amplifications were obtained with the plant material (sample f) using the primers combinations trnL_c - trnL_h (amplicon size about 187 bp) and trnL_g - trnL_h (about 79 bp). The consequent Sanger sequencing of the amplicons had reliable results only with the product of about 187 bp. In fact, the obtained 136 bp sequence had a 98.53% of similarity with several plant species belonging to the family of Apiaceae (MN167277, *Cuminum cyminum*; MN167276, *Anethum graveolens*; MN167272, *Petroselinum crispum*; MH142518, *Glehnia littoralis*; MK688991, *Angelica sinensis*). Using the DNA extracted from the plant sample, a totally of 2287 contigs with a length range of 79–1423 bp were produced by high-throughput sequencing analysis. The Poaceae family was represented by three species: *Digitaria exilis*, *Hordeum vulgare*, and *Triticum aestivum*. The other two detected species, *Brassica juncea* and *Nicotiana tabacum* belonged to the family Brassicaceae and Solanaceae, respectively.

## 4. Discussion

### 4.1. Bacterial Community

The two strategies of analysis (culture-dependent and culture-independent) evidenced a link with the members of the genera *Micrococcus* and *Paenibacillus*. These bacteria were detected by both approaches, but in different samples. They are frequently isolated from historical textile [1,20] and also in the past studies, as well as here, they showed significant hydrolytic properties. Members of the genus *Paenibacillus* produce a broad range of enzymes used in biotechnology processes [21]. In addition to *Micrococcus*, other isolated Actinobacteria with interesting proteolytic and fibroinolytic properties belonged to the genus *Kocuria*. Representatives of this genus are commune in subterranean environment [22] and were also isolated from archaeological textile [23]. Therefore, bacteria of the genera *Paenibacillus*, *Micrococcus*, and *Kocuria* can be considered as dangerous textile degrading microorganisms.

By high-throughput sequencing the genera *Micrococcus* and *Paenibacillus* were dominant in the sample c (tissue covering the stiff reinforcement), where *M. luteus* comprised 8.47% of all reads classified. *Micrococcus luteus* is also one of the species inhabiting the human skin [24]. Members of the genus *Micrococcus* were also previously identified as part of microbiota on mummy surface [2].

Regarding the MinION analysis, three species of the genus *Cutibacterium* (*C. acnes*, *C. granulosum*, and *C. avidum*), formerly known as *Propionibacterium* [25], were detected in all four samples (a, b, c, and d) with the remarkable prevalence of *Cutibacterium acnes*, which is involved in the maintenance of a healthy skin, but it can also act as an opportunistic pathogen in acne vulgaris [26] or can cause implant-associated infections [27].

Genus *Acinetobacter* dominated mostly in sample b (sampled from silk fabric of the bodice), whose probably touched the torso of cadaver. Hyde et al. [28] observed the highest abundance of *Acinetobacter* in samples of male (5–20% of relative abundance) and female (5–15% of relative abundance) cadavers studied during the time, mostly in the late stages of their decomposition.

The genus *Brevibacterium* dominated in the samples b and c, namely *B. casei* (5.63%) and B. pityocampae (4.15%), respectively. *Brevibacterium* was the most frequent genera identified in human archaeological remains by Philips et al. [29], also typically found in a wide range of soils and waters.

The genus *Pseudonocardia* was in our case the most abundant (silk bobbin lace cap) in sample d (7.06%). This genus was previously detected in the caves [30], catacombs [31], tomb walls, as part of the ancient paintings deteriorating microflora [32], or part of the so-called moonmilk [33] and also on the body of a mummy [2].

The reads identified as genus *Pedomicrobium* (Alphabacteria class) predominated in samples b and d (2.67% and 5.42%, respectively). *Pedomicrobium* is known as iron- and manganese-oxidizing and accumulating bacterium often detected in biofilms formed on different surfaces [34,35]. Probably, in these samples *Pedomicrobium* could be responsible for various corrosion phenomena [36] of metal materials present in the structure of the bodice, but also as decoration jewels on the head of girl. The isolates *Roseomonas mucosa* and *Sphingomonas* sp. also belonged to the class of Alphabacteria. Here they contribute to the degradation of silk fibroin and other kind of proteinaceous substrates. These bacteria have the characteristic to possess hydrolytic enzymes active also at low temperature [37,38].

The iron-oxidizing bacteria of the genus *Sediminibacterium* were present with approximately 2% in all samples, except of sample c. We think that their role is connect mainly with the corrosion [39] of metal materials present in these samples.

In the samples a and c, *Brachybacterium* was dominated by reaching 3.73% and 4.19%, respectively. Affiliates of this genus were already isolated from subterranean environment [40] and also from funeral cloths [1] demonstrating significant proteolytic property.

Other bacteria were present exclusively in samples a (2.12%) and c (4.25%), where the members of the genus *Virgibacillus*. In the sample a prevailed *V. marseillensis*, *V. carmonensis*, and *V. halodenitrificans*. In sample c mostly *V. necropolis* and *V. carmonensis* were identified. *V. carmonensis*, *V. necropolis*, and *V. picturae* were isolated from deteriorated mural paintings [41]. In addition, these bacteria are able to produce different extracellular enzymes, such as proteases, cellulases, xylanases, and amylases, which can degrade several substrates [42].

### 4.2. Fungal Community

*Malassezia*, the most abundant taxa detected by 28S rRNA sequencing analysis, significantly prevailed in sample c (51.77%), from which, the *M. globosa* reached a percentage of 21.59%. Although *Malassezia* yeasts are part of the cutaneous microbiota detected mainly on healthy human skin, species of this genus are also associated with mammalian cutaneous disorders such as dandruff, seborrhoeic dermatitis, pityriasis versicolor, psoriasis, folliculitis, and otitis [43]. The ITS analysis placed *Malassezia* in the 16th place of the most occurring taxa (Figure 5), it dominated in sample e (c-0.09%; d-0.25%; e-1.88%).

*Aspergillus* and *Metarhizium* were detected in all samples using both fungal markers; by 28S rRNA the highest occurrence, 20.08% for *Aspergillus* and 8.41% for *Metarhizium*, was recorded in sample b. ITS analysis revealed the highest percentages of these genera in sample d, 1.51% and 31.36% for *Aspergillus* and *Metarhizium*, respectively. The presence of *Metarhizium* is probably due to because members of this genus are known as entomopathogens [44], which infected various insects contributing to human decomposition. In both analysis, *A. fumigatus* was the most dominant species among the other *Aspergillus* spp. This species, in previous studies, was isolated from various organs of cadavers and human remains [45,46], several cultural heritage items made by organic materials [47], and it showed also proteolytic, keratinolytic, and fibroinolytic abilities [1]. Therefore it is not surprising, that the Aspergilli and Penicilli isolated in this study possessed also interesting hydrolytic properties. They were recovered almost exclusively from a portion of silk tissue, which covered the stiff reinforcement made by animal substances. These characteristics evidenced their biodeteioration dangerousness for textile and other proteinaceous materials occurring in various cultural heritage objects.

*Pestalotiopsis* is another abundant taxa in sample b (10.11%) according to 28S rRNA analysis, and their species are known as plant pathogens and as the producers of antifungal active metabolites [48–50]. The ITS analysis did not detect this taxa.

Both taxa *Phycomyces* and *Colletotrichum* were detected by 28S rRNA (*Phycomyces*: a-6.75%, b-0.73%, c-0.10%; *Colletotrichum*: a-1.69%, b-1.02%, c-0.52%) and ITS (*Phycomyces*: c-0.20%, d-0.18%, e-4.94%; *Colletotrichum*: c-1.75%, d-7.05%, e-5.23%) analyses. These fungi are not common colonizers of historical items, in fact, to our knowledge, the first detection of *Phycomyces* in two canvas paintings was very recent [51].

Genus *Serpula*, dominated significantly only in sample a (a-4.44%, b-0.62%, c-0.36%) analyzed by 28S rRNA genes sequencing. All reads assigned to the genus *Serpula* belonged to only one species - *S. lacrymans*, a common wood-decaying dry rot causing deterioration of wood monuments [52] and buildings [53]. The ITS analysis revealed lower occurrence of *Serpula* taxa with highest presence in sample c (0.34%). Other phytopathogens were detected in sample c (ITS) in remarkable abundance, e.g., *Neofusicoccum* with one representative species of this genus *N. parvum* (c-4.56%), known as phytotoxins-producing grapevine canker agent [54] and Botrytis with one representative species of this genus *B. cinerea* (2.50%), well known pathogen causing serious losses in more than 200 crop species worldwide [55]. The genus *Torulaspora* represented 3.56% in sample e (ITS), with one representing species *T. delbrueckii*, which is also typical inhabitant of the grape surface, known as well studied non-Saccharomyces yeasts used in winemaking [56]. Another plant pathogen, *Phaeosphaeria nodorum* was detected mostly in sample d (2.51%). It is known as fungal pathogen of wheat (*Triticum aestivum*). The relevant detection of plant pathogens is due of the wood of coffin, but also of the presence of several types of plant species that usually served to adorn the body of the dead.

*Wallemia ichthyophaga*, as the only representatives of the taxa *Wallemia*, occurring in highest abundance in the sample e (6.27%), is a halophilic fungus with optimally growth on media with salinity above 15% of NaCl (*w/v*) [57]. Another extremophile yeast [58] was the isolate *Sporidiobolus metaroseus* recovered from the silk bodice (sample b). It displayed significant hydrolytic properties including fibroin degradation.

### 4.3. Microscopic Observation

The combination of sampling by nitrocellulose membrane and the consequent SEM screening, can be considered as a suitable method for the microscopic analysis of fragile cultural heritage items. Usually, the adhesive tape sampling is applied in combination of microscopic observation [59], but depending of the fragility of surface, the use of membranes is a valid alternative.

The SEM monitoring permitted to display the presence of an active microbial communities, mainly fungi that are able to degrade various types of substrates. In addition, the observation of ascospores [60] confirmed the results of microbial and sequencing analysis where different Aspergillus members were detected.

The presence of bacteria was evidenced by the observation of bio-mineralization phenomena, such as the occurrence of $CaCO_3$ crystals, probably vaterite or aragonite. These $CaCO_3$ formations are generally produced by various bacterial members belonging to the phylum Proteobacteria (vaterite forming) and Firmicutes (aragonite) [61,62]. Also, these findings are in accordance with the bacteria isolated and detected by both culture-dependent and culture-independent strategies.

### 4.4. Plant and Animal DNA Analysis

Usually, during the XVII century, the stiff materials used as reinforcement of the bodices were whalebone from the jaws of the whale, wood, or horn [63]. Therefore, when we have observed the reinforcement of the bodice by microscope (Figure S1), these busks were particular because they appeared as several fibers bonded, presumably, with animal glue. The restorers supposed that this structure was formed by horsehair, and the molecular analysis confirm such hypothesis. Although, the horsehair was used for different types of textile materials (horsehair interlining; [64]), in literature we did not find any previous study, where this kind of reinforcement material was described, so the analysis of the DNA help us to unambiguously identify the origin of these accessories of the bodice.

DNA of Apiaceae was identified from the small portion of plant material through PCR sequencing analysis. Unfortunately, the small sequence did not permit a specific recognition of the plant species. However, excluding *Glehnia littoralis* which is a typical plant growing at temperate sandy coasts around the North Pacific Ocean [65], and *Angelica sinensis* which is also characteristic of different world regions other than Europe [66]. One of the others species (*Cuminum cyminum*, *Anethum graveolens* and *Petroselinum crispum*) probably were the main ornamental plants on the head of the young girl. These species present nice inflorescence [67], which can create beautiful decorative effects. Moreover, the use of these kinds of plants in burial custom was evidenced in several archaeological studies [68–70].

The Illumina analysis on plant DNA did not confirm the results of PCR and sequencing approach, but other interesting outcomes were produced. In fact, the detection of several sequences is due by the presence of traces of plants, probably pollen belonged mainly to the species *Hordeum vulgare* (barley), *Triticum aestivum* (common wheat), *Brassica juncea* (canola), and *Nicotiana tabacum* (tobacco) which were commonly cultivated also in central Europe. We think that the identification of these plants perhaps indicates somehow the period of the funeral which could be in the late spring or summer. The register of deceased persons has been lost, therefore our molecular investigation provided at least some hypothesis about the seasonal period of the burial ceremony.

## 5. Conclusions

The combination of non-invasive sampling, classical microbiological methods (cultivation, SEM observation, and hydrolytic assays), molecular approaches, and high-throughput sequencing (culture-independent strategy and DNA analysis) permitted to investigate the deteriorating microbiota and to discover various characteristics of diverse funeral accessories. The complete microbiological analysis displayed the occurring microbiota and evidenced its dangerous hydrolytic properties and bio-precipitation abilities. The MinION approach can be optimized in order to perform the analysis in situ. In fact, this high-throughput sequencing platform is portable and in the future the analysis of samples can be done directly in the place of sampling. Moreover, MinION can sequence long sequences, which facilitates the better identification of microbial communities. By the DNA analysis, we observed important information about the use of certain materials (horse-hair reinforcement) in textile manufacturing. Knowledge about the use of herbal and flower ornaments in the 17th century burial customs in central Europe were also revealed. This study evidenced again the importance of a multi-disciplinary approach applied to cultural heritage objects. Such study showed how it is important to develop periodical and precise microbial monitoring strategies which permit prompt decisions for the safeguard of these fragile items against the microbial deterioration and also to obtain valuable data on the uses and customs of a certain historical period.

**Supplementary Materials:** The following are available online at http://www.mdpi.com/2076-3417/10/16/5451/s1; Figure S1: Stereomicroscope visualization of reinforcement material of the bodice.

**Author Contributions:** Conceptualization, D.P., M.B., and A.P.; Methodology, D.P., K.Š., and M.P.; Formal Analysis, Z.K., J.P., M.K., L.K., M.B., A.P., and K.Š.; Investigation, Z.K., J.P., M.K., L.K., M.B., A.P., and M.P.; Resources, Z.K. and D.P.; Writing—Original Draft Preparation, M.P. and D.P.; Writing—Review and Editing, D.P. and Z.K.; Visualization, D.P.; Supervision, D.P.; Project Administration, M.B.; Funding Acquisition, M.B. and K.Š. All authors have read and agreed to the published version of the manuscript.

**Funding:** This research was mainly funded by the project APVV-19-0059 (Colored stains on historical papers: biological and chemical characterization coupled with removal solutions). This publication is also the result of implementation of the projects: Research and Development Operational Programme funded by the ERDF (grant number: ITMS 26240220086) and VEGA 1/0404/19.

**Acknowledgments:** We would like to thank the members of the Academy of Fine Arts and Design in Bratislava (Zuzana Machatová, Sylvia Birkušová, Ingrid Ondrejičková Soboslayová) who permitted us to investigate these interesting samples in the frame of their restoration project and for a photograph (Figure S1). We acknowledge the archaeologists Peter Harčar and Marián Uličný of the Monument Board of the Slovak Republic, the pioneers who started the archaeological research (Uličný, M. - Harčar, P.: Bardejov - Bazilika sv. Egídia, Výskumná dokumentácia z archeologického výskumu vo svätyni baziliky v rokoch 2008 - 2009, Pamiatkový úrad SR, február 2017). We also thank Ľubomír Orovčík (Institute of Materials and Machine Mechanics, Slovak Academy of Sciences) for the opportunity to use the SEM.

**Conflicts of Interest:** The authors declare no conflict of interest.

## References

1. Pangallo, D.; Kraková, L.; Chovanová, K.; Bučková, M.; Puškarová, A.; Šimonovičová, A. Disclosing a crypt: Microbial diversity and degradation activity of the microflora isolated from funeral clothes of Cardinal Peter Pázmány. *Microbiol. Res.* **2013**, *168*, 289–299. [CrossRef] [PubMed]
2. Kraková, L.; Šoltys, K.; Puškárová, A.; Bučková, M.; Jeszeová, L.; Kucharík, M.; Budiš, J.; Orovčík, L.U.; Szemes, T.; Pangallo, D. The microbiomes of a XVIII century mummy from the castle of Krásna Hôrka (Slovakia) and its surrounding environment. *Environ. Microbiol.* **2018**, *20*, 3294–3308. [CrossRef] [PubMed]
3. Sterflinger, K.; Piñar, G. Microbial deterioration of cultural heritage and works of art—tilting at windmills? *Appl. Microbiol. Biotechnol.* **2013**, *97*, 9637–9646. [CrossRef] [PubMed]
4. Gutarowska, B.; Pietrzak, K.; Machnowski, W.; Milczarek, J.M. Historical textiles–A review of microbial deterioration analysis and disinfection methods. *Text. Res. J.* **2017**, *87*, 2388–2406. [CrossRef]
5. Marvasi, M.; Cavalieri, D.; Mastromei, G.; Casaccia, A.; Perito, B. Omics technologies for an in-depth investigation of biodeterioration of cultural heritage. *Int. Biodeterior. Biodegrad.* **2019**, *144*, 104736. [CrossRef]
6. Szostak-Kotowa, J. Biodeterioration of textiles. *Int. Biodeterior. Biodegrad.* **2004**, *53*, 165–170. [CrossRef]

7. Forlani, G.; Seves, A.M.; Ciferri, O. A bacterial extracellular proteinase degrading silk fibroin. *Int. Biodeterior. Biodegrad.* **2000**, *46*, 271–275. [CrossRef]
8. Brzozowska, I.; Bogdanowicz, A.; Szczęsny, P.; Zielenkiewicz, U.; Laudy, A. Evaluation of bacterial diversity on historical silk velvet textiles from the Museum of King John III's Palace at Wilanów, Poland. *Int. Biodeterior. Biodegrad.* **2018**, *131*, 78–87. [CrossRef]
9. Martins, C.; Pereira, C.S.; Plechkova, N.V.; Seddon, K.R.; Wang, J.; Whitfield, S.; Wong, W. Mycobiota of silk-faced ancient Mogao Grottoes manuscripts belonging to the Stein collection in the British library. *Int. Biodeterior. Biodegrad.* **2018**, *134*, 1–6. [CrossRef]
10. MacHugh, D.E.; Larson, G.; Orlando, L. Taming the past: Ancient DNA and the study of animal domestication. *Annu. Rev. Anim. Biosci.* **2017**, *5*, 329–351. [CrossRef]
11. Bieker, V.C.; Martin, M.D. Implications and future prospects for evolutionary analyses of DNA in historical herbarium collections. *Bot. Lett.* **2018**, *165*, 409–418. [CrossRef]
12. Puškárová, A.; Bučková, M.; Habalová, B.; Kraková, L.; Maková, A.; Pangallo, D. Microbial communities affecting albumen photography heritage: A methodological survey. *Sci. Rep.* **2016**, *6*, 20810. [CrossRef] [PubMed]
13. Grivalský, T.; Bučková, M.; Puškárová, A.; Kraková, L.; Pangallo, D. Water-related environments: A multistep procedure to assess the diversity and enzymatic properties of cultivable bacteria. *World J. Microbiol. Biotechnol.* **2016**, *32*, 42. [CrossRef] [PubMed]
14. Lane, D.J. 16S/23S rRNA sequencing. In *Nucleic Acid Techniques in Bacterial Systematics*; Stackenbrandt, E., Goodfellow, M., Eds.; John Wiley & Sons: New York, NY, USA, 1991; pp. 115–148.
15. White, T.J.; Bruns, T.; Lee, S.; Taylor, J. Amplification and direct sequencing of fungal ribosomal RNA genes for phylogenetics. In *PCR Protocols: A Guide to Methods and Applications*; Innis, M.A., Gelfand, D.H., Sninsky, J.J., White, T.J., Eds.; Academic Press: New York, NY, USA, 1990; pp. 315–321.
16. Kurtzman, C.P.; Robnett, C.J. Identification and phylogeny of ascomycetous yeasts from analysis of nuclear large subunit (26S) ribosomal DNA partial sequences. *Antonie Van Leeuwenhoek* **1998**, *73*, 331–371. [CrossRef] [PubMed]
17. Šoltys, K.; Planý, M.; Biocca, P.; Vianello, V.; Bučková, M.; Puškárová, A.; Sclocchi, M.C.; Colaizzi, P.; Bicchieri, M.; Pangallo, D.; et al. Lead soaps formation and biodiversity in a XVIII Century wax seal coloured with minium. *Environ. Microbiol.* **2020**, *22*, 1517–1534. [CrossRef] [PubMed]
18. Taberlet, P.; Coissac, E.; Pompanon, F.; Gielly, L.; Miquel, C.; Valentini, A.; Vermat, T.; Corthier, G.; Brochmann, C.; Willerslev, E. Power and limitations of the chloroplast trn L (UAA) intron for plant DNA barcoding. *Nucleic Acids Res.* **2007**, *35*, e14. [CrossRef]
19. Vilà, C.; Leonard, J.A.; Götherström, A.; Marklund, S.; Sandberg, K.; Lidén, K.; Wayne, R.K.; Ellegren, H. Widespread origins of domestic horse lineages. *Science* **2001**, *291*, 474–477. [CrossRef]
20. Wawrzyk, A.; Gutarowska, B.; Rybitwa, D.; Pietrzak, K.; Machnowski, W.; Wrzosek, H.; Papis, A.; Walawska, A.; Otlewska, A.; Szulc, J.; et al. Vapourised hydrogen peroxide (VHP) and ethylene oxide (EtO) methods for disinfecting historical cotton textiles from the Auschwitz-Birkenau State Museum in Oświęcim, Poland. *Int. Biodeterior. Biodegrad.* **2008**, *133*, 42–51.
21. Grady, E.N.; MacDonald, J.; Liu, L.; Richman, A.; Yuan, Z. Current knowledge and perspectives of Paenibacillus: A review. *Microb. Cell Fact.* **2016**, *15*, 203. [CrossRef]
22. Piñar, G.; Ettenauer, J.; Sterflinger, K. La vie en rose: A review of rosy discoloration of subsurface monuments. *Conserv. Subterr. Cult. Herit.* **2014**, 113–124.
23. Pietrzak, K.; Puchalski, M.; Otlewska, A.; Wrzosek, H.; Guiamet, P.; Piotrowska, M.; Gutarowska, B. Microbial diversity of pre-Columbian archaeological textiles and the effect of silver nanoparticles misting disinfection. *J. Cult. Herit.* **2017**, *23*, 138–147. [CrossRef]
24. Davis, C.P. Chapter 6 normal flora. In *Medical Microbiology*, 4th ed.; Baron, S., Ed.; University of Texas Medical Branch at Galveston: Galveston, TX, USA, 1996; ISBN -100-9631172-1-1.
25. Scholz, C.F.; Kilian, M. The natural history of cutaneous propionibacteria, and reclassification of selected species within the genus Propionibacterium to the proposed novel genera Acidipropionibacterium gen. nov., Cutibacterium gen. nov. and Pseudopropionibacterium gen. nov. *Int. J. Syst. Evol. Microbiol.* **2016**, *66*, 4422–4432. [CrossRef] [PubMed]

26. Dréno, B.; Pécastaings, S.; Corvec, S.; Veraldi, S.; Khammari, A.; Roques, C. Cutibacterium acnes (Propionibacterium acnes) and acne vulgaris: A brief look at the latest updates. *J. Eur. Acad. Dermatol. Venereol.* **2018**, *2*, 5–14. [CrossRef] [PubMed]
27. Gharamti, A.A.; Kanafani, Z.A. Cutibacterium (formerly Propionibacterium) acnes infections associated with implantable devices. *Expert Rev. Anti. Infect. Ther.* **2017**, *15*, 1083–1094. [CrossRef]
28. Hyde, E.R.; Haarmann, D.P.; Petrosino, J.F.; Lynne, A.M.; Bucheli, S.R. Initial insights into bacterial succession during human decomposition. *Int. J. Legal. Med.* **2014**, *129*, 661–671. [CrossRef]
29. Philips, A.; Stolarek, I.; Kuczkowska, B.; Juras, A.; Handschuh, L.; Piontek, J.; Kozlowski, P.; Figlerowicz, M. Comprehensive analysis of microorganisms accompanying human archaeological remains. *Gigascience* **2017**, *6*, 1–13. [CrossRef]
30. Nakaew, N.; Pathom-aree, W.; Lumyong, S. Generic diversity of rare actinomycetes from Thai cave soils and their possible use as new bioactive compounds. *Actinomycetologica* **2009**, *23*, 21–26. [CrossRef]
31. De Leo, F.; Iero, A.; Zammit, G.; Urzı, C. Chemoorganotrophic bacteria isolated from biodeteriorated surfaces in cave and catacombs. *Int. J. Speleol.* **2012**, *41*, 125–136. [CrossRef]
32. Diaz-Herraiz, M.; Jurado, V.; Cuezva, S.; Laiz, L.; Pallecchi, P.; Tiano, P.; Sanchez-Moral, S.; Saiz-Jimenez, C. The actinobacterial colonization of Etruscan paintings. *Sci. Rep.* **2013**, *3*, 1440. [CrossRef]
33. Cirigliano, A.; Tomassetti, M.C.; Di Pietro, M.; Mura, F.; Maneschi, M.L.; Gentili, M.D.; Cardazzo, B.; Arrighi, C.; Mazzoni, C.; Negri, R.; et al. Calcite moonmilk of microbial origin in the Etruscan Tomba degli Scudi in Tarquinia, Italy. *Sci. Rep.* **2018**, *8*, 1–10. [CrossRef]
34. Braun, B.; Richert, I.; Szewzyk, U. Detection of iron-depositing Pedomicrobium species in native biofilms from the Odertal National Park by a new, specific FISH probe. *J. Microbiol. Methods* **2009**, *79*, 37–43. [CrossRef]
35. Zhu, Y.; Wang, H.; Li, X.; Hu, C.; Yang, M.; Qu, J. Characterization of biofilm and corrosion of cast iron pipes in drinking water distribution system with UV/Cl2 disinfection. *Water Res.* **2014**, *60*, 174–181. [PubMed]
36. Lee, J.S.; Little, B.J. A mechanistic approach to understanding microbiologically influenced corrosion by metal-depositing bacteria. *Corrosion* **2019**, *75*, 6–11. [CrossRef]
37. Männistö, M.K.; Häggblom, M.M. Characterization of psychrotolerant heterotrophic bacteria from Finnish Lapland. *Syst. Appl. Microbiol.* **2006**, *29*, 229–243. [CrossRef] [PubMed]
38. Kim, M.S.; Baik, K.S.; Park, S.C.; Rhee, M.S.; Oh, H.M.; Seong, C.N. Roseomonas frigidaquae sp. nov., isolated from a water-cooling system. *Int. J. Syst. Evol. Microbiol.* **2009**, *59*, 1630–1634. [CrossRef] [PubMed]
39. Li, X.; Wang, H.; Zhang, Y.; Hu, C.; Yang, M. Characterization of the bacterial communities and iron corrosion scales in drinking groundwater distribution systems with chlorine/chloramine. *Int. Biodeterior. Biodegrad.* **2014**, *96*, 71–79. [CrossRef]
40. Urzì, C.; De Leo, F.; Bruno, L.; Pangallo, D.; Krakova, L. New species description, biomineralization processes and biocleaning applications of Roman Catacombs-living bacteria. *Conservation Subterr. Cult. Herit.* **2014**, 65–72.
41. Heyrman, J.; Logan, N.A.; Busse, H.-J.; Balcaen, A.; Lebbe, L.; Rodriguez-Diaz, M.; Swings, J.; De Vos, P. Virgibacillus carmonensis sp. nov., Virgibacillus necropolis sp. nov. and Virgibacillus picturae sp. nov., three novel species isolated from deteriorated mural paintings, transfer of the species of the genus Salibacillus to Virgibacillus, as Virgibacillus marismortui comb. nov. and Virgibacillus salexigens comb. nov., and emended description of the genus Virgibacillus. *Int. J. Syst. Evol. Microbiol.* **2003**, *53*, 501–511.
42. Kour, D.; Rana, K.L.; Kaur, T.; Singh, B.; Chauhan, V.S.; Kumar, A.; Rastegari, A.A.; Yadav, N.; Yadav, A.N.; Gupta, V.K. Extremophiles for hydrolytic enzymes productions: Biodiversity and potential biotechnological applications. *Bioprocess. Biomol. Prod.* **2019**, 321–372.
43. Batra, R.; Boekhout, T.; Guého, E.; Cabañes, F.J.; Dawson, T.L., Jr.; Gupta, A.K. Malassezia baillon, emerging clinical yeasts. *FEMS Yeast Res.* **2005**, *5*, 1101–1113. [CrossRef]
44. Chouvenc, T.; Efstathion, C.A.; Elliott, M.L.; Su, N.-Y. Resource competition between two fungal parasites in subterranean termites. *Naturwissenschaften* **2012**, *99*, 949–958. [CrossRef]
45. Martínez-Ramírez, J.A.; Strien, J.; Sanft, J.; Mall, G.; Walther, G.; Peters, F.T. Studies on drug metabolism by fungi colonizing decomposing human cadavers. Part I: DNA sequence-based identification of fungi isolated from postmortem material. *Anal. Bioanal. Chem.* **2013**, *405*, 8443–8450. [CrossRef]
46. Šimonovičová, A.; Kraková, L.; Pangallo, D.; Majorošová, M.; Piecková, E.; Bodoriková, S.; Dörnhoferová, M. Fungi on mummified human remains and in the indoor air in the Kuffner family crypt in Sládkovičovo (Slovakia). *Int. Biodeterior. Biodegrad.* **2015**, *99*, 157–164. [CrossRef]

47. Sterflinger, K. Fungi: Their role in deterioration of cultural heritage. *Fungal Biol. Rev.* **2010**, *24*, 47–55. [CrossRef]
48. Strobel, G.; Ford, E.; Worapong, J.; Harper, J.K.; Arif, A.M.; Grant, D.M.; Fung, P.C.W.; Chau, R.M.W. Isopestacin, an isobenzofuranone from Pestalotiopsis microspora, possessing antifungal and antioxidant activities. *Phytochemistry* **2002**, *60*, 179–183. [CrossRef]
49. Harper, J.K.; Arif, A.M.; Ford, E.J.; Strobel, G.A.; Porco, J.A., Jr.; Tomer, D.P.; Oneill, K.L.; Heider, E.M.; Grant, D.M. Pestacin: A 1,3-dihydro isobenzofuran from Pestalotiopsis microspora possessing antioxidant and antimycotic activities. *Tetrahedron* **2003**, *59*, 2471–2476. [CrossRef]
50. Li, E.; Jiang, L.; Guo, L.; Zhang, H.; Che, Y. Pestalachlorides A–C, antifungal metabolites from the plant endophytic fungus Pestalotiopsis adusta. *Bioorg. Med. Chem.* **2008**, *16*, 7894–7899. [CrossRef]
51. Piñar, G.; Poyntner, C.; Lopandic, K.; Tafer, H.; Sterflinger, K. Rapid diagnosis of biological colonization in cultural artefacts using the MinION nanopore sequencing technology. *Int. Biodeterior. Biodegrad.* **2020**, *148*, 104908. [CrossRef]
52. Blanchette, R.A. A review of microbial deterioration found in archaeological wood from different environments. *Int. Biodeterior. Biodegrad.* **2000**, *46*, 189–204. [CrossRef]
53. Brebbia, C.A. *Structural Studies, Repairs and Maintenance of Heritage Architecture XI*; Wessex Institute of Technology: Southampton, UK, 2009; ISBN 978-1-84564-196-2.
54. Evidente, A.; Punzo, B.; Andolfi, A.; Cimmino, A.; Melck, D.; Luque, J. Lipophilic phytotoxins produced by Neofusicoccum parvum, a grapevine canker agent. *Phytopathol. Mediterr.* **2010**, *49*, 74–79.
55. Williamson, B.; Tudzynski, B.; Tudzynsk, P.; Van Kan, J.A.L. Botrytis cinerea: The cause of grey mould disease. *Mol. Plant Pathol.* **2007**, *8*, 561–580. [CrossRef] [PubMed]
56. Benito, S. The impact of Torulaspora delbrueckii yeast in winemaking. *Appl. Microbiol. Biotechnol.* **2018**, *102*, 3081–3094. [CrossRef] [PubMed]
57. Zajc, J.; Kogej, T.; Galinski, E.A.; Ramos, J.; Gunde-Cimerman, N. Osmoadaptation strategy of the most halophilic fungus, Wallemia ichthyophaga, growing optimally at salinities above 15% NaCl. *Appl. Environ. Microbiol.* **2014**, *80*, 247–256. [CrossRef] [PubMed]
58. Urbina, H.; Aime, M.C. A closer look at Sporidiobolales: Ubiquitous microbial community members of plant and food biospheres. *Mycologia* **2018**, *110*, 79–92. [CrossRef] [PubMed]
59. Urzı, C.; De Leo, F. Sampling with adhesive tape strips: An easy and rapid method to monitor microbial colonization on monument surfaces. *J. Microbiol. Methods* **2001**, *44*, 1–11. [CrossRef]
60. Hubka, V.; Kolařík, M.; Kubátová, A.; Peterson, S.W. Taxonomic revision of Eurotium and transfer of species to Aspergillus. *Mycologia* **2013**, *105*, 912–937. [CrossRef]
61. Silva-Castro, G.A.; Uad, I.; Gonzalez-Martinez, A.; Rivadeneyra, A.; Gonzalez-Lopez, J.; Rivadeneyra, M.A. Bioprecipitation of calcium carbonate crystals by bacteria isolated from saline environments grown in culture media amended with seawater and real brine. *BioMed Res. Int.* **2015**, *2015*, 816102. [CrossRef]
62. Dhami, N.K.; Mukherjee, A.; Watkin, E.L. Microbial diversity and mineralogical-mechanical properties of calcitic cave speleothems in natural and in vitro biomineralization conditions. *Front. Microbiol.* **2018**, *9*, 40. [CrossRef]
63. Waugh, N. *Corsets and Crinolines*; Routledge: Abingdon-on-Thames, UK, 2017.
64. Arnold, J.; Tiramani, J.; Costigliolo, L.; Passot, S.; Lucas, A.; Pietsch, J. *Patterns of fashion 5: The Content, Cut, Construction and Context of Bodies, Stays, Hoops and Rumps c. 1595–1795*; The School of Historical Dress: London, UK, 2018.
65. Yang, H.X.; Chu, J.M.; Liu, X.S. Natural persistence of the coastal plant Glehnia littoralis along temperate sandy coasts. *Sci. Rep.* **2017**, *7*, 42784. [CrossRef]
66. She, M.L.; Pu, F.T.; Pan, Z.H.; Watson, M.F.; Cannon, J.F.M.; Holmes-Smith, I.; Kljuykov, E.V.; Phillippe, L.R.; Pimenov, M.G. Apiaceae. In *Flora of China Editorial Committee (ED), Flora of China*; Missouri Botanical Garden Press: St. Louis, MO, USA, 2005; Volume 14, pp. 1–205.
67. Sayed-Ahmad, B.; Talou, T.; Saad, Z.; Hijazi, A.; Merah, O. The apiaceae: Ethnomedicinal family as source for industrial uses. *Ind. Crop. Prod.* **2017**, *109*, 661–671. [CrossRef]
68. Beneš, J.; Čulíková, V.; Kosňovská, J.; Frolík, J.; Matiášek, J. New plants at Prague Castle and Hradčany in the early modern period: A history of selected species. *Interdiscip. Archaeol.* **2012**, *3*, 103–114. [CrossRef]

69. Tranberg, A. 10 burial customs in the northern Ostrobothnian region (Finland) from the late medieval period to the 20th Century. Plant remains in graves. In *The Archaeology of Death in Post-Medieval Europe*; De Gruyter: Berlin, Germany, 2015; pp. 189–203.
70. Jarosińska, M.; Nowak, S.; Noryśkiewicz, A.M.; Badura, M. Plant identification and significance in funeral traditions exemplified by pillow filling from a child crypt burial in Byszewo (18th/19th centuries). *Analecta Archaeol. Ressoviensia* **2019**, *14*, 187–197. [CrossRef]

applied sciences

MDPI

Article

# Fungal Community Analysis and Biodeterioration of Waterlogged Wooden Lacquerware from the Nanhai No. 1 Shipwreck

**Yin Jia [1,†], Liuyu Yin [1,†], Fengyu Zhang [1], Mei Wang [1], Mingliang Sun [1], Cuiting Hu [1], Zijun Liu [1], Yue Chen [2], Jie Liu [2] and Jiao Pan [1,*]**

1 Key Laboratory of Molecular Microbiology and Technology of the Ministry of Education, Department of Microbiology, College of Life Sciences, Nankai University, Tianjin 300071, China; 117759@zju.edu.cn (Y.J.); 1510836@mail.nankai.edu.cn (L.Y.); 2120171023@mail.nankai.edu.cn (F.Z.); 1510821@mail.nankai.edu.cn (M.W.); 2120181154@mail.nankai.edu.cn (M.S.); 2120160931@mail.nankai.edu.cn (C.H.); 2120150877@mail.nankai.edu.cn (Z.L.)

2 Conservation and Restoration Institute, Chinese Academy of Cultural Heritage, Beijing 100029, China; chenyue@cach.org.cn (Y.C.); liujie@cach.org.cn (J.L.)

* Correspondence: panjiaonk@nankai.edu.cn; Tel.: +86-022-2350-5961

† The two authors have the equal contribution to this work.

Received: 19 April 2020; Accepted: 28 May 2020; Published: 29 May 2020

**Abstract:** To avoid the lacquerware of the Nanhai No. 1 shipwreck from being corroded by microorganisms and to improve the knowledge on microbial ecology of the wood lacquers, we conducted a series of tests on the two water samples storing the lacquerware and colonies on the surface of the lacquerware. The high-throughput sequencing detected dominant fungal communities. After that, the fungal strains were isolated and then identified by amplification of ITS- 18S rRNA. Then the activity of ligninolytic and cellulolytic enzymes was detected on potato dextrose agar (PDA) plates with 0.04% (*v/v*) guaiacol and carboxymethyl cellulose (CMC) agar plates. Finally, we tested the biocide susceptibility of these fungi. *Penicillium chrysogenum* (NK-NH3) and *Fusarium solani* (NK- NH1) were the dominant fungi in the sample collected in April 2016 and June 2017. What is more, both showed activity of ligninolytic and cellulolytic enzymes. Four biocidal products (Preventol® D7, P91, BIT 20N, and Euxyl® K100) inhibited the growth of the fungal species in vitro effectively. In further research, the microbial community and environmental parameters in the museum should be monitored to assess the changes in the community and to detect potential microbial outbreaks.

**Keywords:** Nanhai No.1 shipwreck; wood lacquers; fungi; high-throughput sequencing; carboxymethyl cellulose activity; biocides

## 1. Introduction

In traditional Chinese art, lacquerware refers to a vessel made of bamboo, wood or metal, coated with a hard resin layer on the surface. Arising in the Neolithic Age, lacquering was initially regarded as a form of waterproof protection and later developed into an important decorative art. The application of multiple layers on lacquerware can protect the core material [1]. Various types of items could be lacquered, such as furniture and many other household objects, cosmetics boxes, domestic ware, musical instruments, food-serving implements, even coffins. The cores of the lacquerware were made from various materials—for instance, the core of a lacquered tray was made of wood and fabric. Sometimes other materials such as iron, pottery, rattan, leather or the like were also used as lacquer cores. Chinese lacquerware continues to develop and has a far-reaching impact on Japanese and Korean art as well as Indian culture.

Large quantities of lacquerware have been excavated at different sites in China. For example, there were thousands of pieces of wood lacquerware found in Nanhai No.1 shipwreck in Guangdong province. Unearthed wood lacquers were usually preserved in water to prevent degradation caused by the microbes existing in the air. Soaking in the water for a long time however can reinforce the damage of wood lacquers, including the deterioration, lacquer layer loss, and microbial damage [2]. Plenty of studies have uncovered the fact that many kinds of degradation and microbial successions may occur in waterlogged wooded objects. In waterlogged sites, tunneling bacteria and erosion bacteria have been proved to be able to degrade wooden items in near anaerobic environments [3,4]. In fact, some fungi also can degrade these waterlogged lignocellulosic materials as a part of the carbon cycle [5]. Typical fungi inhabitants of waterlogged wooden objects are the species belonging to the group of Ascomycetes and Deuteromycota such as species in genus *Cladosporium*, *Acremonium*, *Fusarium*, and *Chaetomium*, among others [6,7].

A large percentage of the world's cultural heritage artifacts have been severely and irreversibly damaged and degraded by microorganisms [8,9]. In July 2012 to August 2013, the Chengdu Institute of Cultural Relics and the Jingzhou Cultural Relics Protection Center formed a joint archaeological team who excavated four pits in the wooden tombs of the Western Han Dynasty in Chengdu, and uncovered hundreds of precious bamboo and wood lacquers, including some bamboo cases. In order to avoid the damage caused by microbial degradations during the period of water conservation, the research team collected the microbial samples from the unearthed bamboo cases. *Penicillium* sp., *Aspergillus* sp., and *Trichoderma* sp. were identified by isolation and purification, DNA-ITS, and other biological methods [10]. Therefore, the protection against mildew should be carried out in time [11]. These studies on microbial degradations of cultural heritage artifacts revealed the occurrence of fungi. Moreover, the Nanhai No.1 shipwreck is suffering an outbreak of a white filamentous fungus. To this end, the detection and identification of biodegradative microorganisms are an essential means for the control and prevention of microbial biodeterioration, since only if we understand the biological and physiological properties of organisms can we have the capacity to control the potential harm they cause [12].

The aims of this study were (1): to isolate and identify fungi colonizing the surfaces of the wood lacquers excavated from Nanhai No.1 shipwreck; (2) to detect fungi dwelling in the water storing the lacquerware; (3) to assess the ability of the fungi recovered from the surfaces of the wood lacquers to degrade cellulose and lignin; (4) to evaluate the susceptibility of the fungi recovered from the surfaces of the wood lacquers to four different biocides.

## 2. Materials and Methods

### 2.1. Sample Collection

This research was carried out in June 2017 in the Marine Silk Road Museum of Guangdong. This museum is the world's first underwater archaeology theme museum that shows the excavation, protection, display, and research of Nanhai No.1 shipwreck. It displays dynamic demonstrations of the excavation of the underwater archeology site. The museum is divided into two theme areas: exhibit area of Nanhai No.1 and exhibit area of Maritime Silk Road. Nanhai No.1 shipwreck was wholly salvaged from the seabed more than 20 m deep and moved to the museum in 2007 [13]. In November 2013, the excavation of the Nanhai No.1 shipwreck began and a number of archaeological objects including lacquerware, ornaments, bronze mirrors, wood combs, and even seeds of plants were extracted from the wreck. The map of Nanhai No.1 can be seen in our former study about the microbial community of waterlogged archaeological wood from Nanhai No.1 shipwreck. In this research, the overall microbiological analyses were to identify the fungi present on the wood lacquers extracted from Nanhai No.1 and to reveal fungal communities thriving in water samples of the wood lacquers. So far, the wood lacquers have been stored in a 4 °C fridge and contained with moist gauze to preserve moisture with deionized water as the storage water of the lacquerware.

Lacquerware surface sample collection using non-invasive sampling with PDA medium (Figure 1). The lacquerware surfaces were sampled with sterile swabs and then inoculated in petri dishes with PDA medium in a sterile operation. Three distinct fungal plaque areas on each lacquerware surface were selected for inoculation. Afterwards, these petri dishes were incubated in the laboratory and pure colonies were isolated for identification and investigation. In addition, 50 mL sterile centrifuge tubes were used to collect water samples with one water sample for each lacquerware. Sample NHI935 was collected from lacquerware 2016NHIT0501⑩935 and sample NHI58 was collected from lacquerware 2015NHIT0202③:58. The two water samples were conducted with high-throughput sequencing and purification.

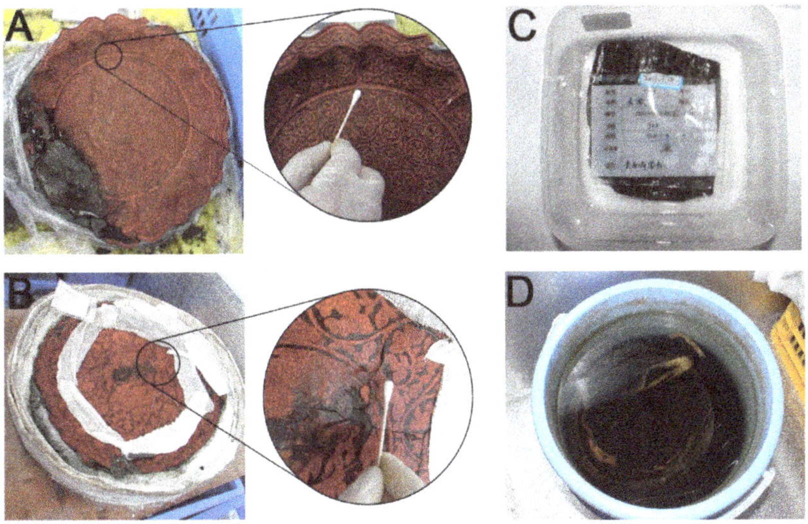


**Figure 1.** Sampling areas for microbiological analyses on the wood lacquers surfaces: (**A**) lacquer NHI566; (**B**) lacquer NHI883; (**C**) water sample NHI935; (**D**) water sample NHI58.

### *2.2. Microbial Isolation, Cultivation, and Identification*

According to the common cultural laboratory practices, the PDA medium plates containing lacquerware surface samples were incubated at 28 °C for 5–30 days. Fungi were isolated and pure cultures obtained after subsampling. The DNA of the pure strains isolated from the surface of lacquerwares were extracted by the hexadecyltrimethylammonium bromide (CTAB) method [14,15]. Two primers, ITS1 (5′-TCCGTAGGTGAACCTGCGG-3′) and ITS4 (5′-CCTCCGCTTATTGATATGC-3′), were used to amplify the fungal ITS region [16]. The PCR reaction consisted of 50 μL mixtures, including genomic DNA (2 μL), 10× Reaction Buffer (5 μL), 2.5 mM deoxy-ribonucleoside triphosphate (dNTP) mix (4 μL), 10 μM forward primer (2 μL), 10 μM reverse primer (2 μL), 5 U/μL Transtaq-T DNA polymerase (0.5 μL) (TransGen Biotech, China), and dd $H_2O$ to 50 μL. The thermocycling program was set as follows: 95 °C for 3 min, followed by 95 °C for 32 cycles of 30 s, 54 °C for 30 s, 72 °C for 20 s, 72 °C for 5 min as a final extension [17]. After detecting the PCR products with 1% agarose gel electrophoresis, an AxyPrep PCR Clean Up Kit (Axygen, California, USA) was used for purification.

The purified PCR products were sequenced by GENEWIZ (Beijing, China). The sequences were used as queries in BLAST searches [18]. The sequences retrieved were deposited in the NCBI GenBank under the accession number SRP219493 for fungi.

### 2.3. High-Throughput Sequencing Analysis

Water samples of lacquers were collected using 50 mL centrifuge tubes. Total DNA of the water samples was detected by the MoBio PowerSoil® DNA Isolation Kit (Mo Bio Laboratories, Carlsbad, CA, USA).

Fungal communities were studied by amplifying internal transcribed spacer 1 (ITS1) fragments using primers ITS5-1737F (5′-GGAAGTAAAAGTCGTAACAAGG-3′)/ITS2-2043R (5′-GCTGCGTTCTT CATCGATGC-3′) combined with adapter sequences and barcode sequences. Amplifications were carried out in a 50 μL mixture including 25 μL of Master Mix (2X), a 0.5 μM final concentration of the forward and reverse primers, 10 ng of template DNA and nuclease-free water to 50 μL. The PCR conditions were 98 °C for 1 min, followed by 30 cycles of 10 s at 98 °C, 30 s at 55 °C for ITS region amplification, and 30 s at 72 °C, with a final extension of 5 min at 72 °C.

After extraction, the total DNA was operated in Novogene (Beijing, China) including amplification by amplicons to analyze the fungal community. The construction of a library used the TruSeq® DNA PCR-Free Sample Preparation Kit (Illumina, San Diego, CA, USA) [13]. The library was sequenced on an Illumina Hiseq2500 PE250 platform. The sequences were processed and analyzed by Novogene (Beijing, China).

### 2.4. Activity of Carboxymethyl Cellulose (CMCase) Production by Fungi

Fungal spores of pure isolates strains illustrated in Section 2.2 at a concentration of $2 \times 10^7$ CFU/mL were added to the liquid fermentation medium. The liquid fermentation medium formulation consisted of 1.0% CMC-Na, 1.0% peptone, 0.5% beef extract, 0.2% $KH_2PO_4$, 0.05% $MgSO_4{\cdot}7H_2O$, 0.2% $(NH_4)_2SO_4$ and 1 L of tap water. Four single variables were set for time, temperature, pH, and $FeCl_3$ concentration. Fungi were cultivated under single variable conditions as above.

For measuring enzyme activity, 2 mL of fermentation brothwas centrifuged at 12,000 r/min, 4 °C for 5 min. Then 0.5 mL of the supernatant was incubated with 0.5 mL of 1% CMC-Na prepared with 0.05 mol/L citric acid buffer (pH 4.8) at 50 °C for 30 min. The reaction was terminated by addition of 3 mL of 3,5-dinitrosalicylic acid reagent. All of the assay tubes were placed in boiling water for 5 min. The absorbance of the contents in the tubes was measured at 530 nm. To the blank control group was added 0.5 mL of the fermentation supernatant after boiling to make sure it would not react with the reagents.

The unit of CMCase (IU) was defined as the amount of enzyme required to produce 1 μmol of glucose per minute and per mL of crude enzyme solution under reaction conditions. The activity of CMCase calculation formula is $X = (A \times V_T \times 1000)/(M \times V \times T)$. X is the CMCase activity ($IU{\cdot}mL^{-1}$) in the sample; A is the glucose content ($mg{\cdot}mL^{-1}$) calculated according to the standard curve; $V_T$ is the total volume of the reaction solution (mL); M is the molar mass of glucose (180.2 $g{\cdot}mol^{-1}$); V is the volume of the crude enzyme solution (mL) added in the reaction; T is the reaction time (min).

The total amount of protein in the supernatant was measured by the Enhanced Lowry Assay Kit (Leagene, Beijing). The specific activity of CMCase is calculated and defined as the unit of enzyme activity per mg of protein (IU/mg) [19].

### 2.5. Biocide Susceptibility of Fungal Strains

The biocide susceptibility test used the disk diffusion method [15]. Four biocidal products were evaluated, Preventol® D 7, P 91, BIT 20 N and Euxyl® K100 [13]. In brief, the tested fungi were dipped with a sterile cotton swab and painted on PDA medium plates. Afterwards five paper discs (radius of 3 mm) were put on the medium and then four were loaded with 30 μL of the corresponding 0.5% biocides, one was soaked with sterile water as a control. After incubating at 28 °C for 4 days, the inhibition zones around the discs were measured and statistical analysis of these measures was conducted.

## 3. Results

### 3.1. Fungi Communities Analyzed by Classical Cultivation Methods and High-Throughput Sequencing

Fungal strains from the surface of two lacquers were isolated. As shown in Table 1, the cultivable fungal strains were identified by molecular identification. Two isolates were obtained from the surface of lacquer NHI556, and three isolates from the surface of lacquer NHI883. These isolates were most closely related to *P. chrysogenum* and *F. solani*.

**Table 1.** Molecular identification of pure strains isolated from lacquerware surfaces.

| Taxonomy | Closet Strain | Similarity (%) | Accession Number (GenBank Database) |
|---|---|---|---|
| NHI566 | | | |
| Isolate 1 | *Penicillium chrysogenum* | 99% | KF986423 |
| Isolate 2 | *Fusarium solani* | 99% | KX349467 |
| NHI883 | | | |
| Isolate 3 | *Penicillium chrysogenum* | 99% | KF986423 |
| Isolate 4 | *Fusarium solani* | 99% | KX349467 |
| Isolate 5 | *Colletotrichum dematium* | 99% | JN998109 |

The fungal community composition of two water samples (NHI935 and NHI58) was analyzed. High-quality reads numbering 171,814 were obtained after filtering. There were five fungal phyla in the two samples (Figure 2A). Ascomycota was the most abundant phylum, accounting for 97.50% in sample NHI935 and 95.33% in sample NHI58. Four additional fungal phyla including Glomeromycota, Basidiomycota, Zygomycota, and Chytridiomycota accounted for a small proportion. In the genus level, high-throughput sequencing revealed the presence of a total of 48 genera in the two water samples. The main fungal genera are summarized in Figure 2B, with "Other" consistently being the largest category. *Fusarium* belongs to the phylum Ascomycota and was the most dominant genus in the samples, where it accounted for 34.4% and 7.6% of the community. *Cadophora* was the second largest genus in NHI935, accounting for 16.7%, but only 0.8% in NHI58. *Rhizopus*, *Pestalotiopsis*, *Cryptococcus*, *Alternaria*, *Aspergillus*, *Talaromyces*, and *Exophiala* were detected in all two samples with average abundance 0.13%, 1.23%, 0.58%, 1.21%, 0.05%, 0.71%, and 0.69%, respectively.

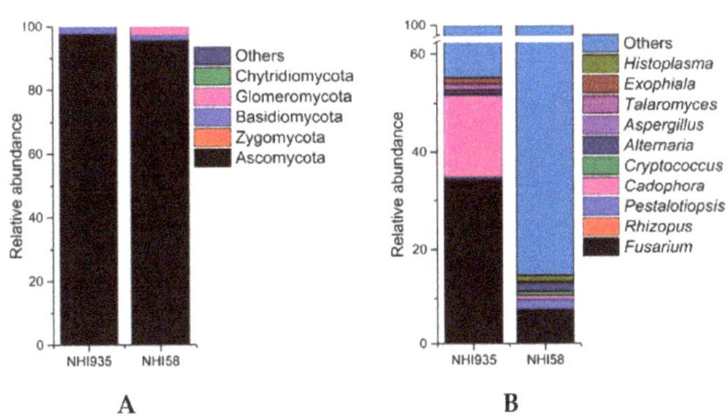


**Figure 2.** Fungal community at phyla (**A**) and genera (**B**) level detected in the two water samples.

### 3.2. Isolation of Dominant Fungi

The plate pictures and photomicrographs of five fungi isolated from the surface of the lacquerware are presented in Figure 3. Meanwhile, as was shown in Table 1, the strain 'isolate 1' from the sample NHI566 and the 'isolate 3' from the sample NHI883 were the same species and the strain *P. chrysogenum* was named NK-NH3(MH392741). The strain 'isolate 2' from the sample NHI566 and the 'isolate 4' from the sample NHI883 were the same species and the same as *F. solani* NK- NH1(KY410238) in the former study on Nanhai No.1 shipwreck in 2015 [13].

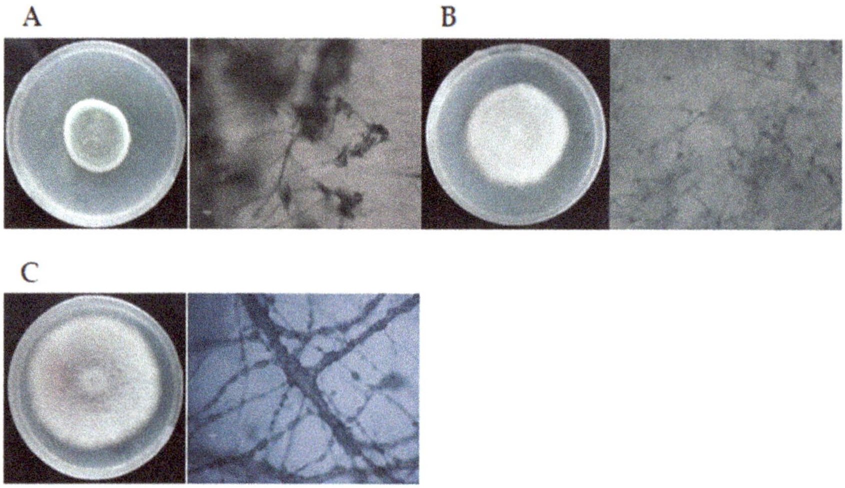


**Figure 3.** The plate pictures (left) and photomicrographs (right) of five fungi isolated and purified from the surface of the lacquerware. (**A**) Isolate 1 and 3: *P. chrysogenum*, the scale bar is 10 μm; (**B**) Isolate 2 and 4: *F. solani*, the scale bar is 5 μm; (**C**) Isolate 5: *C. dematium*, the scale bar is 10 μm.

### 3.3. Activity of CMCase Production by F. Solani NK-NH1 and P. Chrysogenum NK-NH3

During the protection process, the pH value and iron ion concentration of the lacquerware from ocean change, resulting in changes in the activity of CMCase. In the CMCase activity test, *P. chrysogenum* NK-NH3 reached a peak (0.5272 IU/mL and 0.3481 IU/mg, respectively) when the fermentation medium pH was 3.0, the $FeCl_3$ concentration was 0.2 g/L, cultured under 32 °C for 5d. While, *F. solani* NK-NH1 reached the highest CMCase activity and specific enzyme activity (0.11 IU/mL and 0.096 IU/mg, respectively) when the pH of fermentation medium was 5.4, and cultured under 32 °C for 5d. The effect of $Fe^{3+}$ ion concentration on the CMCase activity is not significant compared with when no ferric ion is added. The results are illustrated in Figure 4 and three replicates were conducted per conditions for each activity test. As is shown in Figure 4, CMCase presents different effects on the two strains. The difference in the enzymatic properties of the two strains is related to the strain and its genetic characteristics—we will conduct further study. If the CMCase of the two fungi are separated and purified, the optimal pH value of these enzymes and other physiological and biochemical characteristics can be further determined.

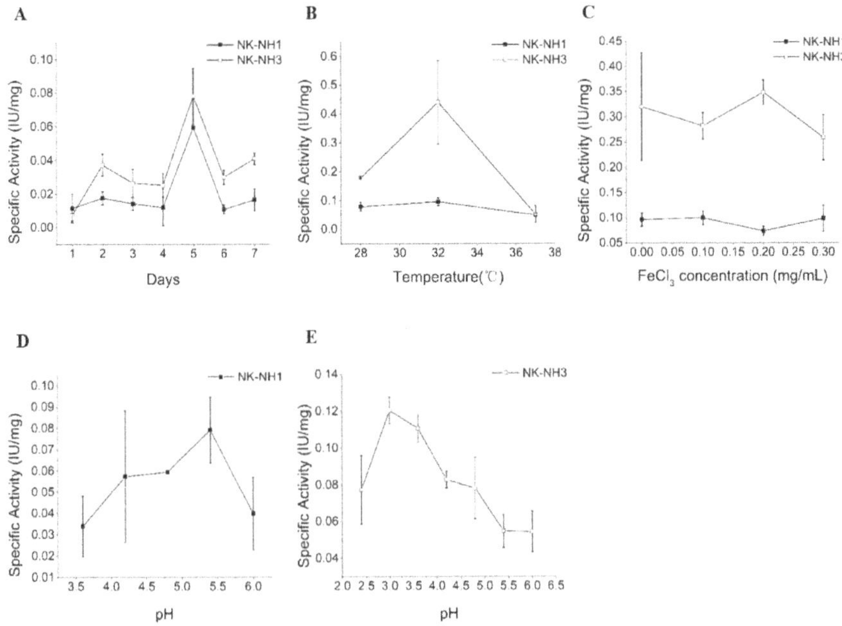


**Figure 4.** Effects of time, temperature, ferric chloride concentration, and pH on the CMCase activity of *F. solani* NK-NH1 and *P. chrysogenum* NK-NH3. (**A**) Trend of CMCase activity of two strains cultured in PDA liquid medium for at 28 °C 1–5 days. (**B**) Trend of CMCase activity of two strains cultured in PDA liquid medium at 28, 32, and 37 °C for 5 days. (**C**) Trend of CMCase activity of two strains cultured in PDA liquid medium (*F. solani* NK-NH1 at pH 5.4 and *P. chrysogenum* NK-NH3 at pH 3.0) containing 0.1, 0.2, and 0.3 mg/mL $FeCl_3$ for five days. (**D**) Trend of CMCase activity of *F. solani* NK-NH1 cultured in PDA liquid medium with pH 3.6–6.0 at 28 °C for five days. (**E**) Trend of CMCase activity of *P. chrysogenum* NK-NH3 cultured in PDA liquid medium at pH 2.4–6.0 at 28 °C for five days.

### *3.4. Efficiency of Biocide Products against Targeted Strains*

Four biocides used in the protection of cultural relics, K100, Preventol® D7, P91, and BIT 20N, were selected to test the susceptibility of strains. *F. solani* NK-NH1 and *P. chrysogenum* NK-NH3 were used for the tests. Four biocides were able to inhibit the fungal growth when applied at 0.5% concentration, which is much lower than the concentration (2%) recommended by manufacturers. In this study, the most effective agent was Preventol® D7, based on isothiazolinone, followed by BIT 20N (isothiazolinone derivative). Biocide products P91 combining bronopol and isothiazolinone had similar efficacy when compared with BIT 20N. Another product K100 based on the combination of benzyl alcohol and isothiazolinone derivative had the lowest efficacy against fungal strains.

Under the treatment of 0.1% and 0.5% concentration bacteriostatic agents, obvious inhibition zones appeared around the filter paper with the bacteriostatic agent added, indicating that all the four bacteriostatic agents had biocide effects on the strains. Under the treatment of 0.1% concentration of bacteriostatic agent, the treatment effects of P91 and K100 were poor, and the treatment differences of the other bacteriostatic agents were not obvious. For strains *F. solani* NK-NH1 and *P. chrysogenum* NK-NH3, under the treatment of 0.5% concentration of bacteriostatic agent, the biocide efficacy of the four bacteriostatic agents could be clearly seen, D7 had the strongest effect, P91 was second, 20N was poor, and K100 was the weakest. The results are shown in Figure 5 and statistical analysis is presented in Figure S1.

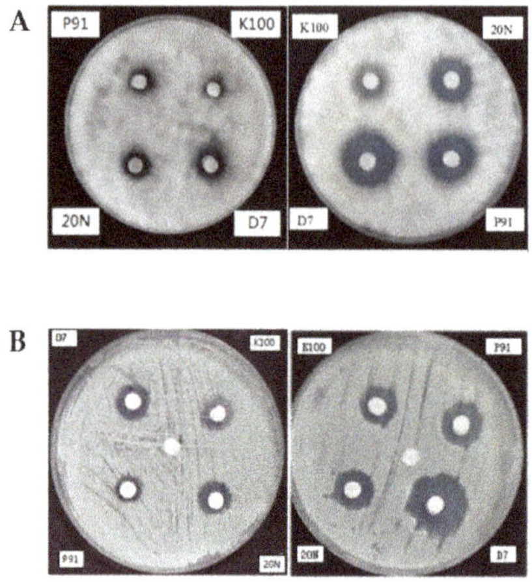


**Figure 5.** The inhibition of fungal growth by four biocides. (**A**) *F. solani* NK-NH1 under the treatment of 0.1%(left) and 0.5%(right) concentration biocides and cultured on PDA plates at 28 °C for 3 days. (**B**) *P. chrysogenum* NK-NH3 under the treatment of 0.1% (left) and 0.5% (right) concentration biocides and cultured on PDA plates at 28 °C for 3 days.

## 4. Discussion

### *4.1. Fungal Communities*

Our previously unpublished results from 2016 showed that *Penicillium* was the main fungus genus in the water sample of lacquer, while results from 2017 in the present article showed that *Fusarium* was the dominant one (Supplementary Tables S1–S3). Moreover, fungi of these two genera, namely *F. solani* NK-NH1 and *P. chrysogenum* NK-NH3, were successfully isolated from the lacquerware surfaces.

This is the first time that *Fusarium* species have been reported to lead a putative microbial deterioration of wooden lacquerware. Another famous case concerning *Fusarium* in the biodeterioration of cultural heritage is the Paleolithic painting of the Lascaux Cave in France [20]. The ground and walls were covered by the dense white hyphae of *F. solani*. They not only destroyed the aesthetics of the paintings, but also changed the material structure of them [21]. In addition, *Fusarium* sp. was isolated from documentary heritage [22]. *Fusarium* species are mainly plant pathogens that cause root and stem rots. They live in the soil after the plant has been harvested or decayed [20]. Some members of the genus produce mycotoxins that can affect human and animal health andare often found in wounds [23]. The genus also can be found in food and indoor environments [24]. Thus in lacquerware surfaces and water *Fusarium* might have been brought from the outside environment by visitors or conservation-restorers.

It is worthy of mention that besides the *Fusarium*, the genus *Cadophora* should not be ignored, which accounts for 16.7% in sample NHI935. Most *Cadophora* members were primarily isolated from plants and soil [25]. For example, isolates related to *C. malorum* cause wood decay of kiwifruit [26]. *Cadophora* species, such as *C. malorum*, *C. luteo-olivacea*, and *C. fastigiata* have also been isolated in extreme environments in Antarctica, causing soft decay of wood huts [25,27,28]. Thus this genus may also be responsible for the deterioration of wooden lacquerware.

Another fungus, *P. chrysogenum*, is known to be a deteriorative agent and was both isolated from ancient documents and the air of indoor environments [12,29,30]. In addition, sequences of the cellulase and hemicellulase producing strain, *P. chrysogenum*, were detected. Therefore, this fungus is a threat to the wooden lacquerware.

### 4.2. Degradation of Lignocellulose by Fungi

Lignocellulose consists of lignin, carbohydrates, pectin, proteins, ash, salts, and minerals. Lignocellulose is highly difficult to degrade by microorganisms. Fungi are mainly responsible for lignocellulose degradation because they have two types of efficient enzymatic systems [31]. A hydrolytic system to degrade polysaccharide by producing hydrolases and an oxidative and extracellular ligninolytic system to open phenyl rings and degrade lignin.

*Fusarium* is an important type of parasitic pathogenic fungi. Cell wall degrading enzymes (CWDEs) are one of the main pathogenic ways by which *Fusarium* invades and infects the host [32]. According to J.Chen, CWDEs secreted by *Fusarium* mainly include pectinase, cellulase, hemicellulase, protease, amylase, and phospholipase [33]. *Penicillium* is known to be able to secrete lignocellulolytic enzymes, such as hemicellulase, cellulases, manganese peroxidase, lignin peroxidase, and laccase [34]. Therefore, *Fusarium* and *Penicillium* may be responsible for the microbial degradation of lacquerware.

According to the CMCase activity test, notably, the activity of CMCase increased to some extent under acidity and iron ions-containing conditions. Studies have shown that marine effluent wood relics usually deposit ferro-iron compounds, which are easily oxidized in effluent, causing degradation of organic matter [35]. Since the wood lacquers were effluent in the ocean, strains isolated from their surface may be adaptable to this environment. Therefore, attention must be paid to control the adverse effects of iron sulfide on wood lacquers for subsequent protection. Considering the further protection of the lacquer, it should undergo de-ironing, desalination, reinforcement, and other treatments. Lacquer serves in marine waterlogged cultural relics, it has a great degree of a problem with acidification. Therefore, during the protection process, the pH value and iron ion concentration in the wood will change. We need to examine the enzyme activity of CMCase at different pH values and iron ion concentrations of these two main fungi. In addition, the storage temperature of the lacquer is also an important factor to be considered in further protection. Therefore, we tested the enzyme activity of CMCase of two fungi at different temperatures.

### 4.3. Biocide Discussion

The results of biocide products provide preliminary knowledge about the susceptibility of lacquerware fungi to four commercial biocides. According to these data, one may think that a biocide application based on an active compound like isothiazolinone could be a useful method to control fungal biodeterioration, and commercial biocide products based on isothiazolinone have been widely used on paper objects and stone cultural heritage monuments [36–39]. However, there are few studies on their long-term efficacy, and the potential recolonization after biocide application has not been monitored [36]. In the worst case, microbial communities may develop resistance to these biocides, and might become even more harmful to the objects [40,41]. In addition, under the selective pressure of biocides, microorganisms previously absent may replace the original microbial communities, leading to new microbial outbreaks [36]. The outbreak of the fungi *F. solani* species complex of Lascaux Cave in France is a precedent. Biocides, such as benzalkonium chloride (BC), were used to control this outbreak between 2001 and 2004 [21]. Four months after the initial application, some black stains progressively invaded the cave, marking a new outbreak. Besides, years of BC application have selected resistance populations, including a mix of *Ralstonia* and *Pseudomonas*, as well as free-living amoebae [20,21,42]. As suggested by Martin-Sanchez et al., it is necessary to weigh the pros and cons of using biocides in each specific environment. Furthermore, any biocide treatment should be carefully designed even if it may be difficult, which includes not also laboratory assays like this study, but also field assays in the real environment [21].

We believe that the choice of biocides is also very important. BC has been used as an antifungal agent against fungi isolated from cultural heritage. The biological activity of BC is ascribed to its quaternary ammonium group [43]. However, microorganisms tend to develop resistance to quaternary ammonium compounds (QACs), resulting in microorganisms colonizing surfaces treated with QACs for a long period [44]. It has been suggested that use of nitrogen-containing biocides should be avoided because they enable several groups of micro- and macro-organisms to use it as a nitrogen source favoring recolonization [45]. The main active ingredients of the biocides we used is isothiazolinone, which is relatively environmentally friendly and efficient. A study isolated some fungi species from the stuccos of the vault of a religious building in Torino (NW-Italy), and evaluated the sensitivity of the fungi species to widely used biocidal products [46]. As a result BC, in the low range of recommended concentrations (0.25%) was demonstrated ineffective against most of the tested fungal strains, while isothiazolinone-containing biocides displayed the best effectiveness against all the strains. Based on the above considerations and our experimental results, we focused on the effect of isothiazolinone-containing biocides. However, its effects on wooden lacquerware still requires evaluation.

In the museum, the fungal contamination of lacquerware is low. To avoid breaking the relatively stable current balance, commercial biocides should not be used at the present stage. For the conservation of these wood lacquers, they should be preserved at low temperature to control fungal deterioration. Moreover, the stored water could be replaced with sterile water to avoid fungal colonization. Alternatively, continuous monitoring of the fungal community is also necessary. Once there is a significant fungal outbreak, physical, not chemical cleaning methods should be considered first, such as careful mechanical removal of fungal biomass and so on.

## 5. Conclusions

We studied the potential fungal biodeterioration of waterlogged wooden lacquerware from the Nanhai No. 1 shipwreck by conventional cultivation methods and modern molecular biotechnology. The high-throughput sequencing of the water samples storing the lacquerware revealed that *Fusarium* was the dominant fungus in 2017. This is the first time that *Fusarium* species have been reported to lead a putative microbial deterioration of wooden lacquerware. Considering the characteristics of *Fusarium*, we suggest that *Fusarium* might have been brought from the outside environment by visitors or conservation-restorers. Even though the corrosion and damage of the lacquerware was not very serious at a relatively stable state, the potential biodeterioration was continuous. The fungal isolates, *F. solani* NK-NH1 and *P. chrysogenum* NK-NH3, may not be directly responsible for the potential biodeterioration, while the CMCase activity test showed that they could be the potential deteriorative agents of the wood and must be regarded as a threat to the lacquerware. The results also indicated that attention must be paid to control the adverse effects of iron sulfide on wooden lacquerware in subsequent protection. The biocide susceptibility assay indicated that isothiazolinones can inhibit fungal isolate growth effectively. However, it is not recommended to use these biocides immediately without considering the actual protection of the lacquerware. In future research, the microbial community and environmental parameters in the museum should be monitored to assess the changes in the community and to detect potential microbial outbreaks at any time.

**Supplementary Materials:** The following are available online at http://www.mdpi.com/2076-3417/10/11/3797/s1, Table S1: Fungal clone library results of NHI58 (2016.4); Table S2: High throughput sequencing results of NHI58 (2017.6); Table S3: High throughput sequencing results of NHI935 (2017.6).

**Author Contributions:** Conceptualization, J.P., Y.J., and L.Y.; Methodology, Y.J., and L.Y.; Validation, Y.J., L.Y., and M.W.; Formal Analysis, Y.J. and L.Y.; Investigation, Y.J., L.Y., and M.W.; Resources, Y.C. and J.L.; Writing—Original Draft Preparation, Y.J. and L.Y.; Writing—Review and Editing, J.P., Y.J., L.Y., M.W., Z.L., F.Z., M.S., and C.H.; Visualization, Y.J. and L.Y.; Supervision, J.P.; Project Administration, J.P.; Funding Acquisition, J.P. All authors have read and agreed to the published version of the manuscript.

**Funding:** This research was funded by the Natural Science Foundation of Tianjin (19JCZDJC33700), the Fundamental Research Funds for the Central Public Research Institutes (2016-JBKY-19 and 2018-JBKY-01).

**Acknowledgments:** We thank other members of our lab who provided valuable and constructive suggestions.

**Conflicts of Interest:** The authors declare no conflict of interest.

## References


1. You, C.; Han, Y.; Wei, Q. Brief Analysis on the Developmental and Creative Approach of Large-Scale Lacquerware Crafts with Ethnic Characteristics. *Art Des. Rev.* **2019**, *7*, 115–123. [CrossRef]
2. Fang, B.; Wu, S. History, status and future of the conservation and restoration of waterlogged bamboo, wood and lacquer. *Sci. Conserv. Archaeol.* **2008**, *20*, 122–130.
3. Blanchette, R.A. A review of microbial deterioration found in archaeological wood from different environments. *Int. Biodeterior. Biodegrad.* **2000**, *46*, 189–204. [CrossRef]
4. Carpetti, C.; Macchioni, N.; Pizzo, B.; Galotta, G.; Giachi, G.; Giampaola, D. The characterization of waterlogged archaeological wood: The three roman ships found in Naples (Italy). *Archaeometry* **2008**, *50*, 855–876.
5. Gjelstrup Björdal, C. Microbial degradation of waterlogged archaeological wood. *J. Cult. Herit.* **2012**, *13*, S118–S122. [CrossRef]
6. Bucher, V.V.C.; Pointing, S.B.; Hyde, K.D.; Reddy, C.A. Production of Wood Decay Enzymes, Loss of Mass, and Lignin Solubilization in Wood by Diverse Tropical Freshwater Fungi. *Microb. Ecol.* **2004**, *48*, 331–337. [CrossRef]
7. Pournou, A.; Bogomolova, E. Fungal colonization on excavated prehistoric wood: Implications for in-situ display. *Int. Biodeterior. Biodegrad.* **2009**, *63*, 371–378. [CrossRef]
8. Sterflinger, K.; Pinzari, F. The revenge of time: Fungal deterioration of cultural heritage with particular reference to books, paper and parchment. *Environ. Microbiol.* **2012**, *14*, 559–566. [CrossRef]
9. Saarela, M.; Alakomi, H.L.; Suihko, M.L.; Maunuksela, L.; Raaska, L.; Mattila-Sandholm, T. Heterotrophic microorganisms in air and biofilm samples from Roman catacombs, with special emphasis on actinobacteria and fungi. *Int. Biodeterior. Biodegrad.* **2004**, *54*, 27–37. [CrossRef]
10. Xiao, L.; Tang, H.; Yang, T. A preliminary study on microbial degradations during water conservation of unearthed bamboo cases. *World Antiq.* **2014**, *3*, 77–80.
11. Du, J.; Chen, Y.; Liu, J. Preliminary analysis of waterlogged lacquerware from the Nanhai NO.1 shipwreck and take the example of round red lacquer plate. *Chin. Cult. Herit.* **2019**, *05*, 46–51.
12. Polo, A.; Cappitelli, F.; Villa, F.; Pinzari, F. Biological invasion in the indoor environment: The spread of *Eurotium halophilicum* on library materials. *Int. Biodeterior. Biodegrad.* **2017**, *118*, 34–44. [CrossRef]
13. Liu, Z.; Fu, T.; Hu, C.; Shen, D.; Macchioni, N.; Sozzi, L.; Chen, Y.; Liu, J.; Tian, X.; Ge, Q. Microbial community analysis and biodeterioration of waterlogged archaeological wood from the Nanhai No. 1 shipwreck during storage. *Sci. Rep.* **2018**, *8*, 7170. [CrossRef] [PubMed]
14. Möller, E.M.; Bahnweg, G.; Sandermann, H.; Geiger, H.H. A simple and efficient protocol for isolation of high molecular weight DNA from filamentous fungi, fruit bodies, and infected plant tissues. *Nucleic Acids Res.* **1992**, *20*, 6115–6116. [CrossRef]
15. Liu, Z.; Zhang, Y.; Zhang, F.; Hu, C.; Liu, G.; Pan, J. Microbial Community Analyses of the Deteriorated Storeroom Objects in the Tianjin Museum Using Culture-Independent and Culture-Dependent Approaches. *Front. Microbiol.* **2018**, *9*, 802. [CrossRef]
16. White, T.J.; Bruns, T.; Lee, S.; Taylor, J. Amplification and direct sequencing of fungal ribosomal RNA genes for phylogenetics. In *PCR Protocols*; Innis, M.A., Gelfand, D.H., Sninsky, J.J., White, T.J., Eds.; Academic Press: San Diego, CA, USA, 1990; pp. 315–322.
17. Sterflinger, K.; Voitl, C.; Lopandic, K.; Piñar, G.; Tafer, H. Big Sound and Extreme Fungi—*Xerophilic, Halotolerant Aspergilli* and *Penicillia* with Low Optimal Temperature as Invaders of Historic Pipe Organs. *Life* **2018**, *8*, 22. [CrossRef]
18. BLAST: Basic Local Alignment Search Tool. Available online: https://blast.ncbi.nlm.nih.gov/Blast.cgi (accessed on 20 May 2020).
19. Tuncer, M.; Kuru, A.; Sahin, N.; Isikli, M.; Isik, K. Production and partial characterization of extracellular peroxidase produced by *Streptomyces* sp F6616 isolated in Turkey. *Ann. Microbiol.* **2009**, *59*, 323–334. [CrossRef]
20. Dupont, J.; Jacquet, C.; Dennetière, B.; Lacoste, S.; Bousta, F.; Orial, G.; Cruaud, C.; Couloux, A.; Roquebert, M.F. Invasion of the French paleolithic painted cave of Lascaux by members of the Fusarium solani species complex. *Mycologia* **2007**, *99*, 526–533. [CrossRef]

21. Martin-Sanchez, P.M.; Nováková, A.; Bastian, F.; Alabouvette, C.; Saiz-Jimenez, C. Use of biocides for the control of fungal outbreaks in subterranean environments: The case of the Lascaux Cave in France. *Environ. Sci. Technol.* **2012**, *46*, 3762–3770. [CrossRef]
22. Lavin, P.; Saravia, S.G.D.; Guiamet, P. *Scopulariopsis* sp. and *Fusarium* sp. in the Documentary Heritage: Evaluation of Their Biodeterioration Ability and Antifungal Effect of Two Essential Oils. *Microb. Ecol.* **2015**, *71*, 1–6. [CrossRef]
23. Zhang, N.; O'Donnell, K.; Sutton, D.A.; Nalim, F.A.; Summerbell, R.C.; Padhye, A.A.; Geiser, D.M. Members of the *Fusarium solani* Species Complex That Cause Infections in Both Humans and Plants Are Common in the Environment. *J. Clin. Microbiol.* **2006**, *44*, 2186–2190. [CrossRef] [PubMed]
24. Samson, R.A.; Hoekstra, E.S.; Frisvad, J.C. *Introduction to Food- and Airborne Fungi*; Centraalbureau Voor Schimmelcultures: Utrecht, The Netherlands, 2004.
25. Travadon, R.; Lawrence, D.P.; Rooney-Latham, S.; Gubler, W.D.; Wilcox, W.F.; Rolshausen, P.E.; Baumgartner, K. *Cadophora* species associated with wood-decay of grapevine in North America. *Fungal Biol.* **2015**, *119*, 53–66. [CrossRef] [PubMed]
26. Di Marco, S.; Calzarano, F.; Osti, F.; Mazzullo, A. Pathogenicity of fungi associated with a decay of kiwifruit. *Australas. Plant Pathol.* **2004**, *33*, 337–342. [CrossRef]
27. Arenz, B.E.; Held, B.W.; Jurgens, J.A.; Farrell, R.L.; Blanchette, R.A. Fungal diversity in soils and historic wood from the Ross Sea Region of Antarctica. *Soil Biol. Biochem.* **2006**, *38*, 3057–3064. [CrossRef]
28. Blanchette, R.A.; Held, B.W.; Jurgens, J.A.; McNew, D.L.; Harrington, T.C.; Duncan, S.M.; Farrell, R.L. Wood-Destroying Soft Rot Fungi in the Historic Expedition Huts of Antarctica. *Appl. Environ. Microbiol.* **2004**, *70*, 1328–1335. [CrossRef] [PubMed]
29. De Filpo, G.; Palermo, A.M.; Munno, R.; Molinaro, L.; Formoso, P.; Nicoletta, F.P. Gellan gum/titanium dioxide nanoparticle hybrid hydrogels for the cleaning and disinfection of parchment. *Int. Biodeterior. Biodegrad.* **2015**, *103*, 51–58. [CrossRef]
30. Anaya, M.; Borrego, S.F.; Gámez, E.; Castro, M.; Molina, A.; Valdés, O. Viable fungi in the air of indoor environments of the National Archive of the Republic of Cuba. *Aerobiologia* **2016**, *32*, 513–527. [CrossRef]
31. Sánchez, C. Lignocellulosic residues: Biodegradation and bioconversion by fungi. *Biotechnol. Adv.* **2009**, *27*, 185–194. [CrossRef]
32. Gao, F.; Yue, H.D.; Qin, X.M.; Lei, Z.H.; Wang, M.L. Research advances on cell wall degrading enzymes produced by pathogenic *Fusarium* causing plant diseases. *Jiangsu J. Agric. Sci.* **2018**, *34*, 955–960.
33. Chen, J. *Modern Plant Pathology Research Methods*; China Agriculture Press: Beijing, China, 2005.
34. Shen, Y.; Hu, T.J.; Zeng, G.M.; Huang, D.L.; Yin, L.; Liu, Y.; Wu, J.J.; Liu, H. Biodegradation of Lignocellulose by *Penicillium simplicissimum* and Characters of Lignocellulolytic Enzymes. *Environ. Sci.* **2013**, *34*, 781–788.
35. Shen, D.W.; Ge, Q.Y.; Yang, M.; Ma, Q.L. Iron sulfide in the conservation of marine archaeological wood. *Sci. Conserv. Archaeol.* **2013**, *25*, 84.
36. Coutinho, M.L.; Miller, A.Z.; Martin-Sanchez, P.M.; Mirão, J.; Gomez-Bolea, A.; Machado-Moreira, B.; Cerqueira-Alves, L.; Jurado, V.; Saiz-Jimenez, C.; Lima, A.; et al. A multiproxy approach to evaluate biocidal treatments on biodeteriorated majolica glazed tiles. *Environ. Microbiol.* **2016**, *18*, 4794–4816. [CrossRef] [PubMed]
37. De los Ríos, A.; Pérez-Ortega, S.; Wierzchos, J.; Ascaso, C. Differential effects of biocide treatments on saxicolous communities: Case study of the Segovia cathedral cloister (Spain). *Int. Biodeterior. Biodegrad.* **2012**, *67*, 64–72. [CrossRef]
38. Mitova, M.M.; Iliev, M.; Nováková, A.; Gorbushina, A.A.; Groudeva, V.I.; Martin-Sanchez, P.M. Diversity and biocide susceptibility of fungal assemblages dwelling in the Art Gallery of Magura Cave, Bulgaria. *Int. J. Speleol.* **2017**, *46*, 67–80. [CrossRef]
39. Polo, A.; Cappitelli, F.; Brusetti, L.; Principi, P.; Villa, F.; Giacomucci, L.; Ranalli, G.; Sorlini, C. Feasibility of Removing Surface Deposits on Stone Using Biological and Chemical Remediation Methods. *Microb. Ecol.* **2010**, *60*, 1–14. [CrossRef] [PubMed]
40. Morton, L.H.G.; Greenway, D.L.A.; Gaylarde, C.C.; Surman, S.B. Consideration of some implications of the resistance of biofilms to biocides. *Int. Biodeterior. Biodegrad.* **1998**, *41*, 247–259. [CrossRef]
41. Sterflinger, K.; Piñar, G. Microbial deterioration of cultural heritage and works of art—Tilting at windmills? *Appl. Microbiol. Biotechnol.* **2013**, *97*, 9637–9646. [CrossRef]

42. Bastian, F.; Alabouvette, C.; Saiz-Jimenez, C. Bacteria and free-living amoeba in the Lascaux Cave. *Res. Microbiol.* **2009**, *160*, 38–40. [CrossRef]
43. Mehta, J.; Patidar, K.; Vyas, N. Development and validation of a precise method for determination of benzalkonium chloride (BKC) preservative, in pharmaceutical formulation of latanoprost eye drops. *J. Chem.* **2007**, *7*, 11–20. [CrossRef]
44. Stupar, M.; Grbić, M.L.; Džamić, A.; Unković, N.; Ristić, M.; Jelikić, A.; Vukojević, J. Antifungal activity of selected essential oils and biocide benzalkonium chloride against the fungi isolated from cultural heritage objects. *S. Afr. J. Bot.* **2014**, *93*, 118–124. [CrossRef]
45. Scheerer, S.; Ortega-Morales, O.; Gaylarde, C. Microbial deterioration of stone monuments—An updated overview. *Adv. Appl. Microbiol.* **2009**, *66*, 97–139. [PubMed]
46. Favero-Longo, S.E.; Brigadeci, F.; Segimiro, A.; Voyron, S.; Cardinali, M.; Girlanda, M.; Piervittori, R. Biocide efficacy and consolidant effect on the mycoflora of historical stuccos in indoor environment. *J. Cult. Herit.* **2018**, *34*, 33–42. [CrossRef]

applied sciences

MDPI

Article

# Frequent Microalgae in the Fountains of the Alhambra and Generalife: Identification and Creation of a Culture Collection

**Fernando Bolívar-Galiano [1,*], Clara Abad-Ruiz [1], Pedro Sánchez-Castillo [2], Maurizio Toscano [1,*] and Julio Romero-Noguera [3]**

[1] Department of Painting, Faculty of Fine Arts (Facultad de Bellas Artes), University of Granada, 18014 Granada, Spain; cabad@ugr.es
[2] Department of Botanic, Faculty of Sciences (Facultad de Ciencias), University of Granada, 18071 Granada, Spain; psanchez@ugr.es
[3] Department of Painting, Faculty of Fine Arts (Facultad de Bellas Artes), University of Seville, 41003 Seville, Spain; juliorn@us.es
* Correspondence: fbolivar@ugr.es (F.B.-G.); maurizio@ugr.es (M.T.)

Received: 4 August 2020; Accepted: 17 September 2020; Published: 21 September 2020

**Abstract:** Cyanobacteria, green algae and diatoms are significant factors in the biodeterioration of stone cultural heritage sites, and specifically fountain monuments, due to the constant presence of water. In this study, samples were taken from different fountains in the Alhambra and Generalife, which are among the Spanish monuments of greatest historical and artistic value and which together were declared a World Heritage Site by UNESCO in 1984. The aim was to identify which species of colonising microalgae are most frequent and to obtain monoalgal cultures from them. From a conservation point of view, it is interesting to identify which algae are growing in these fountains and how they behave in order to develop new methods to control their growth. The most abundant groups of algae in our samples were green algae and cyanobacteria. The most common genera in the former group were *Bracteacoccus*, *Chlorosarcina*, *Chlorosarcinopsis*, *Apatococcus* and *Klebsormidium*. As for cyanobacteria, the most abundant genera were *Phormidium*, *Calothrix*, *Leptolyngbya*, *Chamaesiphon*, *Pleurocapsa* and *Chlorogloea*. Using our collected samples, 10 genera of green algae and 13 genera of cyanobacteria were isolated, thereby constituting the base samples for the creation of a reference collection of living algae from the Alhambra and Generalife contexts, which can be used in subsequent studies to develop new types of treatment against biodeterioration.

**Keywords:** green microalgae; cyanobacteria; diatoms; biodeterioration; Alhambra; Generalife; conservation; information modelling; stone fountains; cultural heritage

---

## 1. Introduction

Biodeterioration is defined as "any undesirable change in the properties of a material caused by the vital activities of organisms" [1]. Microorganisms such as bacteria, fungi and algae are some of the agents that affect cultural heritage. These include microalgae (microscopic algae), which are photoautotrophic organisms capable of growing in media not abundant in nutrients, such as stone, and which proliferate in places with water or high relative humidity levels. In particular, they play an especially relevant role in fountains and ornamental pools, because of the constant presence of water. These organisms are the main agents involved in the biodeterioration of such cultural assets, causing a range of aesthetic, physical and chemical damage, which has been studied over time around the world [2–8].

### 1.1. Biodeterioration Caused by Microalgae

Algae are a polyphyletic group of organisms, i.e., not all of them come from the same evolutionary ancestor. Within this group we can find enormous diversity in all aspects. For decades, cyanobacteria were called blue-green algae and were grouped with other algae, mainly because of their ability to photosynthesise. Today, however, cyanobacteria are no longer considered to be algae, but rather are considered to be prokaryotic organisms belonging to the domain of bacteria, whereas all algae are eukaryotic organisms. Nevertheless, in this work we have included cyanobacteria in the term "microalgae", since both cyanobacteria and eukaryotic microalgae share important similitudes from an ecological point of view—they both carry out oxygenic photosynthesis, are pioneers in stone colonisation and cause biodeterioration through similar processes.

The microalgae that affect the state of conservation of fountains and other stone monuments can be epilithic (growing on the surface of the stone) or endolithic (growing inside it) [9], with the latter being the most damaging and difficult to remove. In any case, the microalgae grow and produce biofilms, which are composed of a set of organisms and the substances they excrete. The microalgae are often embedded in the matrix formed by extracellular polymeric substances (EPS) secreted by the cells themselves. These biofilms affect the stone in different ways.

Firstly, they produce an aesthetic alteration, since they form a patina with different natural colourings (light or dark green, orange, brown, etc.) on the stone. Furthermore, biofilms can retain atmospheric dust, pollutants from the air and substances dissolved in water that give them a darker colour. This phenomenon is known as "soiling" [10].

These biofilms also cause physical deterioration. Many species of algae and cyanobacteria have sheaths that produce mucilage, which allows them to adhere to the substrate. The mucilage is also used for gliding [11] and to retain water, so that they can survive drought conditions. On absorbing water, the sheaths' volumes vary considerably and their continual expansion and contraction causes the gradual destruction of stony material, which progressively loses granules that go on to form part of the biofilm [12]. The water retention is a fundamental aspect—it fosters mechanical damage due to frost weathering with sharp temperature changes and also allows for the growth of other algae with greater water requirements, so that the communities become more complex. It may also help to support other, more damaging organisms such as fungi and bryophytes [10]. If there is an excessive proliferation of algae, they can affect how the fountains work, for example by blocking spouts, which can speed up other kinds of deterioration.

Lastly, microalgae also cause chemical damage as a result of their metabolism. They directly damage the stone by producing organic acids (pyruvic, glycolic, lactic, acetic, succinic or alpha-ketoglutaric) as a result of respiration. They can also give rise to biotransference phenomena by producing polysaccharides capable of reacting with metals and leading to salts [10]. However, the main damage caused by microalgae is the formation of carbonate crusts on stone as a result of photosynthetic metabolism. Due to the withdrawal of carbon dioxide from the medium and to other biochemical processes, the pH increases and fosters the precipitation and aggregation of calcium carbonate, forming layers of mineral crusts that can end up covering the original materials. All of the taxonomical groups of microalgae are capable of producing this phenomenon, but it is more apparent in green algae and cyanobacteria [13]. In addition to causing aesthetic damage, these crusts foster physical biodeterioration due to thermal contraction and expansion to a different extent than the substrate, leading to stresses that weaken it, and also due to frost weathering, since they have a porous structure that increases water retention further [14]. They even end up burying the algae that have generated them, which then become endolithic and protected from environmental adversity and treatments, such as cleaning or biocides, thereby making it much more difficult to remove them.

### 1.2. Microalgae in the Alhambra

The Alhambra and the Generalife, together named a UNESCO World Heritage site in 1984, form one of the monument complexes of greatest historical and artistic value in Spain, meaning their conservation is of great importance. Given the importance of water in Nasrid and Islamic art in general, "*alberca*" reservoirs abound throughout the site, together with channels and fountains (including iconic ones such the Fountain of the Lions), which are affected by colonisation from microalgae (Figure 1a,b). Although the fountains undergo continual maintenance, these microorganisms are not completely eliminated. Despite the continual investment in chemical and mechanical means of control, the natural process of colonisation starts over.

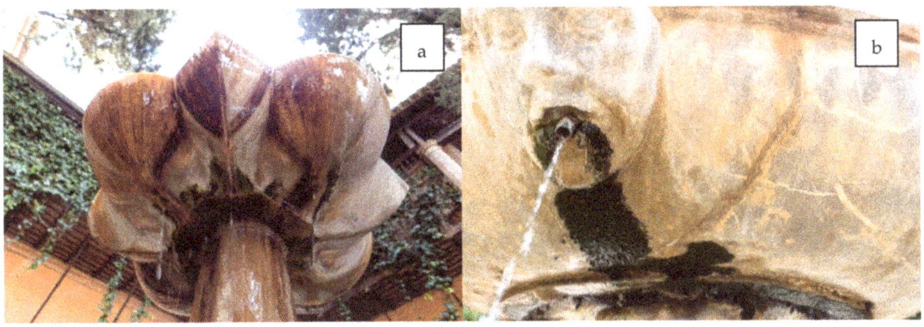


**Figure 1.** (**a**) Two different phycofilms on the *Fuente del Patio de la Reja* at the Nasrid Palaces. (**b**) Algal mat growing on the *Fuente del Patio de la Sultana* at Generalife. Photos by G. Alfano.

From a conservation point of view, an important step will be to learn the structure and functioning of the biofilms—which species they are composed of, which species are the most resistant or harmful and which conditions help them develop—in order to know how to create new methods in order to halt the proliferation of the microorganisms. There are a variety of previous studies on the biodeterioration of fountains in the Alhambra, many of which focus on biodeterioration due to microalgae, their classification and possible treatments [13–18]. Our work was intended to update the knowledge on the types of microalgae growing in different fountains throughout the Alhambra and Generalife today, establishing which genera are the most common and beginning to create a culture collection of algae with species from these sites.

### 1.3. Objectives

The main objectives of this study were as follows:

- To identify the most common types of microalgae in the fountains of the Alhambra and Generalife today.
- To make unialgal cultures of the species found in the Alhambra in order to begin creating a culture collection of living microalgae available for subsequent studies on new kinds of treatment to keep biodeterioration under control.
- To design a data model and develop a database for storing and retrieving sample details and analytical values from the collection.

## 2. Materials and Methods

### 2.1. Sampling

In total, 21 fountains in the Alhambra and Generalife were studied, with 120 microalgae samples being taken (Table 1, Figures 2 and 3). The greatest number of samples was taken from the monumental fountains of the Nasrid Palaces (60%) due to their enormous cultural interest, although fountains from other areas were also selected in order to make a general assessment of the most common microalgae species colonising the monument complex.

**Table 1.** Fountains studied in different areas of the Alhambra and numbers of samples collected. In the last two columns, in brackets, the numbers of samples taken corresponding to different materials and environments are shown.

| Zone | Fountain | No. of Samples | Material | Environment |
|---|---|---|---|---|
| Nasrid Palaces | 1. Fuente del Patio del Cuarto Dorado (Mexuar fountain) | 7 | marble (5), marble and metal (1), mortar (1) | submerged (6), amphibious (1) |
| | 2. Fuente-guitarra N del Patio de los Arrayanes (North "guitar" fountain of the Court of the Myrtles) | 15 | marble (10), metal (5) | submerged (9), amphibious (5), aerial (1) |
| | 3. Fuente-guitarra S del Patio de los Arrayanes (South "guitar" fountain of the Court of the Myrtles) | 6 | marble (5), metal (1) | submerged (4), amphibious (2) |
| | 4. East channel of the Court of the Myrtles | 2 | marble (2) | submerged (1), amphibious (1) |
| | 5. Fuente de los Leones (Fountain of the Lions) | 21 | marble (15), mortar (3), metal (3) | submerged (4), amphibious (13), aerial (4) |
| | 6. Fuente del Patio de la Reja (Fountain of the Court of the Grille) | 11 | marble (11) | submerged (8), amphibious (3) |
| | 7. Fuente del Patio de Lindaraja (Fountain of the Court of the Lindaraja) | 10 | marble (10) | submerged (1), amphibious (5), aerial (4) |
| Partal | 8. Fuente-guitarra junto a la Torre de las Damas | 6 | marble (5), metal (1) | submerged (4), amphibious (2) |
| | 9. Pilar (Basin) debajo del Mirador del Partal | 5 | brick (5) | submerged (1), amphibious (2), aerial (2) |

**Table 1.** *Cont.*

| Zone | Fountain | No. of Samples | Material | Environment |
|---|---|---|---|---|
| Casa del Arquitecto | 10. Inner fountain of the court with oranges | 1 | painted plaster (1) | aerial (1) |
| Alcazaba | 11. Pilar E. del Jardín de los Adarves (East basin of the Garden of the Ramparts) | 3 | marble (3) | amphibious (2), aerial (1) |
| Secano | 12. Fuente de la Glorieta del Secano | 5 | marble (5) | submerged (3), amphibious (2) |
| Generalife | 13. Fuente S de los Jardines Bajos (South fountain of the Lower Gardens) | 2 | marble (2) | submerged (1), amphibious (1) |
| | 14. Fuente central del Patio de la Sultana (Central fountain of the Sultana's Court) | 7 | marble (6), brick (1) | submerged (2), amphibious (2), aerial (3) |
| | 15. Pretil de la Escalera del Agua (The handrail of the Water Stairway) | 4 | glazed tile (2), plaster (2) | submerged (2), amphibious (2) |
| Walkways | 16. Fuente del Tomate (Tomato fountain) | 3 | marble (3) | submerged (3) |
| | 17. Fuente de Ángel Ganivet | 3 | bronze (1), mortar (2) | submerged (2), amphibious (1) |
| | 18. Pilar de Carlos V | 3 | marble (3) | submerged (1), amphibious (2) |
| | 19. Pilar de Washington Irving | 3 | marble (3) | amphibious (2), aerial (1) |
| | 20. Pilar de la Puerta de las Granadas (Basin of the Gate of the Pomegranates) | 2 | marble (2) | aerial (2) |
| Carmen de Bellavista | 21. Fuente derecha de la 2ª terraza (Right fountain of the 2nd terrace) | 1 | marble (1) | amphibious (1) |
| Total samples | | 120 | | |

**Figure 2.** Sampling from the Fountain of the Lions. Photo by G. Alfano.

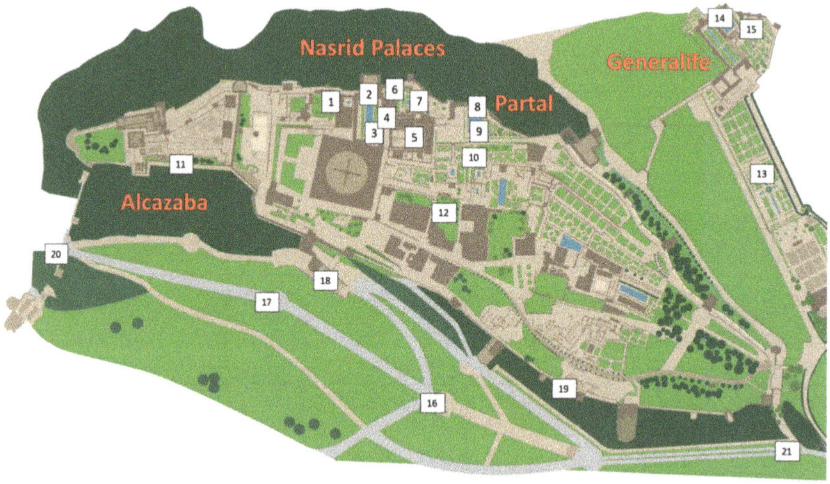


**Figure 3.** Sampling points in the monumental complex of the Alhambra and the Generalife. The labels correspond to Table 1.

The main materials used in the construction of the fountains studied are white marble from Macael (Almería, Spain) and limestone from Sierra Elvira (Granada, Spain). Samples were also taken from the surface of the metal spouts in several fountains, and occasionally from mortar, brick, plaster and glazed tile.

Seasonal samples were taken in spring, summer, autumn and winter throughout 2017 and 2018. The sampling points chosen were the surfaces of fountains, where generally abundant macroscopic growth of epilithic microalgae had been observed (Figures 4a and 5a). For each of them, the appearance of the microalgal communities was taken into account, as well as the environment they belong to, according to the water available—submerged, amphibious or aerial. The submerged zones are those below the water level, as well as those exposed to continually running water, for example inside the spouts or the outer surfaces of basins where the water overflows. The amphibious zones are those with water available intermittently, such as surfaces that are splashed by the spouts. Lastly, we considered aerial zones to be those that do not receive water directly, such as the inner surfaces of basins above the waterline. As for the appearance of the communities, we can distinguish: patinas (sheens) when they are of scant thickness and less structural complexity; mats of greater thickness and complexity with filamentous species; and mineral crusts, which may arise due to any of the aforementioned formations. On taking the different samples, a variety of fountains and environments were chosen to take into account the greatest possible variation. In some cases, samples were also taken in the same zone at different times of year to check if there were variations in the types of species. All of the sampling processes were photographically documented (i.e., fountain, sampling point, sample, etc.), giving a total of around 450 high quality images.

**Figure 4.** (**a**) Sampling a mineralising crust with a scalpel at *Fuente N del Patio de los Arrayanes*. (**b**) Two sheeted crusts from perpendicular angles with buried communities. Photos by G. Alfano.

**Figure 5.** (**a**) South fountain of the Lower Gardens. The base of the basin, where a dark green mat can be seen, was one of the sampling points; (**b**) Taking a sample of mineral crust in the *Escalera del Agua* (Water Stairway); (**c**) Sample of mineral crust taken in the Washington Irving basin. Photos by G. Alfano.

### 2.2. Procedures

The samples were taken with the help of tweezers and a scalpel, taking the utmost care not to damage the original materials (Figures 4a and 5b). The fresh samples were stored in a polyethylene sample collection vial with water from the same fountains from which they were taken (Figure 5c). In the laboratory, they were refrigerated until their observation. After observation and creation of the cultures, the samples were fixed in a glutaraldehyde and 25% glycerol solution (Sigma Aldrich, Madrid, Spain) to be permanently conserved for subsequent study.

The microalgae present in each sample were identified in the phycology lab at the University of Granada, using a binocular Olympus SZX10 microscope and an inverted optical Nikon Eclipse TS100 microscope. The following references were used for the taxonomical identification [19–23]. In each sample, the different microalgal species were identified according to their proportion by assigning an index (+–5) representing the following percentages of the total: + = 0–1%; 1 = 1–10%; 2 = 10–25%; 3 = 25–50%; 4 = 50–75%; 5 = 75–100%.

Microalgal cultures were made in a solid medium in agar plates at 2%. A high concentration of agar (Sigma-Aldrich, Madrid, Spain) was used because we intended to cultivate benthonic species that grow on stone. The media for cultivation used were Bold's Basal Medium (BBM) (Phyto-Tech, Lenexa, KS, USA) for green algae and BG11 (Sigma-Aldrich, Madrid, Spain) for the cyanobacteria. Diatoms were not cultivated since they have stricter nutritional requirements and are the most uncommon group of the three.

The streaking method was used to isolate microalgae on plates using fresh samples, successive isolations and re-streaking until unialgal cultures were obtained. The plates were incubated at 22 °C with a cycle of 12 h of light/darkness using fluorescent lamps that provided a light intensity of 800 lux. On average, the cultures took about two weeks to develop, but the time varied depending on multiple factors, so the re-streaking was carried out depending on the needs in each case.

### 2.3. The Database

From the planning phase of this study, it was clear that the information collected would have to be processed digitally in order to provide the required features of data consistency, automated backups, multimedia, versioning, automatic quantifications and simultaneous user access. For this reason, two of the fundamental steps towards the creation of the algae culture collection were the design of a specific data model (tables, attributes, constraints and relationships) and creation of a multimedia, geographically enabled database to store and retrieve data collected during the sampling process and in the lab.

The main tables of the database are used to register the collected samples, the monuments (fountains, pools, etc.) and the identified microalgae; the sample represent the central table, which is related to both the monuments (one-to-many cardinality) and microalgae (many-to-many cardinality). Several attributes were added in order to register a wide variety of values, such as the date, location, environment, media and description; classes, genera, species, colour, communities, biotype and biotope; applied treatments, cultivation procedures and microphotographs; material and chronology.

The database is simultaneously accessible by several users with individual accounts and different roles. Stored data can be retrieved using a unified search box or through a multifaceted search engine, supporting filters and Boolean operators. Every change in each record is registered in a different version, so individual states can be restored or compared between each other. The database's architecture is modular so that additional workflows can be introduced in the future, for example to continue tracing the evolution of biodeterioration through time and space.

## 3. Results and Discussion

### 3.1. Microalgae Identification

Firstly, it should be noted that the fountains in the Alhambra and Generalife undergo continual maintenance. In addition to general cleaning, all of the studied fountains are treated with algicide. Chlorine tablets are also added to many of them. In the specific case of the Fountain of the Lions, the water flowing through gets special treatment, since the circulating water is chlorinated and purified independently from the rest of the fountains of the monumental complex. This means that the algae colonising these fountains exist in conditions that are far from natural and that they are generally resistant to such treatment, since it does not manage to eliminate them completely.

Fundamentally, as seen in previous studies [15,16], the microalgae colonising the selected fountains largely belong to the large groups of green algae (Chlorophyta and Charophyta), cyanobacteria (Cyanobacteria) and diatoms (Bacillariophyta). Once in a while, some Euglenophyceae have been found in the "guitar" fountain of the Ladies Tower (Torre de la Damas, Granada, Spain), and a Phaeophyceae (*Pleurocladia* sp.) sample was found in the southern "guitar" fountain of the Court of the Myrtles (Patio de los Arrayanes), but those species were only seen in spring in those two fountains. Table 2 shows a list of the fountains studied and the most common genera of microalgae in each.

**Table 2.** The most common genera of microalgae in each of the fountains studied. For each fountain, the set of samples taken was considered, along with the different sampling points, and where applicable the different times of year. The genera that appear most often and in greater abundance were chosen. For each genus, the total number of samples in which it is present (first number) and the number of samples in which its index is 3 or higher (second number; see 2.2. Procedures) are shown in brackets.

| Fountain | | Most Common Genera | | |
|---|---|---|---|---|
| | | Green Algae | Cyanobacteria | Diatoms |
| 1. | Mexuar fountain | *Bracteacoccus* (3-2), *Chlorosarcinopsis* (2-1) | | *Navicula* (3-1), *Nitzschia* (2-2) |
| 2. | North "guitar" fountain of the Court of the Myrtles (*Patio de los Arrayanes*) | *Chlorosarcinopsis* (5-2) | *Dichothrix* (5-3), *Phormidium* (6-5), *Leptolyngbya* (9-6), | |
| 3. | South "guitar" fountain of the Court of the Myrtles | | *Phormidium* (4-2), *Calothrix* (4-1), *Chlorogloea* (1-1), *Pleurocapsa* (3-1) | |
| 4. | East channel of the Court of the Myrtles | | *Leptolyngbya* (2-2), *Pleurocapsa* (2-1) | |
| 5. | Fountain of the Lions | *Klebsormidium* (3-2) | *Phormidium* (7-6), *Leptolyngbya* (9-2), *Cyanosarcina* (5-4) | |

**Table 2.** *Cont.*

| Fountain | Most Common Genera | | |
|---|---|---|---|
| | Green Algae | Cyanobacteria | Diatoms |
| 6. *Patio de la Reja* fountain | *Chlorosarcina* (7-5) | *Symploca* (1-1), *Leptolyngbya* (5-4), *Chamaesiphon* (3-2), *Calothrix* (2-1) | |
| 7. *Patio de Lindaraja* fountain | *Chlorococcum* (4-2), *Chlorosarcinopsis* (4-1), *Klebsormidium* (5-2) | *Chamaesiphon* (6-3), *Phormidium* (2-2) | |
| 8. Ladies Tower (*Torre de las Damas*) | *Pleurastrum* (2-2) | *Symploca* (3-2), *Phormidium* (3-2), *Leptolyngbya* (2-1) | |
| 9. *Mirador del Partal* basin | *Bracteacoccus* (4-4) | | *Navicula* (2-1) |
| 10. Inner fountain of the court with oranges | *Klebsormidium* (1-1) | | |
| 11. *Jardín de los Adarves* east basin | *Bracteacoccus* (3-2), *Choricystis* (1-1) | | |
| 12. *Glorieta del Secano* fountain | *Bracteacoccus* (4-3) | *Chamaesiphon* (2-1), *Cyanosarcina* (1-1) | |
| 13. *Jardines Bajos* South fountain | | *Phormidium* (2-2) | |
| 14. *Patio de la sultana* central fountain | *Gongrosira* (3-1), *Bracteacoccus* (3-3) | *Phormidium* (3-1), *Lyngbya* (1-1) | *Navicula* (4-1), *Cymbella* (3-1) |
| 15. *Escalera del Agua* handrail | *Pseudopleurococcus* (1-1), *Leptosira* (2-1), *Chlorosarcina* (2-2) | *Myxosarcina* (1-1), *Chlorogloea* (1-1) | |
| 16. *Fuente del tomate* | *Bracteacoccus* (2-1) | *Chlorogloea* (1-1), *Chamaesiphon* (1-1), *Phormidium* (2-1) | |
| 17. Ángel Ganivet pool fountain | *Bracteacoccus* (2-2) | *Leptolyngbya* (1-1) | |
| 18. Carlos V basin | *Bracteacoccus* (3-1) | *Chamaesiphon* (2-1) | |
| 19. Washington Irving basin | *Bracteacoccus* (2-1) | | |

**Table 2.** *Cont.*

| Fountain | | Most Common Genera | | |
|---|---|---|---|---|
| | | Green Algae | Cyanobacteria | Diatoms |
| 20. | *Puerta de las Granadas* basin | *Bracteacoccus* (1-1) | *Chroococcopsis* (1-1) | |
| 21. | Right fountain of the 2nd terrace of the *Carmen de Bellavista* | | *Calothrix* (1-1), *Phormidium* (1-1) | |

In general, diatoms are the least common type of microalgae of the three groups studied. The most common genera are *Navicula*, *Nitzschia*, *Cymbella* and *Achnanthes*, all of which are unicellular; in other words, the cells appear individually and do not form aggregates adhering strongly to surfaces. Of these four genera, *Navicula*, *Nitzschia* and *Cymbella* were the ones that generally appeared most often. All of these diatoms appear to a greater or lesser extent as part of more complex communities of cyanobacteria or green algae, except one specific case in which a monospecific community of *Achnanthes* sp. was found on the surface of the basin at the Fountain of the Lions in a submerged environment.

Among the green algae, the unicellular coccoid algae were very widespread in many fountains. The most common genera were *Chlorococcum*, *Bracteacoccus*, *Chlorosarcina* and *Chlorosarcinopsis* (Figure 6a). The latter two are able to form generally compact cellular aggregates; less frequent (but still common) were the *Chlorella* and *Choricystis* genera.

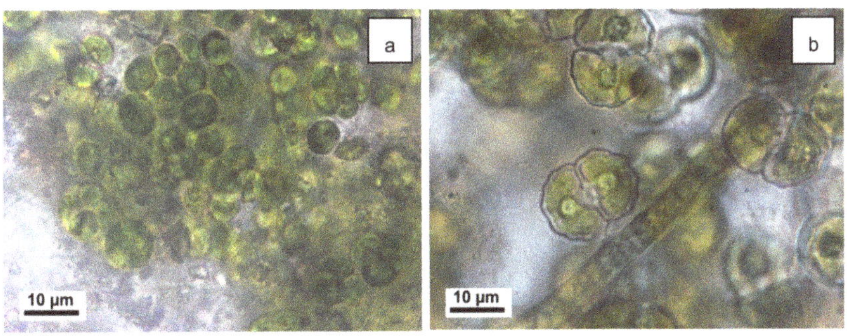


**Figure 6.** Photographs of samples taken using optical microscopy: (**a**) *Chlorosarcinopsis* sp. cells can be seen on the left; (**b**) *Cosmarium* sp. cells can be seen on the right, along with a filament of cyanobacteria.

In addition to the coccoid algae, we also found pseudoparenchymatic green algae—*Pleurastrum* sp. appears on the "guitar" of the Ladies Tower and *Gongrosira* sp. appears in the pool of the central fountain of the Court of the Sultana, both in submerged zones; while *Leptosira* sp. and *Pseudopleurococcus* sp. grow in the Water Stairway's handrail in amphibious and submerged zones. As for filamentous algae, the *Klebsormidium* genus is common, which is also capable of growing in aerial zones. The filamentous alga *Cladophora* sp. has also been found sporadically (in the Lindaraja fountain and the *mirador del Partal* basin), as has *Stigeoclonium* sp. (*Patio de los Arrayanes* southern "guitar" fountain). *Stichococcus* sp. appears in amphibious and aerial zones of the Fountain of the Lions and in the handrail of the *Escalera del Agua*, although not abundantly. This alga is normally unicellular, although it can form lax filaments that disaggregate easily. Other green algae genera that are very widespread, although their proportions in the samples are usually small, were *Scenedesmus* sp., which is characteristic for the formation of coenocytes; *Cosmarium* sp. (unicellular; Figure 6b); and *Apatococcus* sp. (pseudoparenchymatic).

As for the cyanobacteria, we can distinguish between the Chlorococcales order, whose cells group together in colonies (and are occasionally filamentous in appearance), and truly filamentous algae. The most common chlorococcal cyanobacteria among all the fountains studied were *Pleurocapsa* sp. (Figure 7a), *Chamaesiphon* sp. (Figure 7b) and *Chlorogloea* sp., followed by *Chroococcidiopsis* sp., which is less common or abundant. *Cyanosarcina* sp. is very common in amphibious zones in the Fountain of Lions and the fountain of the *Glorieta del Secano*, while *Myxosarcina* sp. grows in crusts on the southern "guitar" of the *Patio de los Arrayanes* and the *Escalera del Agua*. On the Fountain of the Lions and the basin of the *Puerta de las Granadas* there were species of the genus *Chroococcopsis*. Among the filamentous cyanobacteria, the most common genera were *Phormidium* (Figure 8a,b), *Calothrix*, *Leptolyngbya*, *Lyngbya* and *Schizothrix*. The former three appear in a large number of fountains and are relatively abundant. *Symploca* sp. appears abundantly in submerged zones of the "guitar" of the *Torre de las Damas* and in the fountain of the *Patio de la Reja*. In the spout of the northern fountain in the *Patio de los Arrayanes*, where samples have been taken in different seasons, a dark mat forms in which there is a great presence of *Dichothrix* sp.

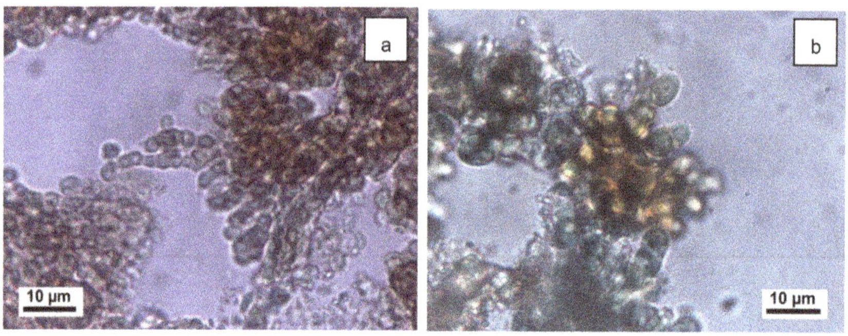


**Figure 7.** Photograph of samples taken using optical microscopy: (**a**) *Pleurocapsa* sp. cells are shown on the left; (**b**) *Chamaesiphon* sp. cells are shown on the right.

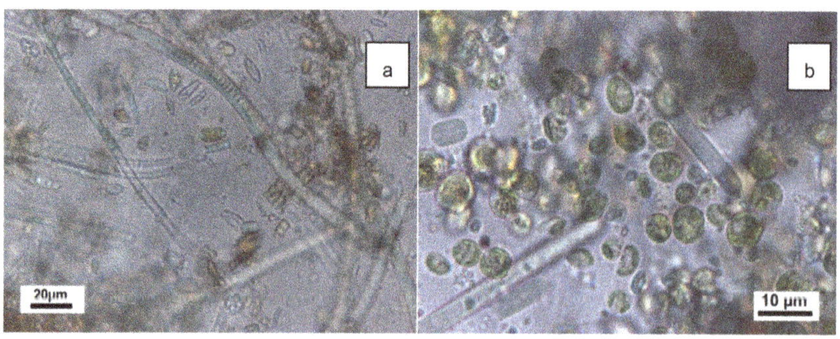


**Figure 8.** Photograph of samples taken using optical microscopy: (**a**) diatoms of the *Cymbella, Navicula and Achnanthes* genera and filaments of *Phormidium* sp. can be seen on the left; (**b**) on the right, there are cells of *Bracteacoccus* sp. and filaments of *Phormidium* sp.

Normally, in each sample we can find generally complex communities made up of two or more different species of microalgae. However, occasionally the sheens (structures of less complexity) are composed of a single species. In the vast majority of cases, these species belong to the unicellular coccoid green algae group, such as *Chlorosarcinopsis* sp. (Figure 6a) and *Bracteacoccus* sp. (Figures 8b and 9a), which form green sheens in submerged environments. On the one hand, this could show that these genera are especially resistant in adverse conditions and capable of surviving in environments with chlorine and biocides; on the other hand, this could indicate that they are the first colonisers of the stone. Mats are usually found in submerged and amphibious environments and are typical of the presence of filamentous algae, especially cyanobacteria. *Phormidium*, *Leptolyngbya* (Figure 9c,d) and *Calothrix* are genera that are often the majority species in mats. On some fountains, these mats also have a high proportion of filamentous green algae, such as *Gongrosira* sp. and *Klebsormidium* sp. (Figure 9b). As for the mineral crusts, we found variability; they are usually made up of complex communities of various species of green algae and cyanobacteria (which sometimes form separate strata), but sometimes they are monospecific, made up of *Chlorosarcina* sp. and other indeterminate coccoid green algae.

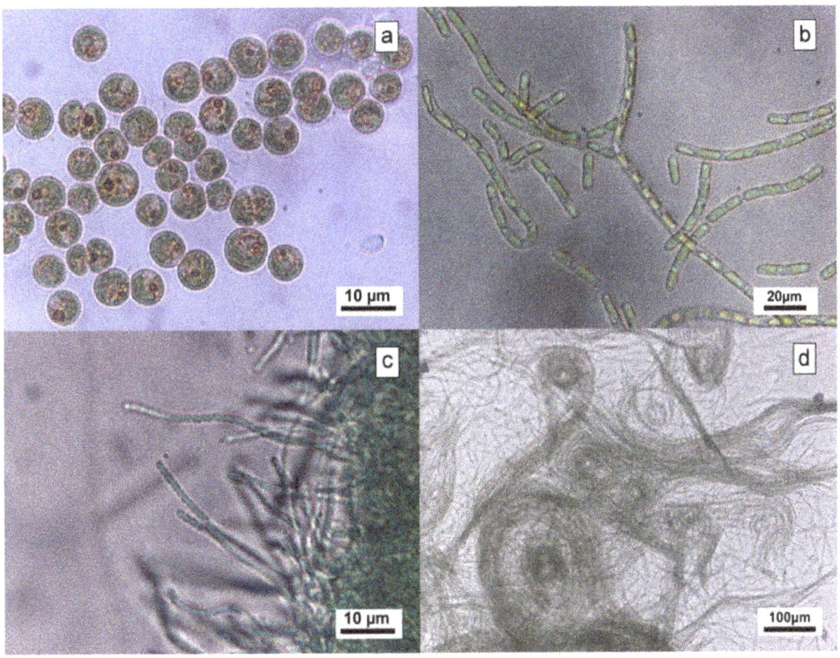


**Figure 9.** Photographs of the algal cultures *Bracteacoccus* sp. (**a**) and *Klebsormidium* sp. (**b**) and two different species of *Leptolyngbya* (**c**,**d**).

Most of the samples belong to amphibious or submerged environments. Only a few were taken in aerial zones far from water, although sometimes with high levels of relative humidity. In such zones, genera such as *Leptolyngbya*, *Klebsormidium*, *Stichococcus*, *Chlorococcum*, *Bracteacoccus*, *Cosmarium*, *Phormidium*, *Navicula* and *Chroococcopsis* were observed, forming coloured sheens and sometimes crusts. These same genera appear in other submerged and amphibious zones, so they are not exclusive to aerial environments. In the sampling points where samples were taken in different seasons, it was observed that there may be small variations between the observed species of algae, however the general composition is largely maintained and the predominant algae are usually the same. The small variations may be due to seasonal environmental factors or cleaning and maintenance work on the fountains.

If we compare the results with previous studies carried out on the fountains of the Alhambra [14,15,17,18], only the genera *Bracteacoccus*, *Klebsormidium*, *Choricystis*, *Stichococcus*, *Chroococcopsis*, *Cyanosarcina* and *Dichothrix* had never been recorded before. The rest of the observed microalgae had already been recorded in the past. It could be considered that new genera have appeared, but it has to be remembered that the same fountains have not always been sampled. In the thesis by Bolívar-Galiano [15] only seven fountains in the Nasrid Palaces were sampled, while in the study by Cuzman et al. [18] only samples from the Lindaraja fountain and the central fountain of the *Patio de la Sultana* were studied. In addition to some of the genera currently identified, in the previous works there are also some different ones. Specifically, we can observe a much greater number of species in the thesis by Bolívar-Galiano in 1994 [15], even though only the Nasrid Palaces' fountains were considered. One of the explanations for this phenomenon is that the presence of microalgae in these fountains was much greater 26 years ago, because today much more attention is paid to these microorganisms and the fountains undergo much stricter maintenance (Figure 10).

**Figure 10.** Presence of microalgae on one of the lions in the Fountain of the Lions: (**a**) an image from 1994 by F. Bolivar is shown on the left; (**b**) a modern image by G. Alfano in which one can see the effect of the cleaning treatment and lack of large sheens on the surface is shown on the right.

The genera we have identified are also common in other monumental fountains in Spain and Italy, where the following genera have been found: *Phormidium*, *Calothrix*, *Leptolyngbya*, *Lyngbya*, *Symploca*, *Schizothrix*, *Myxosarcina*, *Chamaesiphon*, *Chroococcidiopsis*, *Pleurocapsa*, *Chlorogloea*, *Pseudopleurococcus*, *Pleurastrum*, *Apatococcus*, *Chlorella*, *Chlorosarcina*, *Chlorosarcinopsis*, *Cosmarium*, *Chlorococcum*, *Scenedesmus*, *Stichococcus*, *Navicula*, *Nitzschia*, *Achnanthes* and *Cymbella* [18,24,25]. In addition to fountains, microalgae are also pioneers in colonising other stone monuments and are significant agents in their biodeterioration. In fact, algae and cyanobacteria are the majority components in the biofilms found in stone buildings in Europe, while cyanobacteria are the majority components in those found in Latin America [26]. In a study carried out on the most common green algae and cyanobacteria in monuments of different types of stone in the entire Mediterranean basin [27], all of the genera we found in the Alhambra were present. In fact, *Klebsormidium* sp., which did not appear in the fountains of the Alhambra in the past, is among the most common species in stone monuments. All of this information makes it clear that the microalgae growing in the Alhambra and Generalife are species that are very widespread and which are commonly found in stone monuments.

### 3.2. Obtaining Unialgal Cultures

Using some of the fresh samples, and before fixing them permanently, agar plates were inoculated and cultures were made in them to isolate the greatest possible number of microalgae species (Figure 11). The media used to isolate green algae were BBM (Phyto-Tech, Lenexa, KS, USA), while BG11 (Sigma-Aldrich, Madrid, Spain) media were used for cyanobacteria. From the beginning, the cultivation of diatoms was discarded, since they have greater nutritional demands and are a less common group. Finally, it was possible to isolate a total of 23 genera—10 green algae and 13 cyanobacteria (Table 3).

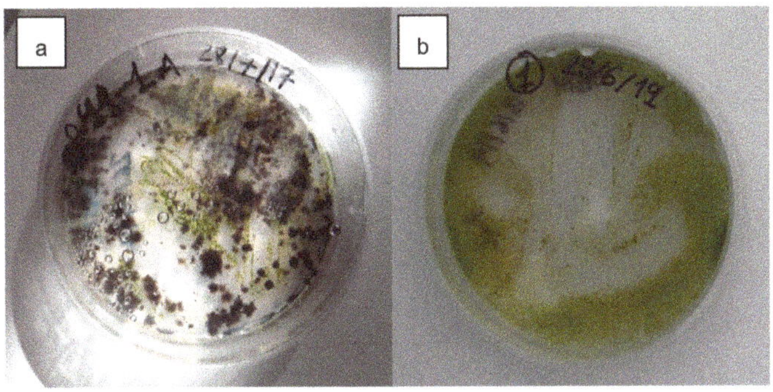


**Figure 11.** Microalgal cultures: (**a**) an initial culture using a sample is shown on the left; (**b**) a unialgal culture of *Neochloris* sp is shown on the right.

**Table 3.** Species obtained from unialgal cultures using samples collected in the Alhambra and Generalife. The fountain numbers (Tables 1 and 2) from which the cultures came are indicated in the left column.

| Fountains | Green Algae | Fountains | Cyanobacteria |
|---|---|---|---|
| 4 | *Neochloris* sp. | 2 | *Dichothrix* sp. |
| 14 | *Stigeoclonium* sp. | 4, 8, 15, 17 | *Leptolyngbya* spp. |
| 14, 8 | *Chlorococcum* sp. | 20, 7 | *Chroococcopsis* sp. |
| 9, 12, 16, 19 | *Bracteacoccus* sp. | 2, 8 | *Pseudophormidium* sp. |
| 4, 8, 14 | *Scenedesmus* spp. | 13 | *Pseudanabaena* sp. |
| 15 | *Pseudopleurococcus* sp. | 6, 21 | *Calothrix* sp. |
| 5, 10 | *Klebsormidium* sp. | 3 | *Pleurocapsa* sp. |
| 11, 13 | *Choricystis* sp. | 3 | *Phormidium* sp. |
| 3 | *Apatococcus* sp. | 5 | *Cyanosarcina* sp. |
| 4 | *Gloeocystis* sp. | 5 | *Schizothrix* sp. |
| | | 5 | *Nostoc* sp. |
| | | 5 | *Chroococcidiopsis* sp. |
| | | 4 | *Ammatoidea* sp. |

The set of data is not completely representative of the site's microbiota, since it is known that the different species have different capacities for developing within in vitro cultures. Some species do not prosper in cultivation, even if they are predominant in the sample. Likewise, species that are not at all abundant in the samples may grow when cultivated, including species that are not observed in the sample, as has occurred with *Neochloris* sp., *Pseudanabaena* sp., some species of *Nostoc* sp., *Gloeocystis* sp. and *Leptolyngbya* sp. The latter genus grows very well when cultivated (Figure 9c,d), with up to 4 different strains having been isolated that are unidentified at the species level. This shows that there is a great diversity within this genus in the Alhambra and Generalife study areas. Moreover, in communities with great diversity it can be difficult to isolate different algae, especially filamentous cyanobacteria that form mats.

It is useful to have isolated cultures of species, firstly for identification, since the microalgae's life cycles can be observed, as well as for other different characteristics that are often difficult to appreciate in samples. Furthermore, through this process a culture collection of living algae has been initiated that may be used in future studies. All of the microalgae from the Alhambra and Generalife are to a greater or lesser extent resistant to the treatments currently used to eliminate them (for example, many are resistant to chlorine), which makes them very interesting for use in trials with new treatments aiming to halt the proliferation of these organisms in research projects currently in progress.

## 4. Conclusions

The fountains of the Alhambra and the Generalife are affected mainly by the growth of green algae, cyanobacteria and diatoms, which grow in spite of the physical and biocide treatments that are periodically carried out. Within these groups, the green algae are the most abundant, as is the case in many other examples of cultural heritage sites involving water and ornamental stone. In most cases different algae grow, forming complex communities and giving rise to sheens, mats or crusts. Coccoid unicellular green algae such as *Chlorococcum* sp., *Bracteacoccus* sp., *Chlorosarcina* sp. and *Chlorosarcinopsis* sp. are especially common, which sometimes form monospecific sheens. Other common genera of green algae are *Klebsormidium, Scenedesmus, Chlorella, Choricystis, Pleurastrum, Gongrosira, Leptosira* and *Pseudopleurococcus*. As for cyanobacteria, the most notable genera are *Pleurocapsa, Chamaesiphon, Chlorogloea, Chroococcidiopsis, Cyanosarcina, Myxosarcina* and *Chroococcopsis* (chroococcal); and *Phormidium, Calothrix, Leptolyngbya, Lyngbya* and *Schizothrix* (filamentous). The species of microalgae found here hardly vary with seasonal changes, and the most common microalgae are always the same ones. All of the genera found in this study correspond to widespread algae that are common in the colonisation of monumental fountains and stone cultural assets in general.

Ten green algae genera and thirteen cyanobacteria genera have been isolated in cultures (with different species of some of them), thereby initiating the creation of a collection of algal cultures from the Alhambra and Generalife. Although the identification of each species among them is still in progress, this will gradually be carried out as new observations are made in the cultures. This collection may be used in the future to carry out trials of new treatments or to study the synthesis of certain products using microalgae in the study area.

**Author Contributions:** Conceptualization and work design, F.B.-G. and P.S.-C.; methodology and analyses, F.B.-G, C.A.-R. and P.S.-C.; software, M.T.; writing—original draft preparation, F.B.-G., C.A.-R. and J.R.-N.; writing—review and editing, J.R.-N., M.T. and P.S.-C.; data curation, C.A.-R. and M.T.; project administration, funding acquisition, F.B.-G and M.T. All authors have read and agreed to the published version of the manuscript.

**Funding:** This research was funded by the following projects: "Desarrollo de Nuevas Sinergias Arte-Ciencia aplicadas a la Conservación y Restauración de los Palacios y Jardines de la Alhambra y el Generalife" (VIRARTE), MINECO, with reference no. HAR2016-79886-P; "Métodos sinérgicos Arte-Ciencia-Tecnología para la Conservación-Restauración de la Alhambra y otros Bienes Culturales" (VIRARTE II), MICINN, with reference no. PID2019-109713RB-I00; "La aplicación de las algas procedentes de la Alhambra y el Generalife en técnicas artísticas y de conservación-restauración" (FICOARTE), Universidad de Granada, with reference no. A-HUM-279-UGR18. And the contract "Aplicación de nuevas técnicas de diagnosis y tratamiento de las microalgas presentes en la arquitectura del agua de la Alhambra, mediante la combinación de disciplinas", Patronato de la Alhambra y el Generalife, with reference no. Adenda 1/2017.

**Acknowledgments:** We would like to thank the Alhambra and Generalife Trust for their logistical support, without which it would not have been possible to do this work. We would also like to thank the Vice-Rector for Research and Knowledge Transfer, Enrique Herrera-Viedma, for his support on this research and for his initiative in creating the Unit of Excellence "Science in the Alhambra", as well as the anonymous reviewers for their constructive criticism and suggestions.

**Conflicts of Interest:** The authors declare no conflict of interest.

## References


1. Hueck, H.J. The biodeterioration of materials as a part of hylobiology. *Mater. Org.* **1965**, *1*, 5–34.
2. Anagnostidis, K.; Economou-Amilli, A.; Roussomoustakaki, M. Epilithic and chasmolithic microflora (Cyanophita, Bacillariophyta) from marbles of the Parthenon (Acropolis-Athens, Greece). *Nova Hedwigia* **1983**, *38*, 227–287.
3. Ortega-Calvo, J.J.; Sánchez-Castillo, P.M.; Hernández-Marine, M.; Sáiz-Jiménez, C. Isolation and characterization of epilithic chlorophytes and cyanobacteria from two Spanish cathedrals (Salamanca and Toledo). *Nova Hedwigia* **1993**, *57*, 239–253.
4. Bolivar-Galiano, F.; Peraza-Zurita, Y.; Romero-Noguera, J.; Sanchez-Castillo, P. L'Alhambra a Granada:15 anni di indagini sulla conservazione delle fontane. In *L'Acqua, le Pietre i Bronzi. le Fontane Monumentali*; Palombi Editori: Roma, Italy, 2010.
5. Vázquez-Nion, D.; Rodriguez-Castro, J.; López-Rodríguez, M.; Fernández-Silva, I.; Prieto, B. Subaerial biofilms on granitic historic buildings: Microbial diversity and development of phototrophic multi-species cultures. *Biofouling* **2016**, *32*, 657–669. [CrossRef] [PubMed]
6. Ibarra-Gallardo, C.E.; Novelo, E. Algas y cianoprocariontes epilíticos de la Zona Arqueológica de Yaxchilán, Chiapas, México. *Rev. Mex. Biodivers.* **2018**, *89*, 590–603. [CrossRef]
7. Mohammadi, P.; Gholami-Nejad, P.; Asghari-Daryasari, R.; Asgarani, E. The study of microbial communities of rudkhan castle. *Geomicrobiol. J.* **2019**, *37*, 119–129. [CrossRef]
8. Favero-Longo, S.E.; Viles, H. A review of the nature, role and control of lithobionts on stone cultural heritage: Weighing-up and managing biodeterioration and bioprotection. *World J. Microbiol. Biotechnol.* **2020**, *36*, 1–18. [CrossRef] [PubMed]
9. Golubic, I.F.S. The lithobiontic ecological niche, with special reference to microorganisms. *J. Sediment. Res.* **1981**, *51*, 475–478. [CrossRef]
10. Galiano, F.C.B.; Castillo, P.M.S. Biodeterioro del patrimonio artístico por cianobacterias, algas verdes y diatomeas. *Rev. PH* **1998**, *24*, 52. [CrossRef]
11. Hoiczyk, E. Gliding motility in cyanobacteria: Observations and possible explanations. *Arch. Microbiol.* **2000**, *174*, 11–17. [CrossRef] [PubMed]
12. Saiz-Jimenez, C. Biogeochemistry of weathering processes in monuments. *Geomicrobiol. J.* **1999**, *16*, 27–37. [CrossRef]
13. Bolívar-Galiano, F.; Sánchez-Castillo, P.M. Biomineralization processes in the fountains of the Alhambra, Granada, Spain. *Int. Biodeterior. Biodegrad.* **1997**, *40*, 205–215. [CrossRef]
14. Peraza Zurita, Y. Biodeterioro por Microalgas en Fuentes de Mármol. Ph.D. Thesis, University of Granada, Granada, Spain, 2004.
15. Bolívar-Galiano, F.C. Diagnosis y Tratamiento del Deterioro por Microalgas en los Palacios Nazaríes de la Alhambra. Ph.D. Thesis, University of Granada, Granada, Spain, 1994.
16. Sánchez Castillo, P.M.; Bolívar Galiano, F.C. Caracterización de comunidades algales epilíticas en fuentes monumentales y su aplicación a la diagnosis del biodeterioro. *Limnetica* **1997**, *13*, 31–46.
17. Sarró, M.I.; García, A.M.; Rivalta, V.M.; Moreno, D.A.; Arroyo, I. Biodeterioration of the lions fountain at the Alhambra Palace, Granada (Spain). *Build. Environ.* **2006**, *41*, 1811–1820. [CrossRef]
18. Cuzman, O.A.; Ventura, S.; Sili, C.; Mascalchi, C.; Turchetti, T.; D'Acqui, L.P.; Tiano, P. Biodiversity of phototrophic biofilms dwelling on monumental fountains. *Microb. Ecol.* **2010**, *60*, 81–95. [CrossRef] [PubMed]
19. Bourrelly, P. *Les Algues d'eau Douce. Algues Bleues et Rouges*; Éditions N. Boubée and Cie: Paris, France, 1970.
20. Komárek, J.; Anagnostidis, K. *Süßwasserflora von Mitteleuropa, Bd. 19/1: Cyanoprokaryota. 1st Part: Chroococcales*; Springer Spektrum: Berlin/Heidelberg, Germany, 1998.
21. Galiano, F.C.B.; Castillo, P.M.S. Claves de identificación de microalgas frecuentes en monumentos. *Rev. PH* **1999**, *26*, 93. [CrossRef]
22. Komárek, J.; Anagnostidis, K. *Süßwasserflora von Mitteleuropa, Bd. 19/2: Cyanoprokaryota. 2nd Part: Oscillatoriales*; Elsevier GmbH: Munich, Germany, 2005.
23. Komárek, J. *Süßwasserflora von Mitteleuropa, Bd. 19/3: Cyanoprokaryota. 3nd Part: Heterocytous Genera*; Springer Spektrum: Berlin/Heidelberg, Germany, 2013.

24. Peraza, Y.; Cultrone, G.; Sánchez-Castillo, P.; Bolívar-Galiano, F. Il biodeterioramento delle fontane dei Reales Alcazares di Siviglia e dell'Alhambra di Granada (Spagna). *Sci. Tech. Cult. Heritage* **2002**, *11*, 111–118.
25. Peraza Zurita, Y.; Cultrone, G.; Sánchez Castillo, P.; Sebastián, E.; Bolívar, F.C. Microalgae associated with deteriorated stonework of the fountain of Bibataubín in Granada, Spain. *Int. Biodeterior. Biodegrad.* **2005**, *55*, 55–61. [CrossRef]
26. Gaylarde, C.C.; Gaylarde, P.M. A comparative study of the major microbial biomass of biofilms on exteriors of buildings in Europe and Latin America. *Int. Biodeterior. Biodegrad.* **2005**, *55*, 131–139. [CrossRef]
27. Macedo, M.F.; Miller, A.Z.; Dionisio, A.; Saiz-Jimenez, C. Biodiversity of cyanobacteria and green algae on monuments in the Mediterranean Basin: An overview. *Microbiology* **2009**, *155*, 3476–3490. [CrossRef] [PubMed]

*applied sciences*

MDPI

*Review*

# Biological Risk for Hypogea: Shared Data from Etruscan Tombs in Italy and Ancient Tombs of the Baekje Dynasty in Republic of Korea

**Giulia Caneva [1], Daniela Isola [1,2,*], Hyun Ju Lee [3] and Yong Jae Chung [3]**

1 Department of Sciences, University Roma Tre, Viale Marconi 446, 00146 Rome, Italy; giulia.caneva@uniroma3.it

2 Department of Ecological and Biological Sciences (DEB), University of Tuscia, Largo dell'Università snc, 01100 Viterbo, Italy

3 Department of Heritage Conservation and Restoration, Graduate School of Cultural Heritage, Korea National University of Cultural Heritage, Buyeo, Chungcheongnam-Do 33115, Korea; smile2581@naver.com (H.J.L.); iamchung@nuch.ac.kr (Y.J.C.)

* Correspondence: daniela.isola@uniroma3.it

Received: 30 July 2020; Accepted: 27 August 2020; Published: 2 September 2020

**Abstract:** Biological growth represents one of the main threats for the conservation of subterranean cultural heritage. Knowledge of the conditions which favour the various taxonomic groups is important in delineating their control methods. Combining our experience regarding hypogea in Italy and the Republic of Korea, we aim to perform a critical review and comparison of the Biodeterioration Patterns (BPs) found, the materials used, and the conservative treatments applied. For this purpose, we focused on Etruscan tombs (Italy, 7th to 3th century BC) and the ancient tombs of the Baekje Dynasty (Republic of Korea, 6th to 7th centuries AD), most of which have been designated UNESCO World Heritage Sites, collecting original and bibliographic data as well as official documents. Results highlight the rich biodiversity of the bacterial and fungal species. Phototrophs were observed only in niches with sufficient light and the development of roots was also detected. Changes in humidity and temperature, the nature of the soil, nutrient accumulation, and vegetation above the hypogea along with human activities explain the different BPs. The effects of biocide treatments are also discussed, such as the emergence of dangerous fungal species. The shared data also enhance the role of overlaying tumuli and vegetation as well as protective barriers to reduce biological risk.

**Keywords:** cultural heritage conservation; eco-friendly biocides; ecology of biodeterioration; mural paintings; stone biodeterioration

## 1. Introduction

Subterranean sites such as caves and catacombs are generally characterised by limited air circulation, a limited variation in temperatures throughout the year, a high level of humidity, and at times, pronounced gradients of light [1]. Hypogea show similarities but also differences with caves due to their more limited size which leads to greater fragility also on the basis of the variability of materials used by human activities [2,3]. In fact, they were often embellished by wall paintings, stuccos, ceramics, bricks, and mosaics for decorative and augural purposes, as well as by statues, sarcophagi, and funerary objects which need to be preserved as testimony of their cultural heritage [4,5]. Moreover, the tombs which often maintained an environment cut off from the exterior world for hundreds of years, undergo substantial environment changes after excavations and consequent weathering [6,7].

Literature on biological colonisation in caves is becoming highly relevant, with hundreds of papers describing the microbial diversity and the relationship with environmental conditions, whereas much less is available for tombs. The most common inhabitants of subterranean sites are oligotrophs (bacteria, represented mainly by *Actinobacteria*, and fungi), accustomed to long periods of starvation and, at times, phototrophs (cyanobacteria, green algae, diatoms, mosses, ferns) that have adapted to scarce light [4–6,8–13]. The current knowledge of the ecological succession of microbial communities in hypogea in relation to environmental factors, such as the efficiency or risks of different control methods seems insufficiently addressed. The peculiarities of hypogea, which vary according to different cultures also deserve special attention, as they give rise to distinct biodeterioration phenomena. There are relatively few studies on the biodeterioration of tombs. The first tombs were in Egypt [14,15] and Japan [16–19], then in Italy and other European countries, and more recently in Korea and China [6,7,13,20–22].

Environmental conditions of subterranean tombs give rise to physic-chemical weathering especially when water evaporation occurs which leads to efflorescence (carbonation, nitrification, sulfatation) or when it reaches condensation, causing a loss of pigment and lithic effects of materials [2,23,24]. They also greatly increase the biological risk since the levels of humidity, temperature, input of nutrients, light, and the nature of the substratum, have long been considered facilitating factors for microbial colonisation [16,18,25–28]. Knowledge of microbial biodiversity and their metabolic and ecological profile, such as the environmental conditions which favor the various taxonomic groups, are important in defining the risk for artefacts and in creating both indirect (preventive) and direct (biocide treatments) control methods [4,29]. Knowledge of the tolerance range of each limiting factor for the various species has a great practical relevance in stopping or reducing undesirable growth [4,30]. Such information is also relevant when biocide treatment is taken into consideration, due to the need to choose the most efficient one against specific microorganisms [31].

Within an international cooperation project between Italy and Korea, we selected Etruscan tombs (mainly in Tarquinia but also at other sites of ancient Etruria), along with the ancient tombs in Songsan-ri (Gongju) and in Neungsan-ri (Buyeo), respectively. They are of interest also because most of them constitute relevant UNESCO World Heritage Sites. The aim of this paper is to analyse and provide a critical review and comparison of recurrent biodeterioration phenomena observed in the Etruscan tombs in Italy and the ancient tombs in Republic of Korea, as well as the respective conservative treatments carried out at these sites, with a special focus on preventive intervention and biocides. Combining experience from countries with significant differences in bioclimatic conditions and historical traditions, we also discuss the ecological relationships and successions among the communities thus making the results more relevant on an international scale.

## 2. Materials and Methods

In order to assess recurrent biodeterioration phenomena observed in Etruscan tombs in Italy and the ancient tombs in the Republic of Korea, we carried out an extensive study among peer-reviewed literature and other bibliographic source in particular from ICCROM (International Centre for the Study of the Preservation and Restoration of Cultural Property) as well as from the Italian ICR named later ISCR (Istituto Superiore per la Conservazione e il Restauro) and from the Korean KNUCH (Korea National University of Cultural Heritage), CHA (Cultural Heritage Administration), NRICH (National Research Institute of Cultural Heritage), and KNU (Kongju National University, Research Institute for Basic Science).

Further information was gathered from our recent field studies carried out on selected tombs in both countries to assess the microflora associated to some deterioration phenomena [32–34]. In this context, culture methods were used to isolate microorganisms and sequencing of some DNA barcode markers for their identification. Biochemical screening test were also performed to assess the detrimental potential of isolates [32–34]. We also created a taxonomical database on the species/genera/families that were identified, which took into consideration both qualitative and quantitative data, if present, along with documentation on the surrounding ecological conditions. As this work covers over 40 years of research, the identification of microorganisms has different level of taxonomical details being performed partly by morphological and partly by molecular approaches.

Such data were also examined to understand the ecological relationships and successions among the communities. Any data on conservative treatment (mainly biocides) were also considered and discussed.

### 2.1. Etruscan Tombs in Italy

The practice of building hypogea and decorating them with mural paintings was developed in several Etruscan cities especially in northern Latium, Tuscany, Umbria and Romagna covering a period from the 7th to the 3rd century BC. Since 2004, both the Necropolises of Cerveteri and Tarquinia (Latium) have become UNESCO World Heritage sites in consideration of their painted tombs which have an extraordinary importance as they reflect the daily life and burial practices of the Etruscans and bear witness to the achievements of their culture [35,36].

Our data set collected information on biodeterioration phenomena in 23 Etruscan tombs (Figure 1) present in three necropolises in Latium (in Tarquinia: Lotus, Jugglers, Mercareccia, Pulcella, Hunting and Fishing, Bulls, Shields, Ogre, Moretti, Blue Demons, Lions, Animas, Hunters, Elderly Men, Painted Vases, Giustiniani, Bartoccini, Lionesses, Caronti, in Cerveteri: Triclinium, Well, Reliefs) [25,26,28,32,37–43], and in Veio (Tomb of the Ducks) [44,45]. A further two tombs that were studied belong to a necropolis in Chiusi, Tuscany (Tombs of Hill and of the Monkey) [46–48].

**Figure 1.** Location of the Italian study cases and details of two tombs belonging to the UNESCO World Heritage sites of the Necropolis of Tarquinia (Monterozzi) (**a**) external view, (**b**) mural paintings, and of Cerveteri (Banditaccia): (**c**) external view, (**d**) mural paintings. (**b**,**d**: Courtesy of Amici delle Tombe di Tarquinia).

These tombs were dug in different geological substrata to obtain the sepulchral chambers, which lie at depths that vary from 2 m to 8 m. In Tarquinia, the geological features were calcarenites banks (Macco), which is an organogenic limestone rich in small shells. This bank constituted the top of the "Monterozzi" hill which was in contact with the underlying Pliocene clays that were more pliable and erodible. Both in Cerveteri and Veio, the characteristic tuffs, containing fragments of pumice arising from ignimbrite deposits, were preferred by sculptors and builders thanks to their softness and low permeability [49]. In the Chiusi necropolises, the tombs are situated in a lithological complex made from Pliocene sands (quartz, feldspars and a smaller amount of calcite) with clay cement [44].

From a bioclimatic point of view, the Cerveteri, Tarquinia and Veio sites fall within the Mediterranean macrobioclimate, with lower mesomediterranean thermotype and a lower subhumid ombrotype, whereas the Chiusi area has temperate characteristics [50]. In most of the Etruscan tombs of this study, the microclimate was found to be quite stable, with temperatures ranging from 10–15 °C to 17–20 °C and relative humidity (RH) between 90% and 100%. Such values sometimes led to dew point temperature, with water condensation on surfaces, especially in the cooler tombs [23].

The necropolises were often characterised by the presence of a large number of tumuli (circular tombs covered by earth mound, varying in diameter and height) (Figure 1) and sometimes of different shapes such as "dadi" (cube tombs). In the various necropolises, the original tumuli were often dismantled and only sometimes rebuilt for the purpose of protection, without considering the original shape [36,51].

The materials of Etruscan mural paintings were made utilising several mineral pigments, such as hematite, Egyptian blue and charcoal black which were then deposited on a previously prepared thin clay layer on the surface [46,51].

Several weathering phenomena of these tombs have accentuated both the physic-chemical weathering as well as biological issues. Each tomb has a specific history and problem, and we will subsequently describe the recurrent phenomena as well as some peculiar ones.

### 2.2. *The Ancient Tombs of Baekje Dynasty in Republic of Korea*

Many of the tombs with wall paintings were built by the people of Goguryeo in the Korean Peninsula. Goguryeo tombs are found throughout North Korea and in the Jian region, China. Sixty-three of these tombs have been designated as UNESCO World Heritage Sites since 2004. There are approximately ten tombs of this type in Republic of Korea, constructed from the 5th to 15th centuries. They have been managed as historical sites for their cultural importance and currently, access to these tombs is restricted due to damage caused by exposure to the external environment and influx [52].

During the Three Kingdoms period (Goguryeo: 37 BC–668 AD; Baekje: 18 BC–660 AD; and Silla: 57 BC–668 AD), three tomb styles developed in the Baekje Kingdom: "(1) a stone mound tomb (tomb made by piling up the stone) in the Hansung period, (2) a brick chamber tomb (a tomb style made by laying bricks for the tomb's corridor and the main chamber and constructing a tumulus on top of them) in the Ungin period, (3) a corridor-style stone chamber tomb (a tomb style where the corridor and main chamber were made of stone then covered with soil on top) in the Ungin/Sabi period" [53,54]. In lieu of previous reports and investigations [55–62], we selected Ancient Tomb No. 6 in Songsan-ri, Gongju, a brick-chamber tomb, and Ancient Tomb No. 1 in Neungsan-ri (Buyeo), a corridor-style stone chamber tomb, in Baekje Dynasty, listed as UNESCO World Heritage Sites for this study (Figure 2).

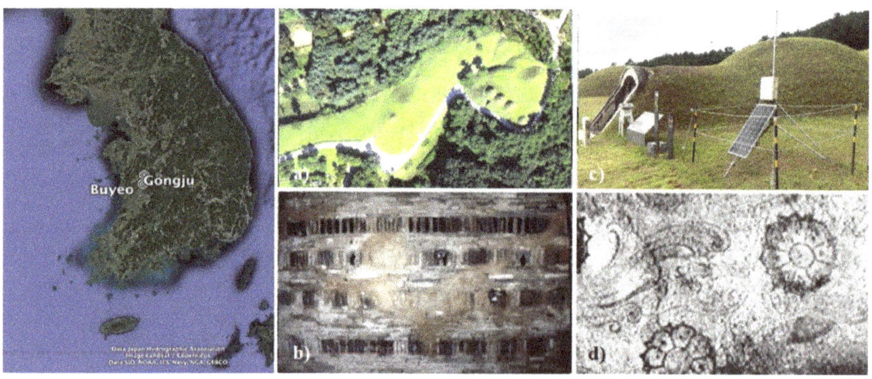


**Figure 2.** Location of South Korean study cases and details of the UNESCO World Heritage sites of the Ancient Tombs in Songsan-ri, Gongju (**a**) external view, (**b**) mural paintings [63], and the Ancient Tombs in Neungsan-ri, Buyeo: (**c**) external view [64], (**d**) mural paintings [64].

Tomb No. 6 was built in the mid-6th century as a royal tomb pertaining to the Baekje Dynasty. It has double corridors under an arch-shaped ceiling and a long rectangular brick chamber. It was discovered in 1933 and opened to the public after its excavation. Four guardian deities (east: blue dragon; west: white tiger; south: red phoenix; and north: black tortoise) are painted on the four walls. In addition, the "sun and moon" were also painted on the southern wall. Plastering the white lime upon the black ground on surface of the bricks, the only white pigments were painted on top of the ground layer made with soil [63]. The mural paintings were executed utilising only white pigment which was obtained either from oyster shells ($CaCO_3$) or chalk ($CaCO_3$) [65]. Today, only traces of these mural paintings remain, hence it is difficult to discern the original figures [52,63,66].

Tomb No. 1 was constructed between the late 6th and 7th centuries. It is considered the King and the Royal Family's tomb and the outer shape consists of a circular tumulus. It is a corridor-style stone chamber tomb dug from the ground with the east, west, and north walls made of well-trimmed stone slabs [52,67–69]. Four types of rock were used for main chamber: "(1) augen gneiss: on the northern and eastern wall, (2) two-mica granite: on the western wall, (3) hornblende schist: on the floor, and (4) granodiorite: on the ceiling" [70,71]. Four guardian deities are painted on the four walls of the burial chamber inside the tomb, with lotuses and clouds engraved on the ceiling. It was excavated in 1916 and the layout of the murals was clear at that time [61]. After excavation, it was opened to the public, but the murals gradually faded leaving only a faint trace, hence it was closed in 1971 to preserve the wall paintings.

Republic of Korea is geographically located in the mid-latitude temperate climate zone with four distinctive seasons (spring, summer, autumn, and winter: the annual average temperature is 10–15 °C, the highest temperature ranges 23–26 °C, and the lowest temperature of minus 3–6 °C, with an annual relative humidity of 60–75% [72]. The indoor temperature of the two tombs differed from the outdoor temperature, as there was little heat transfer due to the tumuli [73,74]. The indoor temperature of Tomb No. 6 ranged from 13.6 to 20.6 °C, and the annual average temperature was 17.0 °C, with the lowest values in April, and the highest in September and October (environmental data from April 2018 to March 2019) [75]. The indoor temperature of Tomb No. 1 ranged from 12.9 to 18.0 °C and the annual average temperature was 15 °C, with the lowest values in May and the highest in November (environmental date from January to December 2019) [34]. The indoor relative humidity of the two ancient tombs was maintained at approximately 100% throughout the year. The shape of the tombs was reassembled when repair and restoration work took place [76–79]. Moreover, these two tombs were passively controlled without a specific environmental control system.

## 3. Results

Shared data on main BPs found in both countries, such as the phenomenology of biodeterioration and species found, arising from 23 Etruscan tombs in Italy and from two tombs in Republic of Korea are listed in Tables 1 and 2.

**Table 1.** Phototrophic and bacterial deteriogens associated to biodeterioration patterns found in Italian [25,26,28,32,37–48,80–82] and South Korean tombs [33,34,55–60,62,73,83–87].

| | | | Biodeterioration Patterns | | | | | | |
|---|---|---|---|---|---|---|---|---|---|
| | | | Green | White Patina | White Fluffy | Whitish-Gray | Gray | Rose/ Purple | Brown-Blackish |
| Phototrophs | | Phototrophs | +++ | | | | | | |
| | Algae | Diatoms | +/+ | | | | | | |
| | | *Apatococcus* sp. | + | | | | | | |
| | | *Chlorella* sp. | + | | | | | | |
| | | *Chlorella vulgaris* | + | | | | | | |
| | | *Muriella terrestris* | + | | | | | | |
| | | *Stichococcus* sp. | + | | | | | | |
| | | *Pseudococcomyxa simplex* | + | | | | | | |
| | | *Trebouxia* sp. | + | | | | | | |
| | *Cyanobacteria* | *Chroococcus* sp. | + | | | | | | |
| | | *Gloeocapsa* sp. | + | | | | | | |
| | | *Leptolyngbya* sp. | + | | | | | | |
| | | *Lyngbya* sp. | ++ | | | | | | |
| *Actinobacteria* | | *Actinobacteria* | | ++/++ | + | + | | + | |
| | | *Arthrobacter* sp. | | + | | | | | |
| | | *Cellulosimicrobium* sp. | | + | | | | | |
| | | *Dermacoccus* sp. | | + | | | | | |
| | | *Isoptericola* sp. | | + | | | | | |
| | | *Microbacterium* sp. | | + | | | | | |
| | | *Micrococcus* sp. | | + | | | | | |
| | | *Mycobacterium* sp. | | + | | | | | |
| | | *Nocardia* sp. | | + | | | | | |
| | | *Pseudonocardia* sp. | | + | | | | | |
| | | *Rhodococcus* sp. | | ++/++ | | | | | |
| | | *Sinomonas* sp. | | + | | | | | |
| | | *Streptomyces* sp. | + | ++/+ | | ++ | | | |

**Table 1.** *Cont.*

| | | | Biodeterioration Patterns | | | | | | |
|---|---|---|---|---|---|---|---|---|---|
| | | | Green | White Patina | White Fluffy | Whitish-Gray | Gray | Rose/Purple | Brown-Blackish |
| Other bacteria | | Other Bacteria | ++ | | ++ | ++ | + | | |
| Other bacteria | *Firmicutes* | *Firmicutes* | | ++ | | | | | |
| Other bacteria | *Firmicutes* | *Bacillus* sp. | | ++/++ | | | | | |
| Other bacteria | *Firmicutes* | *Bacillus cereus* | | + | | | | | |
| Other bacteria | *Firmicutes* | *Bacillus megaterium* | | + | | | | | |
| Other bacteria | *Firmicutes* | *Bacillus simplex* | | + | | | | | |
| Other bacteria | *Firmicutes* | *Paenibacillus* sp. | | ++/++ | | | | | |
| Other bacteria | *Proteobacteria* | *Proteobacteria* | | +++ | | | | | |
| Other bacteria | *Proteobacteria* | ***Alphaproteobacteria*** | | + | | | | | |
| Other bacteria | *Proteobacteria* | ***Rhizobiales*** | | ++ | | | | | |
| Other bacteria | *Proteobacteria* | *Bradyrhizobium* sp. | | +/+ | | | | | |
| Other bacteria | *Proteobacteria* | *Hyphomicrobium* sp. | | ++ | | | | | |
| Other bacteria | *Proteobacteria* | *Pedomicrobium* sp. | | + | | | | | |
| Other bacteria | *Proteobacteria* | *Phyllobacterium* sp. | | + | | | | | |
| Other bacteria | *Proteobacteria* | *Prosthemicrobium* sp. | | + | | | | | |
| Other bacteria | *Proteobacteria* | *Ochrobactrum* sp. | | +/+ | | | | | |
| Other bacteria | *Proteobacteria* | ***Betaproteobacteria*** | | | | | | | |
| Other bacteria | *Proteobacteria* | *Crenobacter* sp. | | + | | | | | |
| Other bacteria | *Proteobacteria* | *Cupriavidus* sp. | | + | | | | | |
| Other bacteria | *Proteobacteria* | *Burkholderia* sp. | | + | | | | | |
| Other bacteria | *Proteobacteria* | *Variovorax* sp. | | + | | | | | |
| Other bacteria | *Proteobacteria* | ***Gammaproteobacteria*** | | + | | | | | |
| Other bacteria | *Proteobacteria* | *Dyella* sp. | | | | + | | | |
| Other bacteria | *Proteobacteria* | *Lysobacter* sp. | | | + | | | | |
| Other bacteria | *Proteobacteria* | *Pseudomonas* sp. | | | | + | + | | |
| Other bacteria | *Proteobacteria* | *Serratia* sp. | | + | | | | | |
| Other bacteria | *Proteobacteria* | *Stenotrophomonas* sp. | | +/+ | | | | | |
| Other bacteria | | *Acidobacteria* | | + | | | | | |

(+) Italian and (+) South Korean records. Frequencies are expressed as + = cited; ++ = recurrent; +++ = very common. Heterotrophic bacteria are grouped in four phyla (left). In *Proteobacteria* were highlighted also classes (bold underlined) and orders (bold).

**Table 2.** Fungal deteriogens associated to biodeterioration patterns found in Italian [25,26,28,32,37–48, 80–82] and South Korean tombs [33,34,55–60,62,73,83–87].

| | | Biodeterioration Patterns | | | | | |
|---|---|---|---|---|---|---|---|
| | | Green | White Patina | White Fluffy | Whitish-Gray | Gray | Brown-Blackish |
| *Mucoromycota* | *Mortierella* sp. | | | ++/+ | | | |
| *Mucoromycota* | *Mortierella alpina* | | | | + | | |
| *Mucoromycota* | *Mortierella ramanniana* | | | + | | | |
| *Mucoromycota* | *Mucor* sp. | | | +/+ | + | | |
| *Mucoromycota* | *Mucor racemosus* | | | ++ | + | | |
| *Mucoromycota* | *Rhizomucor* sp. | | | | | + | |
| *Mucoromycota* | *Umbelopsis* sp. | | | | + | | |

**Table 2.** *Cont.*

| | | | Biodeterioration Patterns | | | | | |
|---|---|---|---|---|---|---|---|---|
| | | | Green | White Patina | White Fluffy | Whitish-Gray | Gray | Brown-Blackish |
| *Ascomycota* | *Sordariomycetes* | *Acremonium* sp. | + | | | + | | |
| | | *Acremonium*-like fungi | | | | | | + |
| | | *Chaetomium* sp. | | | | | + | |
| | | *Doratomyces* sp. | | | | | ++ | |
| | | *Engyodontium* sp. | | | | + | | |
| | | *Engyodontium album* | | | | + | | |
| | | *Fusarium* sp. | | | ++ | ++ | | |
| | | *Fusarium oxysporum* | | | | + | | |
| | | *Fusarium solani* | | | | + | | |
| | | *Glyomastix cerealis* | | | | | +++ | |
| | | *Hypocrea* sp. | | | | + | | |
| | | *Isaria farinosa* | | | + | | | |
| | | *Lecanicillium* sp. | | | + | | | |
| | | *Microascus* sp. | | | | | + | |
| | | *Purpureocillium* sp. | | | | + | | |
| | | *Scopulariopsis* sp. | + | | | + | | |
| | | *Scopulariopsis brevicaulis* | | | | + | | |
| | | *Torrubiella* sp. | | + | | | | |
| | | *Trichocladium asperum* | | | | + | | |
| | | *Trichoderma* sp. | + | | | | + | + |
| | | *Verticillium* sp. | | | | + | | |
| | | *Verticillium bulbillosum* | | | +++ | | + | |
| | *Eurotiomycetes* | *Aspergillus* sp. | | | | + | | |
| | | *Aspergillus creber* | | | | + | | |
| | | *Aspergillus flavus* | | | | | | + |
| | | *Aspergillus versicolor* | | | | | + | |
| | | *Eurotium* sp. | + | | | | | |
| | | *Eurotium herbariorum* | | | | | | + |
| | | *Exophiala* sp. | | | | | | + |
| | | *Exophiala angulospora* | | | | | | + |
| | | *Paecilomyces* sp. | | | + | + | ++ | |
| | | *Penicillium* sp. | | | + | +/+ | ++ | + |
| | | *Penicillium brevicumpactum* | | | | + | | |
| | | *Penicillium communis* | | | | + | | |
| | | *Penicillium miczynszii* | | | | + | | |
| | | *Penicillium rugulosum* | | | | + | | |
| | *Dothideomycetes* | *Cladosporium* sp. | | | | | + | ++ |
| | | *Cladosporium cladosporioides* | | | | | + | + |
| | | *Paraphaeosphaeria* sp. | | | | | | + |
| | | *Preussia terricola* | | | | | | + |
| | | *Epicoccum* sp. | | | | | | + |
| | | *Leotiomycetes* | | | | | | |
| | | *Geomyces pannorum* | | + | | | | |
| | | *Saccharomycetes* | | | | | | |
| | | *Geotrichum* sp. | | | | + | | |

(+) Italian and (+) South Korean records. Frequencies are expressed as + = cited; ++ = recurrent; +++ = very common. Fungi are organised in two divisions (-mycota) and *Ascomycota* in five classes (-mycetes).

## 3.1. *The Etruscan Tombs*

### 3.1.1. Recurrent Biodeterioration Phenomena

Research which commenced in the 1980s indicate that the Etruscan tombs are mainly populated by *Actinobacteria* (belonging mainly to the order Actinomycetales) which give rise to biodeterioration phenomena of the painted layers. Firmicutes, in particular Bacillaceae, and Alphaproteobacteria (order Rhizobiales) also seem to play an important role.

Modern biomolecular tools utilised for the Tarquinia and Chiusi tombs have permitted a more precise identification of these microbiomes, even if some species do not seem to play a part in the biodeterioration processes. In the case of the Chiusi tombs, microbial colonisation was observed on both painted and unpainted wall surfaces, and a total of 456 non-chimeric bacterial sequences were obtained from different areas and pigments, mostly belonging to the Alphaproteobacteria (order Rhizobiales) and Firmicutes groups, such as *Rhodococcus* sp., *Streptomyces* spp., *Bacillus* spp., *Paenibacillus* sp., and also of *Nocardia* and *Pseudonocardia* genera [46–48]. In the Mercareccia tombs (Tarquinia), 142 different heterotrophic colony morphotypes were identified, 16% of which had not been described in relation to artistic heritage up to now. The most representative bacteria belonged to Actinomycetales and Bacillales, and the most common genera were *Bacillus* (24 strains) and *Streptomyces* (18 strains), together with *Rhodoccoccus* (seven strains), *Paenibacillus* (two strains) and *Pseudomonas* (one strain) [39]. In one Tarquinia tomb (Shields) [42], were also detected moonmilk deposits, which consist of very fine crystals of calcium carbonate, made up of calcite needle fibres (1–2 µm) and nanofibres (less than 1 µm). This phenomenology arises from the progressive accumulation of calcite which also traps microorganisms in the crystal matrix, until they are completely encompassed in the crystals [88,89]. The patina is similar to talcum or chalk powder when dry, and the white layers having different consistencies. One is softer while the other is thick and hard, thus more difficult to remove. The microbial community was made up of Proteobacteria, representing approximately 50% of the population, followed by the phyla *Acidobacteria* and Actinobacteria [42]. More precisely, the class Alphaproteobacteria predominated the community (25% of the population), followed by Betaproteobacteria, Gammaproteobacteria and Actinobacteria. The most representative genera were *Pseudonocardia* and *Hyphomicrobium*. In this case, the high affinity in composition highlighted how representative they were of microbiota contained in the tomb [42].

A comparison of the taxa using biomolecular techniques often shows that microbial communities can vary greatly, in fact, the two tombs of Chiusi have very few in common [47,48].

The presence of fungi on the mural paintings displayed a variable, though not negligible, aspect. In the Chiusi tombs, amplification of archaea and fungi was unsuccessful from all samples taken from the whitish alterations [47,48]. On the contrary, in the Tarquinia, Cerveteri and Veio tombs, a certain amount of fungi was recorded. For example, in the Mercareccia Tomb (Monterozzi Necropolis), fungi occurred in both the dromos and inner chamber. The taxa identified were Sordariomycetes, Eurotiomycetes, and Dothideomycetes, along with several species of *Penicillium* (*P. rugulosum, P. communis, P. brevicompactum*) and *Aspergillus* (*A. candidum, A. versicolor, A. flavus*). Other species that were detected included *Cladosporium cladosporioides, Preussia terricola, Trichocladium asperum, Scopulariopsis brevicaulis, Engyodontium album, Eurotium herbariorum*, and *Isaria farinosa* [39]. Our present study on the Moretti and Blue Demons Tombs in Tarquinia highlighted melanised fungi as responsible for the dark pattern deterioration found in both tombs (Figure 3). *Acremonium*-like fungi (three strains), *Cladosporium* sp. (one strain), and *Exophiala angulospora* (two strains) were identified from the Moretti Tomb, while *Cladosporium* sp. (two strains) and *Exophiala* sp. (one strain) were isolated the Blue Demons Tomb [32]. Tracks of rhizomorphs arising from Basidiomycetes, which utilized the remains of dead wooden roots, were also conjectured for some Tarquinia Tombs [80].

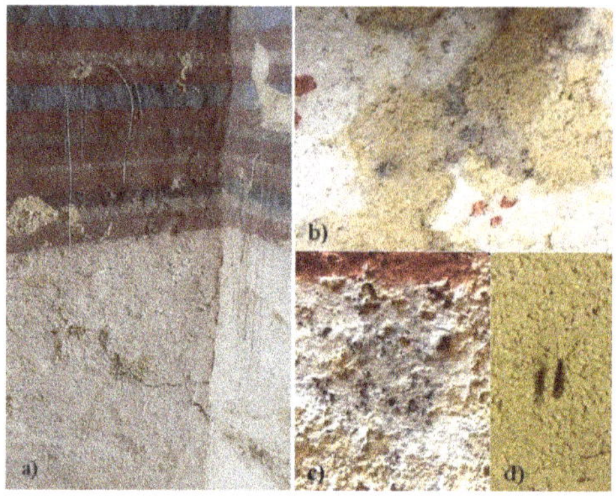


**Figure 3.** Biodegradation patterns in the Italian tombs from the Monterozzi Necropolis in Tarquinia. (**a**) Roots—Lotus flower Tomb; (**b**,**c**) fungal dark discoloration—Moretti Tomb; (**d**) *Limonia* cf. *nubeculosa*—Blue Demons Tomb.

Phototrophic microbial groups, such as algae and cyanobacteria are very rarely reported in Etruscan tombs. Few data available on the Tarquinia tombs (Leonesse, Bartoccini) indicate that they occur inside the tombs where a certain amount of sunlight is present, such as around the lighting systems or close to the entrance [80]. The occurring species were cyanobacteria (*Leptolynbya*), green algae (*Muriella terrestris*), and to a lesser extent *Pseudococcomyxa simplex* and *Chlorella vulgaris*, and Diatoms [80].

As observed in several tombs in Cerveteri and Tarquinia [41,51], the roots of wooden and herbaceous plants were also common deteriorative agents in many Etruscan sites. In the case of herbaceous roots in hypogea (Figure 3a), a precise identification of the occurring species has not yet been carried out [80].

Moreover, insects, slugs, and miriapods (*Scutigera coleoptrata*) were detected in such sites, though their presence did not merit particular attention, despite their potential role in spore dissemination and in utilising and releasing organic materials on surfaces [81]. The occurrence of termites was also detected in the Giustiniani Tomb (Tarquinia), where typical tunnels of *Reticulitermes lucifugus* colonisation were observed [81].

### 3.1.2. Conservation Treatments

Variations in temperature and humidity inside the burial chambers of Etruscan tombs over lengthy periods of time have created significant deterioration [2,23,82]. Indirect control methods were also carried out, stabilising the microclimatic conditions after a partial reconstruction of overlaying tumuli and the creation of a system of double doors to close off the burial chambers, hence preventing entrance to visitors [51]. Glass barrier protection systems were put in place to prevent condensation at the tomb entrance and to allow for good visibility of the interior [23]. A cold light illumination system that does not modify the internal temperature of the chamber was also installed to prevent microclimatic changes and growth of photoautotrophic organisms [51]. In the case of Tarquinia, tests on the possible use of new vegetation cover to reduce the effects of summer heat are still underway [51].

To keep biodeterioration phenomena in check, the most common biocide compounds utilised were quaternary ammonium salts. The product that was initially used was Desogen, and then Neodesogen (both based on benzalkonium chloride), and more recently Preventol R80 (Bayer) Preventol® RI 80 (Lanxess) (dodecyl dimethyl dichlorobenzyl ammonium chloride) [25,47,81]. However, the disinfection of tombs with these biocidal compounds occasionally gave rise to a certain extent of interference with materials and to unexpected problems. In fact, after treatment with Desogen (10%) in the Cerveteri tombs (in particular, a painted relief in the Banditaccia Necropolis), a huge growth of a white fluffy mycelium was observed due to the development of *Mortierella alpina*, *Mucor racemosus*, *Verticillium bulbillosum*, *Mortierella ramamniana*, *Penicillium* sp., and *Gliomastix cerealis*. This last mentioned was darker than others and the most difficult to eradicate [25]. Instead, species belonging to *Acremonium*, *Geotrichum* and the *Penicillium* genera were found in the Tomb of the Triclinium at the same necropolis, as a recolonisation effect following this kind of treatment [27]. Similarly, in the Tomb of the Ducks (Veio), the newly formed white mycelium was associated to *Penicillium miczynskii*, *Fusarium* sp., and *Paecilomyces* sp. [27]. Such a phenomenon was interpreted as the consequence of treatment in limiting the competition of these fungi with more sensitive bacteria. After several tests, only treatment with orto-phenylphenol was effective in blocking their growth [25]. In tests sites, the use of Metatin 58/I0 (Acima Chemicals) a broad-spectrum biocide, resulted as the most effective method of eliminating all microorganism species definitively [27]. The authors suggested to improve tests and not removing the residuals of biocides as usual methods during restoration [27].

In fact, in several further ISCR reports [80,81], treatment with benzalkonium chlorides compounds was suggested since the secondary effect of provoking further colonisation were not observed in several cases. Restorers also consider such treatment as a common practice in case of microbiological attacks, using them at doses of 2% [51]. Some recent studies also looked into identifying microorganisms capable of surviving biocide treatment with ammonium quaternary salts (Preventol RI80) by producing biofilms and/or spores [40]. An analysis carried out one month from the biocide treatment in various places of the tomb showed that it was effective on most bacteria, but not so on species capable of producing resistant forms of life like spore-forming bacteria, such as *Bacillus* sp. and moulds which were still alive [40].

Furthermore, microwave heating treatment was also applied as a new methodology to control biodeterioration phenomena [46]. The prototype system used a 2.45 GHz microwave generator with an adjustable output power of up to 1 kW and an optimal application dose of 65 °C for 3 min. The authors indicate that they checked the possible interference of microwaves on substratum and pigments with a micro photogrammetric system. The efficiency of the treatment, which was evaluated with the Plant Cell Viability Kit on the collected biological samples and on a white spotted area, gave some positive results [46].

In the case of root development, biocide treatments directly on the occurring roots were also carried out in the past, using glyphosate (Rodeo Gold) and hexazinone (Velpar L, Dupont) as herbicides [80].

### 3.2. *The Ancient Tombs of the Baekje Dynasty*

#### 3.2.1. Recurrent Biodeterioration Phenomena

Considerable biodeterioration phenomena took place in both of the analysed tombs with changes occurring after the opening and with further intervention procedures.

In the conservation history of Tomb No. 6, heavy rainfall in Gongju in 1995 led to leakage, which caused blue-green algae, such as *Lyngbya* spp. and *Gloeocapsa* spp. which spread throughout the tomb [83,84]. Since blue-green algae causes colouration and erosion of the structure, they were removed and subsequently no secondary damage caused by the algae was confirmed by a detailed ecosystem study after its removal [90]. In 1997, the Ancient Tombs in Songsan-ri, Gongju were permanently closed to preserve the original state of the mural paintings. Research on microbial distribution of Tomb No. 6 was first conducted in 2010 and twelve species of fungi were identified (e.g., *Acremonium* spp., *Aspergillus* spp., *Epicoccum* spp., *Cladosporium* spp., *Penicillium* spp., *Fusarium* spp., were found in the main chamber [55]. Bacteria belonged to Actinobacteria (three strains), Firmicutes (seven strains) and Proteobacteria (two strains). *Bacillus* sp. resulted as the dominant strain and *B. simplex* was also discovered on the wall surface. After some negative results were produced with microclimatic conditioning, studies were conducted to monitor the microbial distribution of the ancient tombs. Before entering the main chamber of Tomb No. 6, where the wall paintings are present, a person (i.e., visitor, operator) must pass through three buffer zones. The microorganism distribution tended to decrease upon entering the main chamber, due to the buffer activity of such areas caused by the inflow of outdoor air. *Aspergillus, Cladosporium, Penicillium, Bacillus,* and *Pseudomonas* genera were isolated in the air and *Aspergillus* was also identified on the bricks and walls. Soil microorganisms of the genus *Bacillus*, which seem to have flowed inside the tomb from the soil environment, were the dominant bacteria [56]. In a further study, strains collected in the air (*Doratomyces, Engyodontium, Fusarium, Aspergillus* genera) were identified in all four cardinal points of the walls [85].

Since 2008, numerous in-depth studies have been conducted on Tomb No. 1 which was closed in 1971 to preserve the wall paintings, along with an analysis of the biological distribution focusing on discoloured areas on the walls of the main chamber as well as the soil. As a result, 16 bacteria species were identified, including *Actinomycetes* and a total of 15 fungal species (four species of *Penicillium*, three species of *Aspergillus*, one species of *Fusarium*, one species of *Trichoderma*, and six unidentified species) were found. The green specimens collected at the entrance of Tomb No. 1 were microalgae (*Apatococcus, Chlorella, Trebouxia,* and *Stichococcus*), and cyanobacteria (*Chroococcus*) [86]. Results of microbial distribution monitoring performed from 2016 to 2018 highlighted a decrease in microbial numbers upon entering the ancient tomb (entrance > outside > antechamber > corridor > main chamber). The highest airborne microbial distribution was found to be at the entrance as insects inhabiting the space and the air inflow while entering the tomb stationed there for a long time. *Cladosporium, Penicillium,* and *Bacillus* were dominant airborne microorganisms and *Bacillus, Cupriavidus,* and *Streptomyces* were identified on all the walls of the main chamber. *B. cereus, B. megaterium, Dermacoccus nishinomiyaensis, Mucor* sp., *Rhizomucor* sp., and *Penicillium* sp., identified on the walls were equally represented in the air. Accordingly, it was concluded that microorganisms in the air can harm the wall paintings when they settle and grow on the walls [59,60].

Patinas presumed to be microorganisms were observed on the walls of the main chamber in Tomb No. 1 (Figure 4). They were distributed in the middle and bottom sections of the eastern and northern walls and, based on colour, were classified either white or ivory, with white patinas first being mentioned in 2008. At the time, they were thought to be Actinomycetes [86]. Later, studies in 2016 and 2018 confirmed that the patinas continued to exist in the same location. They were presumed to be Actinomycetes based on SEM-EDS analysis, but accurate determination was difficult [60]. Metagenomic sequencing analysis was attempted to accurately identify these patinas, and further studies are currently underway.

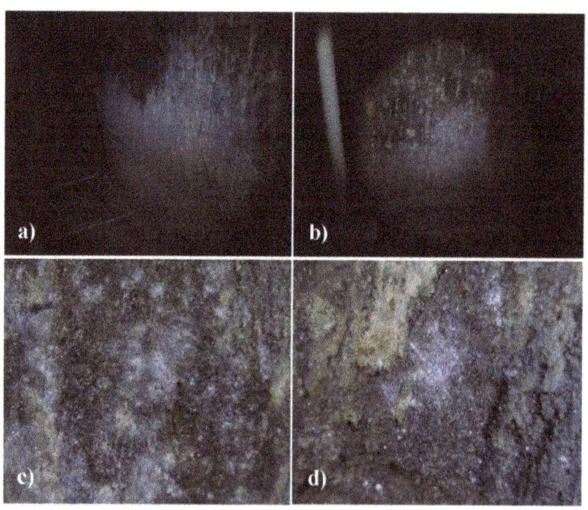


**Figure 4.** Biodeterioration patterns (whitish-grey patinas) on the wall in the main chamber of the Ancient Tombs in Neungsan-ri, Buyeo; (**a**,**b**) general view of the walls, (**c**,**d**) close view of biological patinas.

As regards to roots, photographs and reports showed that only Tomb No. 6 had such a problem. In the early phases of excavation, tree roots were found growing between gaps in the brick work [91]. After maintenance procedures were carried out, no further events have been recorded. Despite the wide distribution of pitch pine (*Pinus rigida* Mill.) around the tumuli, no microbial activity nor root intrusion has been reported for Tomb No. 1 [92].

#### 3.2.2. Conservation Treatments

In Republic of Korea, indirect methods have been used to control the environment to prevent problems caused by microorganisms in ancient tombs. Since 2003, the indoor temperature and humidity in Tomb No. 6 has been maintained with an air conditioning system at 18 °C and 90% humidity on average. However, air flow from the system dried the air too much creating problems for the wall paintings hence the system was shut down in 2011. An analysis of the microclimatic conditions was carried out and buffer zones were set up to assist prolonged conservation of the hypogea. Research was also conducted on the biochemical characteristics and environment influences (temperature, moisture, and nutrients) of microorganisms over the seasons [58]. On the basis of previous investigations, internal microclimatic conditions, and thermal preferences of isolates, three risk periods for microorganism occurrence in ancient tombs were classified [56] as follows: (1) Safe: $T < 18$ °C, (2) Warning: $T = 18$–$22$ °C, (3) Risk: $T > 22$ °C [6]. In addition, the risk periods for microorganism occurrence in the two tombs were respectively: "safe period for microorganism occurrence" from January to July for Tomb No. 6 and from January to October for Tomb No. 1 while the "warning period for microorganism occurrence" was from August to December for Tomb No. 6 and from November to December for Tomb No. 1. Such differences in the risk periods for microorganism occurrence in the two ancient tombs is due to the different structures and environments.

Further research was also conducted to select biocides to remove the blue-green algae (*Gloeocapsa* spp. and *Lyngbia* spp. [83,84] that occurred in Tomb No. 6 and the surroundings tombs. Reagent K101 was selected (composition: EDTA, $(NH_4)_2CO_2$, $Na_2CO_3$, sodium hypochlorite) which was developed by differentiating the composition of AC322 [93] and B57 solutions [83]. After spraying the biocides to the walls, poultices were applied, and ultraviolet rays were subsequently irradiated for 24 h [83,85]. The treatment did not give rise to recolonisation phenomena and secondary damage [90]. A thermo-hygrostat was installed in 2003 to maintain stable environmental conditions but unfortunately it caused the drying of murals and walls, so that its use has been stopped in the Tomb No. 6 since 2011. Then, a treatment was conducted to prevent possible damage being fungal growth started on the facing when the thermo-hygrostat was stopped. Benzalkonium chloride ($C_{22}H_{40}CN$) 1% was used to remove the fungi, and the rayon paper that was applied was later removed using tepid distilled water [94]. After that, there have been no cases in which biocides have been used directly inside ancient tombs and on the wall paintings.

Since heterotrophic microorganisms and algae may grow if there are suitable nutritional and environmental conditions, the development and application of efficient eco-friendly biocides that do not damage the murals is necessary [87]. Several studies have been conducted to select and field test substances from medicinal plants and tree extracts capable of managing microorganisms that occur on such sites [95–101]. Research has also been conducted on ingredients and extracts of herbal medicine to remove biofilms from external stone cultural heritage elements [102]. Such studies have led to selecting a biocide (Stone Keeper) based on eugenol with an eco-friendly emulsifier [103]. This biocide was applied to fungi taken from ancient tombs on the basis of a study showing that the natural substances anethole and eugenol have antibacterial and antifungal properties [87]. The mixture of anethole and eugenol (1:2) significantly inhibited the growth of *Aspergillus creber* and *Aspergillus versicolor*. A simulation also tested the effects of natural ingredients, thereby verifying their potential to inhibit and kill microorganisms that occurred on ancient tomb murals [85,87].

## 4. Discussion

### 4.1. Biodeteriogenic Taxa, Phenomenology of Alteration and Biodeteriogenic Processes

As shown in Tables 1 and 2, in both Italian and Korean sites, a great number of bacterial and fungal species were found in the analysed tombs. The influence of different bioclimatic conditions, such as of the use of different traditional materials (stone, mortar, brick and pigment) explain the significant number of taxa involved. In Italian sites, hundreds of bacterial species (mainly Actinobacteria, but also Firmicutes (Bacillaceae family) and Alphaproteobacteria) and fungi (principally Sordariomycetes and Eurotiomycetes) were detected by molecular methods [39,40,42,46,48]. In Korean sites, *Actinomycetes* and *Bacillus* sp., which seem to have entered the tomb via the soil were the dominant bacteria [56]. Fungal species of different genera (*Aspergillus*, *Cladosporium*, *Doratomyces*, *Engyodontium*, *Fusarium*, *Geomyces*) were also widespread in the air and on the walls of the various tombs [85]. *Geomyces pannorum* and *Fusarium solani* were also identified, and the latter has also been described on the Lascaux cave paintings in France as well as in the Kitora Tomb and Takamatsuzuka Tomb in Japan [19,104]. The presence of some fungal species (e.g., *A. niger and A. flavus*) can also be dangerous to humans due to their production of mycotoxin and allergens [104]. A peculiar situation occurred in the case of burial remains inside the tombs, which determined a specific selection of microbial colonisation [17].

New studies on old Chinese tombs confirmed the important role of bacterial and fungal colonisation in deterioration processes. In particular, bacteria of the genera *Bacillus*, *Massilia* and *Brevibacillus* were identified in the bricks and were considered possible contributors to brick weathering [20]. Such sites contain previously unknown species; new bacterial species of (*Paenibacillus tumbae* sp. nov.) were isolated from the Tomb of the Emperor Yang of the Sui Dynasty [20]. Fungi were also found in great number, the most common genera being *Cordyceps, Fusarium, Harpochytrium, Emericellopsis, Volutella, Cladosporium, Stachybotrys, Trichoderma, Cochlonema* [105,106]. Further studies of interesting and unknown fungal species are underway in Etruscan tombs [32].

In both Italian and Korean sites, algae and cyanobacteria were observed only in conditions of high levels of humidity and sufficient lighting. Similar genera were identified, principally being *Apatococcus, Chlorella, Trebouxia, Stichococcus, Pseudococcomyxa*, and *Muriella* for green algae, and *Chroococcus* and *Leptolynbya* for cyanobacteria. It is possible that vicariant species most likely occur in both countries though further in-depth taxonomical studies are needed. The development of roots was sometimes a frequent occurrence as study results illustrated at the Italian and Korean sites, even if the importance of such phenomenon seems to vary between the two countries, probably due to the different management procedures of the areas. The maintenance of herbaceous cover and regular cutting can reduce the risk of damage [43], even if data from Tarquinia reported depth penetration from non-wooden plants. Recent studies have addressed root identification utilising molecular tools. Insects, Arthropoda, and slugs were also detected, in both countries, even if they have not been sufficiently investigated.

The role that some identified species may have needs to be carefully evaluated via biomolecular analysis, since many species may simply be part of the soil ecosystem, without being responsible for causing damage. On the contrary, other bacterial and fungal species can be more problematic, since their potential growth produces staining and chemical and physical interference with the materials. As observed for the Mogao Grottoes (Dunhuang, China), biodeterioration of wall paintings suggests that only a small fraction of the rich microbial community may in effect grow and cause damage, whereas most of these microorganisms are dormant or metabolically slow [105]. Hence it is of the utmost importance to study the metabolic traits of isolates to associate them to a risk. For example, moonmilk which is present in Etruscan tombs, as in karst caverns or in other hypogeal environments [107] does not seem to greatly detrimental, and *Streptomyces* species may actually exert a protective effect, since restorers have reported that under the patina the wall paintings were well preserved [42].

The most evident effect of microbial colonisation are the different chromatic alterations on the surfaces, often producing whitish or grayish patinas and spots which resemble salt efflorescence or carbonate deposits [26,38,80]. As shown in Tables 1 and 2, the same taxa can at times be associated with quite different biodeterioration patterns, probably due to the complexity of the microbial communities and subterranean trophic networks as recorded for example in *Acremonium* sp. found as a minority component of a green phototrophic patina as well as in a whitish-gray patina (Table 2). Moreover, as observed in other subterranean environments, apparently similar alterations which were often attributed to different species of *Streptomyces*, which arise from the growth of different kinds of bacteria [12]. At times, purple pigmented patinas also due to Actinobacteria colonisation have been detected in the tombs of Tarquinia as well as in other Roman tombs [26,108]. The phenomena of dark spots caused by various fungal colonisations have also been documented [19,32,104,105]. Microbial patinas may not be easily observable at times, probably due to their growth in underground layers, though their potential aggressiveness should not be neglected.

The physico-chemical interaction of the identified species is often underestimated, despite the fact that some species can give rise to biomineralisation phenomena as well as acid metabolite production, such as chelating compounds which can extract ions from the substrata and produce carbonate dissolution, penetrating inside crystal matrices [4,13,32].

Several authors also stress the role of microorganisms in biodeterioration processes, since different actinobacteria are involved in the bio-precipitation of minerals. In the Mercareccia Tomb, a relevant number (71) of bacterial strains capable of precipitating carbonates have been detected [39]. Mechanical interaction, such as the capacity of several fungi to dissolve minerals and mobilise metals, is well known. When colonising rock substrata, fungal hyphae can induce a chemical deterioration secreting siderophore-like compounds, whereas the greatest damage is due to mechanic activity. In fact, hyphae of some black fungi demonstrated their ability to penetrate silicate and carbonate rocks exerting strong mechanical pressure up to 12.39 bar gouging cavities at depths ranging from few hundreds of microns to several millimeters and producing pulverisation of the substrate through chemical processes [109,110]. Recent studies showed that dark spots occurring in the Tomb of Tutankhamun were caused by *Penicillium chrysogenum* colonisation, which would suggest their detrimental nature, due to the ability to produce malic acid [111]. On the light of some interesting metabolic features showed by soil microorganisms, the use of subterranean strains for biotechnological applications could be a possibility, but further analyses are required.

In Etruscan tombs, a fair amount of mechanical damage was caused by root penetration and its development, even up to several meters in depth, which has led to the disintegration of the plaster and painted layers. In some cases, roots systems of different sizes and aggressiveness were highly developed. In the case of herbaceous species, distances of more than 3 m in depth were observed, also for small-sized herbaceous species. Such data were also confirmed in other studies carried out in several hypogeal archaeological sites in Rome, such as the *Domus Aurea* [112], where the root system of an old specimen of *Pinus roxbourgii* was found to be 15 m in depth and 25 m wide. Another example can be found at the Jewish catacombs of Villa Torlonia, where *Ficus carica* was able to grow underground for more than 50 m [113]. In any case, roots should be considered detrimental since they favour water penetration and have been shown to play a conditioning role on the microbial community [47]. In the South Korean ancient tombs, conservation activities included the removal of previously overlying trees. Then, surrounding landscape has been maintained, and burial mounds have been restored after excavation and maintenance activities. Such new situation resulted safe for avoiding root damages [67].

Finally, with respect to insects and slugs, several observations indicate their negative impact due to the mechanical damage they cause, creating micro tunnels under the painted surface and staining light-coloured surfaces [51,80], or supporting the presence and growth of entomophilous/entomopathogenic fungi, such as for example *Isaria farinosa* and *Engyodontium album* (Table 2). These issues have yet to be sufficiently analysed.

### *4.2. Environmental and Biotic Influences*

In both countries, microorganisms that had entered the ancient tombs and microorganisms that originally inhabited the tombs were able to grow in particular environmental conditions. The microbial composition varies in relation to environmental changes, which in turn increase the risk of a biological attack as observed from the very first studies conducted [28]. The microbial colonisation of the hypogeal surfaces is the result of selective, competitive and inhibitory dynamics among different microorganisms from rocks and from the air under diverse and changing environmental conditions. It is also the result of human activity and intervention which form part of the "history of the site". Furthermore, every process of colonisation is different from another, since each community is conditioned by the pioneer species which will in turn influence further processes [12]. The study conducted at Mogao Grottoes also showed some interesting results in that the diversity index of the fungal community was positively correlated with the building period of the caves, which had a greater impact than temperature and relative humidity of the caves [105]. Without disregarding the role of environmental factors, it seems possible that the opening time of the tombs to outdoor exchanges and human presence has a significant influence on its biodiversity.

In fact, several studies which show the great impact that environmental and edaphic factors have in facilitating different taxa are discussed and summarised in Figure 5.

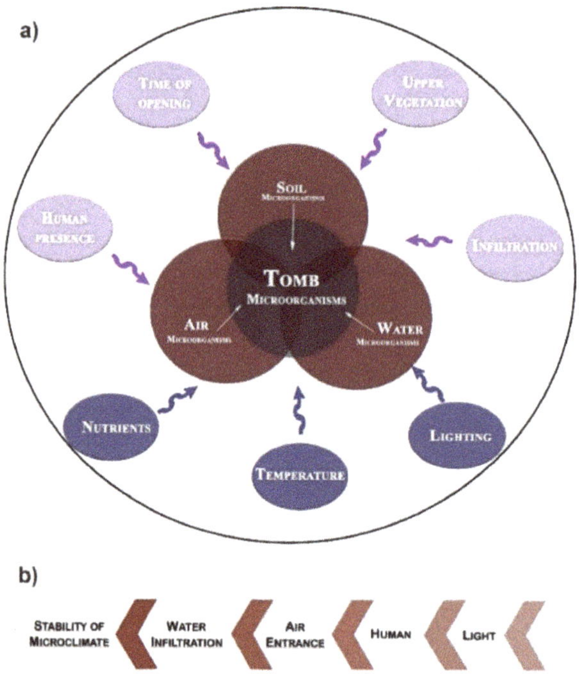


**Figure 5.** Microorganisms occurring in hypogeal tombs and factors influencing their qualitative and quantitative composition (**a**); main factors affecting the microclimate stability, this last is the main goal to be achieved (**b**). (Original elaboration by the authors).

Water originating from percolation, infiltration and capillarity, and at times from human presence along with flooding was found to be the most important factor. The level of humidity greatly affects the various microbial communities as in any archaeological context [30,114]. Such an influence was also recently confirmed by studies conducted in Chinese tombs (Emperor Yang, 1600 years old), where humidity showed the highest degree of microbiological variance (19.2%) than all other environmental factors, followed by illumination (18.3%) and height (12.8%) [106]. In general, fungi were found to be more resistant to dryness, with respect to bacteria.

Lighting conditions were also found to be a very important factor, especially for the Korean sites because, in such humid conditions, the development of phototrophic organisms arises if the amount of radiation exceeds <2 μmoll photons $m^{-2}$ $s^{-1}$, which is approximately three orders of magnitude less than full sunlight [1]. In the areas adjacent to the opening or near artificial sources of light, photoautotrophs grow as they are also aided by the high concentration of $CO_2$, (sometimes ten times higher than those in the atmosphere) [8]. Data from other hypogea, such as catacombs, show that the spectral composition of the light selects different taxonomic groups. Due to their different photosynthetic pattern, green algae and diatoms are favoured by emissions peaking in the blue and red parts of the spectrum, while cyanobacteria are favoured by the emissions in the green and orange-red ones [115].

Furthermore, temperature also greatly influenced the biological risk and for this reason, a risk assessment of microbial colonisation in the Korean sites suggested establishing three ranges of temperature and the risk conditions that occur when temperatures are above 22 °C [6]. In the Italian sites, we observed a higher risk at the end of summer, when the thermal heating reaches below ground, but no precise evaluations have been made. Inside a hypogean, the various walls can be at different risks of biological colonisation. When analysing Korean Tomb No. 1, the interval between the highest and lowest monthly temperature was found to be in the passageway, followed by antechamber and the main chamber [34], due to heat transfer into the tomb. In this case, whitish-grey patinas were observed in the middle and bottom sections of the eastern and northern walls of main chamber, possibly caused by dew and moisture condensation that permeates the tomb.

The edaphic conditions and the porous nature of the substrata were also found to be the primary causes of the development and proliferation of bacteria and fungi, because for these heterotrophic communities, an increase in organic nutrients can change the oligotrophic conditions of mural paintings. The nature of bedrock and its different absorbance potential for organic macromolecules and porosity play an important role. In particular, in one of the Chiusi tombs (Tomb of the Monkey), the abundant actinobacterial colonisation was bolstered by the presence of organic matter and clays on the walls, along with iron oxide that represents an additional factor in promoting growth [47]. Organic matter is also influenced by vegetation cover and root penetration, as observed in the Chiusi tombs [48]. The differences among the microbial communities in the two tombs could not be explained simply by considering the bedrock or bioclimatic conditions, but by considering the influence of differential root penetration as well [48].

Roots from different plant cover were considered responsible for specific microbial communities in response to nutrient availability in the rhizosphere, and in the case of the Tomb of the Hill, there was an abundance of Rhizobiales. Instead, the most common Actinobacteriales were found to be consistent with data on nitrogen-fixing bacteria and rhizosphere studies [48]. The authors suggested that roots had a role in introducing microorganisms into the tomb as well as in influencing organic carbon through root litter and root exudates. Even a small variability of organic substances from biodegraded lignin and humic substances seems to have an important influence on subterranean microbial communities and tomb colonisation by *Nocardia* and *Pseudonocardia* connected to agricultural and/or livestock activities, among other macromolecules on the topsoil [48].

The nature of the pigments utilised in the tombs themselves produced variable effects. In one tomb in Chiusi, the red pigment hematite utilised for the wall paintings played a discriminatory role, since the identified *Nocardia* need iron oxides for the optimal growth [47]. However, in the case of other tombs in Tarquinia, a direct effect of the pigmented minerals displayed no influence on the selection of communities, as indicated by the similarity of the bacterial phyla identified in samples with red, black and ochre pigments. Moreover, in the case of moonmilk, the patina that covered the entire surface of the pictorial layer and its variable thickness showed no relation to the underlying pigments [42].

High values in airborne spore concentration were found to be a risk factor, and preliminary data collected in two Etruscan tombs of Veio on the relationship between biodeterioration of wall frescoes and microbial concentration of the air reported the absence of a qualitative or quantitative parallelism between the microorganisms isolated from the walls and the ones in the air [45]. Ventilation and the consequent risk of contamination/dissemination is also an important factor to take into consideration, hence aerobiological studies were also carried out to assess which areas could be more sensitive [28,116]. Furthermore, the importance of paying great attention during restoration activities was also stressed, as the normal cleaning operations can transfer microorganisms from the altered surfaces to the air, giving rise to further colonisation [38]. The materials used in restoration can sometimes be colonised by several microorganisms as shown in data from the Koguryo Tomb in China with cases of hypocrealean fungi forming "white mould spots" on the acrylic varnish coatings of the murals [22].

Inside the tomb ecosystem, insects and other animals have an additional detrimental effect because they can carry microorganisms which spread inside the tomb, as well as leave behind organic residue as already reported for caves and catacombs [117]. In fact, the highest microbial distribution ascertained at the entrance and the reduced distribution of microorganisms upon entering the ancient Korean tomb may be due to the fact that contamination levels increased as the insects inhabiting the space and the air at the entrance of the tomb had long remained stationary.

### 4.3. Indirect Control Methods and Biocide Treatments

The maintenance of the stability of microclimatic conditions reducing and setting times for the lighting systems, through a decrease in the indoor temperature, and avoiding condensation are goals to be achieved for the tombs conservation [23,28,55,118,119], as shown in Figure 5b. For Etruscan tombs, the presence of a dromos appears to have a positive effect in reducing the risk of microbial development because it helps to stabilise the microclimatic conditions [28]. Further control methods included the reconstruction of overlaying tumuli over the tombs that had been destroyed, and water infiltration prevention by maintaining an herbaceous vegetation layer capable of absorbing humidity and limiting overheating in the hottest seasons [51,58,120]. Conversely, the use of air conditioning systems was found to be counterproductive also because they stabilise the thermo-hygrometric conditions and the air flow can lead to loss of moisture of the wall paintings, hence passive systems are considered preferable. The construction of draught-proof doors and buffer zones inside the hypogea to prevent uncontrolled air circulation was found to be the most important and effective indirect control method [55,56,58,90]. Being human presence a disturbing factor, visits to these hypogea should be planned, the entrance to visitors should be greatly limited, and protective clothing and shoes should be worn (as usually worn by restorers. See Figure 6).

When a biological attack occurs, direct control methods are undoubtedly necessary. The practical use of microwaves as suggested by preliminary testing [46] seems very critical in these fragile environments. The use of biocides firstly needs a careful evaluation of the BPs and the risks entailed [121], and their effective use implies a thorough knowledge of the chemical features of the products as well as the spectrum of efficacy on the resident community [31], this last acting sometimes as a limiting factor for fungal spreading [27,121,122]. With this approach, harmless colonisations could be maintained [42], while the most dangerous ones can be treated. Resistance to certain biocide treatments by spore-forming bacteria, such as *Bacillus* sp. and fungi, also needs to be carefully evaluated. We do not concord with the conclusion that biocide treatments are useless and even detrimental because particular bacteria or fungi can still proliferate on the frescoes [40]. We believe that if the colonisation and the potential microbiomes are well known, efficient treatments can be found with indirect control methods carried out to remove the causes of the proliferation.

In fact, testing new biocides with low toxicological values that do not interfere with materials is still needed. Therefore, it is necessary to develop and apply effective eco-friendly biocides that do not damage the murals and that are harmless for humans and the environment [85,123,124]. In the field of cultural heritage conservation science in South Korea and Italy, studies conducted on such compounds and the efficiency of mixed eugenol with an eco-friendly emulsifier seem promising [125], such as the potential of allelopathic substances derived from lichens [126].

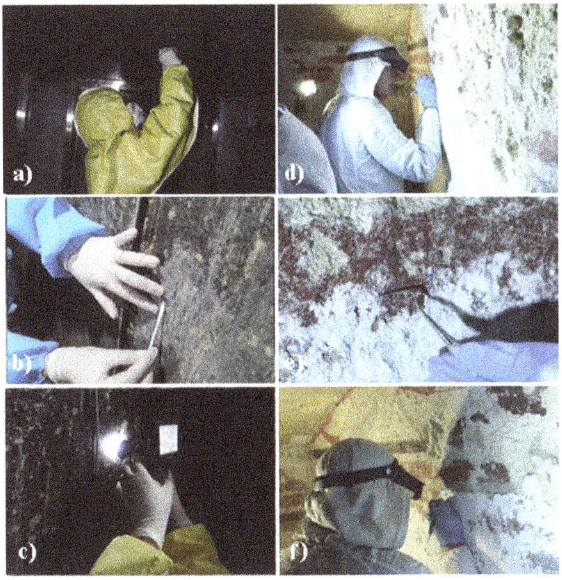


**Figure 6.** Restoration activities carried out in Korean (**a–c**) and Italian sites (**d–f**) respectively. (**d–f** Courtesy of Amici delle Tombe di Tarquinia).

## 5. Conclusions

The collected data show the complexity of the microbial communities in hypogea, as the result of variable environmental values and edaphic conditions. Events that have occurred over time regarding the tombs (opening, flooding, upper vegetation, entrance of visitors, conservation treatments) also play an important role in the biodeterioration processes. Considering the absence of a standardization of the detected sampling methods, and the complexity of the influencing factors, this study shows some limitations in the prevision of the potential occurring biodeterioration phenomena. However, we can observe, as in very humid conditions, cyanobacteria and algae prevail, though only if lighting is sufficient, whereas fungi and bacteria play the greatest role. A possible transition of colonisation among different taxa of fungi and bacteria (Actinomycetes) has been observed, also in consideration of the nutrients and temperatures values. The similarity of phenomena observed in the two countries indicate the importance of these studies. Calibrating indirect control methods and developing eco-friendly biocides are needed.

**Author Contributions:** Conceptualization, G.C. and Y.J.C.; methodology, G.C., D.I., H.J.L., and Y.J.C.; investigation, G.C., D.I., H.J.L., and Y.J.C.; data curation, G.C., D.I., H.J.L., and Y.J.C.; writing—review and editing, G.C., D.I., H.J.L., and Y.J.C.; supervision, G.C. and Y.J.C.; project administration, G.C. and Y.J.C.; funding acquisition, G.C. and Y.J.C. All authors have read and agreed to the published version of the manuscript.

**Funding:** This research was funded by a grant from the Italian Ministry Affairs and International Cooperation Program "Research of conservation environment and eco-friendly damage control of cultural heritage Korea and Italy", grant number PGR06686 and supported by the International Research & Development Program of the National Research Foundation of Korea (NRF) funded by the Ministry of Science and ICT, grant number 2019K1A3A1A25000303.

**Acknowledgments:** The authors wish to thank Adele Cecchini and Massimo Legni (Ass. Amici Tombe di Taquinia) and the Direzione Regionale Musei Lazio (Daniela De Angelis, Direttrice del Sito UNESCO di Cerveteri e Tarquinia) and the Soprintendenza Archeologia Belle Arti e Paesaggio per l'Area Metropolitana di Roma, la Provincia di Viterbo e l'Etruria Meridionale (Daniele Federico Maras), Korea National University of Cultural Heritage (Seonhye Jeong and Haejin Park), Baekje World Heritage Center for sharing information and pictures of restoration activities. We also would also thank Mery Lombardi for English revision.

**Conflicts of Interest:** The authors declare no conflict of interest. The funders had no role in the design of the study; in the collection, analyses, or interpretation of data; in the writing of the manuscript, or in the decision to publish the results.

## References


1. Albertano, P. Epilithic algal communities in hypogean environments. *Plant Biol.* **1993**, *127*, 386–392. [CrossRef]
2. Paribeni, M. *Cause di Deperimento e Metodi di Conservazione Delle Pitture Murali Delle Tombe Sotterranee di Tarquinia*; Rapporto scientifico C.N.R.—Istituto di Fisica Tecnica di Roma, Edizioni Sistema: Rome, Italy, 1970.
3. Sekino, M. A review on the conservation of ornamental tumuli. In *International Symposium on the Conservation and Restoration of Cultural Property. Conservation and Restoration of Mural Paintings, Proceedings of the Congress, Tokyo, Japan, 17–21 November 1983*; Emoto, Y., Miura, S., Eds.; Tokyo National Research Institute of Cultural Properties: Tokyo, Japan, 1984; pp. 217–228.
4. Caneva, G.; Nugari, M.P.; Salvadori, O. *Plant Biology for Cultural Heritage: Biodeterioration and Conservation*; Getty Publications: Los Angeles, CA, USA, 2008.
5. Saiz-Jimenez, C. *The Conservation of Subterranean Cultural Heritage*; CRC Press Taylor & Francis Group: London, UK, 2014.
6. Chung, Y.J.; Lee, H.J.; Kim, H.M.; Kim, D.W. Study on the microbe growth characteristics by environmental condition of ancient tomb. In Proceedings of the 41th Conference of the Korean Society of Conservation Science for Cultural Heritage, Seoul, Korea, 27 March 2015; pp. 27–30.
7. Lee, M.Y.; Kim, D.W.; Chung, Y.J. Conservation environmental assessment and microbial distribution of the Songsan-ri ancient tombs, Gongju, Korea. *J. Conserv. Sci.* **2014**, *30*, 169–179. [CrossRef]
8. Albertano, P.; Urzì, C. Structural interactions among epilithic cyanobacteria and heterotrophic microorganisms in Roman hypogea. *Microb. Ecol.* **1999**, *38*, 244–252. [CrossRef] [PubMed]
9. Sanchez-Moral, S.; Luque, L.; Cuezva, S.; Soler, V.; Benavente, D.; Laiz, L.; Gonzalez, J.M.; Sáiz-Jiménez, C. Deterioration of building materials in Roman catacombs: The influence of visitors. *Sci. Total Environ.* **2005**, *349*, 260–276. [CrossRef]
10. Urzì, C.; De Leo, F.; Bruno, L.; Albertano, P. Microbial diversity in Paleolithic caves: A study case on the phototrophic biofilms of the Cave of Bats (Zuheros, Spain). *Microb. Ecol.* **2010**, *60*, 116–129. [CrossRef]
11. Urzì, C.; De Leo, F.; Bruno, L.; Pangallo, D.; Krakova, L. New species description, biomineralization processes and biocleaning applications of Roman Catacombs-living bacteria. In *The Conservation of Subterranean Cultural Heritage*; Saiz-Jimenez, C., Ed.; CRC Press Taylor & Francis Group: London, UK, 2014; pp. 65–72.
12. De Leo, F.; Iero, A.; Zammit, G.; Urzì, C. Chemoorganotrophic bacteria isolated from biodeteriorated surfaces in cave and catacombs. *Int. J. Speleol.* **2012**, *41*, 125–136. [CrossRef]
13. Ma, W.; Wu, F.; Tian, T.; He, D.; Zhang, Q.; Gu, J.D.; Duan, Y.; Ma, D.; Wang, W.; Feng, H. Fungal diversity and its contribution to the biodeterioration of mural paintings in two 1700-year-old tombs of China. *Int. Biodeterior. Biodegrad.* **2020**, *152*, 104972. [CrossRef]
14. Giacobini, C.; De Cicco, M.A.; Tiglie, I.; Accardo, G. Actinomycetes and biodeterioration in the field of fine arts. In *Biodeterioration 7*; Houghton, D.R., Smith, R.N., Eggins, H.O.W., Eds.; Elsevier Applied Science: London, UK; New York, NY, USA, 1988; pp. 418–423.
15. Arai, H. On microorganisms in the tomb of Nefertari. *Sci. Conserv. (Tokyo)* **1988**, *27*, 13–20.
16. Arai, H. Microbial study on a virgin tumulus. *J. Archaeol. Soc. Nippon.* **1974**, *59*, 328–336.
17. Arai, H. Antimicrobial factors found in virgin tumuli. In *Biodeterioration 6, Proceedings of the 6th International Biodeterioration Symposium, Washington, DC, USA, 5–10 August 1984*; Barry, S., Houghton, D.R., Eds.; C.A.B. International: Slough, UK, 1986; pp. 363–368.
18. Emoto, Y.; Emoto, J. Microbiological investigation of ancient tombs with paintings: Ozuka tomb in Fukuoka and Chibusan tomb in Kumamoto. *Sci. Conserv. (Tokyo)* **1974**, *12*, 95–102. Available online: https://www.koreascience.or.kr/article/JAKO201719554996749.page (accessed on 28 August 2020).

19. Kigawa, R.; Sano, C.; Ishizaki, T.; Miura, S.; Sugiyama, J. Biological issues in the conservation of mural paintings of Takamatsuzuka and Kitora tumuli in Japan. In *Study of Environmental Conditions Surrounding Cultural Properties and Their Protective Measures, Proceedings of the 31th International Symposium on the Conservation and Restoration of Cultural Property, Tokyo, Japan, 5–7 February 2008*; Sano, C., Ed.; National Research Institute for Cultural Properties: Tokyo, Japan, 2009; pp. 43–50.
20. Huang, Z.; Zhao, F.; Li, Y.H. Isolation of *Paenibacillus tumbae* sp. nov., from the tomb of the emperor Yang of the Sui dynasty, and emended description of the genus *Paenibacillus*. *Antonie Leeuwenhoek* **2017**, *110*, 357–364. [CrossRef] [PubMed]
21. Huang, Z.; Zhao, F.; Li, Y.; Zhang, J.; Feng, Y. Variations in the bacterial community compositions at different sites in the tomb of Emperor Yang of the Sui Dynasty. *Microbiol Res.* **2017**, *196*, 26–33. [CrossRef]
22. Sun, J.Z.; Ge, Q.Y.; Zhu, Z.B.; Zhang, X.L.; Liu, X.Z. Three dominating hypocrealean fungi of the 'white mold spots' on acrylic varnish coatings of the murals in a Koguryo tomb in China. *Phytotaxa* **2019**, *397*, 225–236. [CrossRef]
23. Massa, S.; Nichi, D. Conservazione e fruizione: Analisi ambientale sulle tombe dipinte di Tarquinia. In *Etruria Meridionale: Conoscenza, Conservazione, Fruizione, Proceedings of the Atti del Convegno Viterbo, Italy, 29 November–1 December 1985*; Casa Editrice Quasar: Rome, Italy, 1988; pp. 139–143.
24. De Carli, P.; Massa, S.; Nichi, D. Conservation control of an archaeological excavation: The tomb of "Demoni azzurri" of Tarquinia. In *Science, Technology and European Cultural Heritage, Proceedings of the European Symposium, Bologna, Italy, 13–16 June 1989*; Baer, N.S., Sabbioni, C., Sors André, I., Eds.; Butterworth-Heinemann Ltd.: Oxford, UK, 1991; pp. 406–411.
25. Agarossi, G.; Ferrari, R.; Monte, M.; Gugliandolo, C.; Maugeri, T. Changes of microbial system in an Etruscan tomb after biocidal treatments. In Proceedings of the Vlth International Congress on Deterioration and Conservation of Stone, Torun, Poland, 12–14 September 1988; Nicolaus Copernicus University Press: Torun, Poland, 1988; pp. 82–91.
26. Agarossi, G. Biodeterioramento in ambienti ipogei: Esperienze e considerazioni. In *Studi e Ricerche sulla Conservazione delle Opere d'Arte Dedicati alla Memoria di Marcello Paribeni*; Guidobaldi, F., Ed.; CNR: Rome, Italy, 1994; pp. 1–18.
27. Bettini, C.; Agarossi, G.; Ferrari, R.; Monte, M. Fenomeni di biodeterioramento in ambienti ipogei dipinti: Esperienze di controllo di alcune specie microbiche. In Proceedings of the 2nd International Conference on Non-destructive Testing, Microanalytical Methods and Environment Valuation for Study and Conservation of Works of Art, Perugia, Italy, 17–20 April 1988; ICR: Rome, Italy, 1988; pp. 4.1–4.14.
28. Monte, M.; Ferrari, R.; Massa, S. Biodeterioration of Etruscan tombs: Aerobiology and microclimate. In *Aerobiology, Proceedings of the 5th International Conference, Bangalore, 10–15 August 1994*; Agashe India, S.N., Ed.; Science Publishers: Boca Raton, FL, USA, 1997; pp. 333–346.
29. Toreno, G.; Isola, D.; Meloni, P.; Carcangiu, G.; Selbmann, L.; Onofri, S.; Caneva, G.; Zucconi, L. Biological colonization on stone monuments: A new low impact cleaning method. *J. Cult. Herit.* **2018**, *30*, 100–109. [CrossRef]
30. Caneva, G.; Bartoli, F.; Savo, V.; Futagami, Y.; Strona, G. Combining statistical tools and ecological assessments in the study of biodeterioration patterns of stone temples in Angkor (Cambodia). *Sci. Rep.* **2016**, *6*, 32601. [CrossRef] [PubMed]
31. Pinna, D. *Coping with Biological Growth on Stone Heritage Objects: Methods, Products, Applications, and Perspectives*; Apple Academic Press, CRC Press: Oakville, ON, Canada, 2017; 360p.
32. Isola, D.; Zucconi, L.; Cecchini, A.; Caneva, G. Melanized biodeteriogenic fungi in Etruscan tombs: New data on their diversity and favouring conditions. *Fun. Biol.* under review.
33. Lee, H.J.; Jeong, S.H.; Chung, Y.J. Changes in the microbial distribution and the growth characteristics of microorganisms in ancient tombs at Neungsan-ri tomb No. 1, Buyeo. In Proceedings of the New Trends in Cultural Heratage Biodeterioration, Coimbra, Portugal, 5–7 September 2018; pp. 28–29.
34. KNUCH. *Research of Conservation Environment and Eco-Friendly Damage Control of Cultural Heritage Korea and Italy*; National Research Foundation of Republic of Korea: Seoul, Korea, 2020; p. 83.
35. Pallottino, M. *La Peinture Étrusque*; Flammarion: Geneva, Switzerland, 1952.
36. Marzullo, M. *Grotte Cornetane. Materiale e Apparato Critico per lo Studio Delle Tombe Dipinte di Tarquinia. Tarchna Suppl. 6*; Ledizioni Ledi Publishing: Milano, Italy, 2016.

37. Agarossi, G.; Ferrari, R.; Monte, M. Biocides in the control of biodeterioration. In *La Conservazione dei Monumenti nel Bacino del Mediterraneo, Proceedings of the Atti del 10 Simposio Internazionale Bari, Italy, 7–10 June 1989*; Zezza, F., Ed.; Grafo Edizioni: Brescia, Italy, 1990; pp. 511–517.
38. Monte, M.; Ferrari, R. Biodeterioration in subterranean environments. *Aerobiologia* **1993**, *9*, 141–148. Available online: https://link.springer.com/content/pdf/10.1007/BF02066255.pdf (accessed on 28 August 2020). [CrossRef]
39. Sprocati, A.R.; Alisi, C.; Tasso, F.; Vedovato, E.; Barbabietola, N.; Cremisini, C. A microbiological survey of the Etruscan Mercareccia tomb (Italy): Contribution of microorganisms to deterioration and restoration. In *Art 2008, Proceedings of the 9th International Conference on NDT of Art, Jerusalem, Israel, 25–30 May 2008*; NDT of Art: Jerusalem, Israel, 2008; p. 9.
40. Tomassetti, M.C.; Cirigliano, A.; Arrighi, C.; Negri, R.; Mura, F.; Maneschi, M.L.; Gentili, M.D.; Stirpe, M.; Mazzoni, C.; Rinaldi, T. A role for microbial selection in frescoes' deterioration in Tomba degli Scudi in Tarquinia, Italy. *Sci. Rep.* **2017**, *7*, 1–8. [CrossRef]
41. Caneva, G.; Benelli, F.; Bartoli, F.; Cicinelli, E. Safeguarding natural and cultural heritage on Etruscan tombs (La Banditaccia, Cerveteri, Italy). *Rend. Lincei Sci. Fis. Nat.* **2018**, *29*, 891–907. [CrossRef]
42. Cirigliano, A.; Tomassetti, M.C.; Di Pietro, M.; Mura, F.; Maneschi, M.L.; Gentili, M.D.; Cardazzo, B.; Arrighi, C.; Mazzoni, C.; Negri, R.; et al. Calcite moonmilk of microbial origin in the Etruscan Tomba degli Scudi in Tarquinia, Italy. *Sci. Rep.* **2018**, *8*, 15839. [CrossRef]
43. Cicinelli, E.; Benelli, F.; Bartoli, F.; Traversetti, L.; Caneva, G. Trends of plant communities growing on the Etruscan tombs (Cerveteri, Italy) related to different management practices. *Plant Biol.* **2019**, *154*, 158–164. [CrossRef]
44. Bettini, C.; Giacobini, C. Problemi microbiologiche dei dipinti di due tombe etrusche della Necropoli di Veio. In *Conservazione dei Monumenti, Proceedings of the Atti Della Sezione II Dell'associazione Termotecnica Italiana, Firenze, Italy, 25–27 September 1974*; Barbieri, A., Ed.; BIAS: Milan, Italy, 1974; pp. 229–234.
45. Bettini, C.; Giacobini, C.; Marabelli, M.; Oddone, G.; Rossi, M. Gli ipogei dipinti della necropoli di Veio: Indagini sullo stato di conservazione e sulle tecniche pittoriche. *St Etr* **1977**, *45*, 239–257.
46. Bracci, S.; Cuzman, O.A.; Ignesti, A.; Del Fa, R.M.; Olmi, R.; Pallecchi, P.; Riminesi, P.; Tiano, P. Multidisciplinary approach for the conservation of an Etruscan hypogean monument. *Eur. J. Sci. Theol.* **2013**, *9*, 91–106. Available online: http://www.ejst.tuiasi.ro/Files/37/9_Braccietal.pdf (accessed on 28 August 2020).
47. Diaz-Herraiz, M.; Jurado, V.; Cuezva, S.; Laiz, L.; Pallecchi, P.; Tiano, P.; Sanchez-Moral, S.; Saiz-Jimenez, C. The actinobacterial colonisation of Etruscan paintings. *Sci. Rep.* **2013**, *3*, 1440. [CrossRef] [PubMed]
48. Diaz-Herraiz, M.; Jurado, V.; Cuezva, S.; Laiz, L.; Pallecchi, P.; Tiano, P.; Sanchez-Moral, S.; Saiz-Jimenez, C. Deterioration of an Etruscan tomb by bacteria from the order *Rhizobiales*. *Sci. Rep.* **2014**, *4*, 3610. [CrossRef]
49. Mattias, P.P.; Ventriglia, U. La regione vulcanica dei Monti Sabatini e Cimini. *Mem. Soc. Geol. Ital.* **1970**, *9*, 331–384.
50. Pesaresi, S.; Galdenzi, D.; Biondi, E.; Casavecchia, S. Bioclimate of Italy: Application of the worldwide bioclimatic classification system. *J. Maps* **2014**, *10*, 538–553. [CrossRef]
51. Cecchini, A. *Le Tombe Dipinte di Tarquinia. Vicenda Conservativa, Restauri e Tecnica di Esecuzione*; Nardini Editore: Firenze, Italy, 2012; pp. 15–102.
52. NRICH. *The Tomb Murals in South Korea*; NRICH: Daejeon, Korea, 2019; p. 329.
53. Lee, S.Y.; Leidy, D.P. *Silla: Korea's Golden Kingdom*; Metropolitan Museum of Art: New York, NY, USA, 2013; p. 219.
54. Kim, A.R. Contents Analysis and Example of a Class Plan for Ancient Tombs in Three Kingdoms Period in High School History Textbook. Master's Thesis, Hankuk University of Foreign Studies, Seoul, Korea, 2018; p. 70.
55. KNUCH. *Study on the Conservation of Mural Painting in Tomb No. 6 at Songsanri*; Gongju City & KNUCH: Gongju-Buyeo, Korea, 2010; pp. 90–101.
56. KNUCH. *Study on the Preservation Methods of Mural Painting in the Ancient Tombs in Songsan-ri*; Gongju: Gongju City, Korea, 2012; pp. 108–156.
57. KNUCH. *Environment Monitoring of Ancient Tomb No. 1 in Neungsan-ri, Buyeo*; Buyeo-gun: Buyeo, Korea, 2013; pp. 1–29.
58. KNUCH. *Development of Air Distribution System of the Ancient Tombs in Songsan-ri, Gongju*; Gongju City & Kongju National University: Gongju, Korea, 2014; pp. 150–169.

59. KNUCH. *Monitoring on Conservation Environment of the Ancient Tomb No. 1 in Neungsan-ri*; Buyeo-gun & World Heritage Center: Buyeo-Daejeon, Korea, 2016; pp. 56–74.
60. KNUCH. *Continuous Monitoring of the Ancient Tomb No. 1 in Neungsan-ri*; Buyeo-gun Baekje World Heritage Center: Buyeo, Korea, 2018; pp. 61–102.
61. Lee, N.S. *Baekjae kingdom in Neungsan-ri tomb at Buyeo*; The Research Institute of Baekjae Culture: Gongju City, Korea, 2000; pp. 1–24.
62. KNUCH. *The Investigation on the Conservation Environment of the Ancient Tomb No. 1 in Neungsan-ri*; Buyeo-gun: Buyeo, Korea, 2008.
63. Han, K.S. Making technique studies of mural paintings in the No. 6 tomb of Songsanri, Gonju. *J. Conserv. Sci.* **2011**, *27*, 451–458. Available online: https://www.koreascience.or.kr/article/JAKO201120241359979.pdf (accessed on 28 August 2020).
64. Lee, S.O.; Bae, G.W.; Namgung, H.; Nam, D.H.; Choi, Y.G.; Chung, K.Y. Conservation State of Mural Paintings of Royal Tombs in Neungsan-ri, Korea. *J. Conserv. Sci.* **2018**, *34*, 333–343. [CrossRef]
65. Lee, H.S. Manufacturing technology and material characteristic of ancient murals in Korea. 1st KNUCH and ISCR bilateral Workshop. In *Science and Technology Applied to Paintwork of Architecture: Preservation and Restoration*; National Museum of Korea: Seoul, Korea, 17–19 October 2017; pp. 141–151.
66. Cho, H.J.; Lee, T.J.; Lee, H.S.; Chung, Y.J. Study on physical change in the earthen finish layer of tomb murals due to drying. *Kor. J. Cult. Her. Stud.* **2017**, *50*, 148–165. [CrossRef]
67. Jung, H.S. A study on the mural painting tomb of Baekje. *J. Korean Anc. Hist.* **2011**, *61*, 301–336.
68. Seo, H.J. New Achievements and historical interpretations of the excavation of royal tombs from the Sabi period of Baekje. *J. Korean Anc. Hist.* **2017**, *88*, 47–95. [CrossRef]
69. CHA. *Standard Specification for Repairing Cultural Heritages*; CHA: Daejeon, Korea, 2020; p. 391.
70. Lee, S.H. The stone consist of Donghachong. In *The Investigation on the Conservation Environment of Ancient Tomb No. 1 in Neungsan-ri*; Buyeo-gun: Buyeo, Korea, 2008; pp. 91–98.
71. Park, J.H.; Lee, G.H.; Lee, C.H. Consideration for historical application of Augen Gneiss and petrographic characteristics for rock properties of Donghachong tomb from royal tombs of Neungsanri in Buyeo, Korea. *Econ. Environ. Geol.* **2019**, *52*, 91–106. [CrossRef]
72. KMA. Data of Korean Weather-Korean Climate. Available online: http://www.weather.go.kr/ (accessed on 21 June 2020).
73. Chung, Y.J. Conservation environment of Korean mural tombs in the period of the Three Kingdoms. In Proceedings of the International Conference on Conservation of Stone and Earthen Architectural Heritage ICOMOS-ISCS, Gongju, Korea, 20–23 May 2014.
74. Kim, S.H.; Lee, C.H. Interpretation on internal microclimatic characteristics and thermal environment stability of the royal tombs at Songsanri in Gongju, Korea. *J. Conserv. Sci.* **2019**, *35*, 99–115. [CrossRef]
75. Kim, S.H. A Study on Behavior Monitoring and Conservation Environment of Gongju Songsanri Royal Tombs in Korea. Ph.D. Thesis, Kongju National University, Gongju, Korea, 2019; p. 299.
76. Kongju National University the Institute of Paekche Culture. *Tomb of King Muryeong of Baekje: Chapter 1. King Muryeong of Baekje Thought, Tomb of King Muryeong of Baekje*; Research Institute of Baekje Culture: Gongju City, Korea, 1991; pp. 5–57.
77. Kang, W.P. Survey on ancient tombs in Neungsanri, Buyeo, Mounds of Baekje: Documentation in gelatin dry plates. *Nation. Mus. Kor.* **2015**, *191*, 98–167.
78. Jung, S.K. A study on the Songsan-ri Tombs, Gongju based on the data during the Japanese occupation of Korea. *J. Cent. Inst. Cult. Herit.* **2012**, *10*, 249–292.
79. Jeong, H. The recent results and survey on Tomb No. 1 (Donghachong) at Neungsan-ri (Buyeo) in Japanese occupation period. In *Academic Symposium within the Special Exhibition at the Center of Ancient Tomb's Culture in Sabi Baekje, the Ancient Tomb No. 1 in Neungsan-ri*; The Research Institute of Baekjae Culture: Buyeo, Korea, 2019; pp. 3–23.
80. Nugari, M.P.; Pietrini, A.M.; Altieri, A. *Relazione del Laboratorio di Indagini Biologiche: Tarquinia (VT)—Necropoli. Tombe Dipinte—Consulenza Scientifica*; ICR Report: Rome, Italy, 2001.
81. Altieri, A.; Bartolini, M.; Pietrini, A.M. *Relazione del Laboratorio di Indagini Biologiche: Tarquinia Necropoli dei Monterozzi—Consulenza Scientifica*; ISCR Report: Rome, Italy, 2013.
82. Bettini, G.; Massa, S.; Baer, N.S.; Sabbioni, C.; Sors André, I. Preservation problems, visitors and deterioration on the painted Etruscan tomb. In *Science, Technology and European Cultural Heritage, Proceedings of the*

*European Symposium, Bologna, Italy, 13–16 June 1989*; Butterworth-Heinemann Publishers: Guildford, UK, 1991; pp. 761–769.
83. KNU-Research Institute for Basic Science. *In-Situ Status and Conservational Strategy of the Ancient Tombs in Songsan-ri, Gongju*; Research Institute for Basic Science, Kongju National University: Gongju-si, Korea, 1997; pp. 382–398.
84. Suh, M.C.; Lee, N.S.; Choi, S.W.; Kim, G.H.; Jeong, S.M.; Lee, K.B. In-situ status and conservational strategy of the Muryong royal tomb, the Songsanri tomb No. 5 and the Songsanri tomb No. 6 of Baekje dynasty. *J. Nat. Sci.* **1998**, *7*, 147–161.
85. Lee, M.Y.; Park, H.M.; Chung, Y.J. Biochemical characteristics and growth control for fungi isolated from mural painting of tomb No. 6 at Songsan-ri, Gongju. *J. Conserv. Sci.* **2015**, *31*, 227–241. [CrossRef]
86. Yun, Y.G. Biological distribution of Donghachong. In *The Investigation on the Conservation Environment of the Ancient Tomb No. 1 in Neungsan-ri*; Buyeo-gun: Buyeo, Korea, 2008; pp. 157–180.
87. Lee, M.Y. Biochemical Characteristics for Fungi Isolated from Mural Painting of Tomb No. 6 at Songsan-ri. Master's Thesis, Graduate School Chungnam National University, Daejeon, Korea, 2013; p. 68.
88. Baskar, S.; Baskar, R.; Routh, J. Biogenic evidences of moonmilk deposition in the Mawmluh cave, Meghalaya, India. *Geomicrobiol. J.* **2011**, *28*, 252–265. [CrossRef]
89. Portillo, M.C.; Gonzalez, J.M. Moonmilk deposits originate from specific bacterial communities in Altamira Cave (Spain). *Microb. Ecol.* **2011**, *61*, 182–189. [CrossRef]
90. KNU-Research Institute for Basic Science. *Conservation Status of the Ancient Tombs in Songsan-ri*; Gongju-City: Gongju, Korea, 2000; pp. 1–232.
91. CHA. *Report on the Excavation of the Royal of Tomb of King Muryeong*; Ministry of Cultural and Public Information: Daejeon, Korea, 1973; pp. 2–12.
92. NRICH. *Mural Ancient Tombs at Sunheung Eupnae-ri*; Cultural Heritage Administration: Daejeon, Korea, 1986; p. 118.
93. Yamamoto, K.; Kurumazuka, N. Preventive measures for lichens and algae on stone objects. In *Scientific Papers on Japanese Antiques and Art Crafts*; Association of Scientific Research on Historic and Artistic Works of Japan: Tokyo, Japan, 1986; Volume 31, pp. 92–100.
94. Konkuk University. *Study on the Preservation Methods of Mural Painting in the Ancient Tombs in Songsan-ri*; Konkuk University: Gongju, Korea, 2012; pp. 75–86.
95. Chung, Y.J.; Kang, S.Y.; Choi, Y.A. Analysis on antifungal activity of Paulownia-Wood storage box and application of natural biocide for the activity enhancement. *J. Conserv. Sci.* **2008**, *24*, 75–83. Available online: https://www.koreascience.or.kr/article/JAKO200825247232782.page (accessed on 10 April 2020).
96. Lim, J.A.; Choi, Y.A.; Jung, M.H.; Chung, Y.J. Antifungal and insecticidal activity of methanol extract form 11 Korean wood species. *J. Conserv. Sci.* **2008**, *23*, 95–102. Available online: https://www.e-jcs.org/journal/view.php?number=287 (accessed on 28 August 2020).
97. Lim, J.A.; Choi, Y.A.; Jung, M.H.; Kang, S.Y.; Chung, Y.J. Antifungal and insecticidal activity of methanol extracts from 17 medicinal plants against biological deterioration of organic cultural heritage. *NRICH Conserv. Stud.* **2008**, *29*, 126–136. Available online: https://www.koreascience.or.kr/article/JAKO200856853106923.pdf (accessed on 10 April 2020).
98. Kang, S.Y.; Choi, Y.A.; Chung, Y.J. Studies of biological activities of tree extracts for conservation of wooden cultural properties. *NRICH Conserv. Stud.* **2007**, *28*, 60–73.
99. Choi, J.E.; Kim, Y.H.; Hong, J.Y.; Jung, M.H.; Jo, C.W. Research on antifungal and insecticidal activity of natural extracts for protecting organic cultural heritages. *NRICH Conserv. Stud.* **2010**, *31*, 89–102. Available online: https://www.koreascience.or.kr/article/JAKO201056853107136.pdf (accessed on 10 April 2020).
100. Hong, J.Y.; Kim, Y.H.; Jo, C.W.; Choi, J.E. Antifungal activities on organic heritage fungi and antioxidative effect of *Phellodendron amurense* extractives. *J. Korean Tech. Assoc. Pulp Pap. Ind.* **2010**, *42*, 54–61. Available online: https://www.koreascience.or.kr/article/JAKO201015537945329.pdf (accessed on 10 April 2020).
101. Kang, S.Y.; Chung, Y.J.; Lim, J.A. Antifungal and insecticidal activity of essential oil from *Asarum sieboldii* against wood contaminant fungi and *Lasioderma serricorne* L. *J. Conserv. Sci.* **2012**, *28*, 395–401. [CrossRef]
102. Jeong, S.H. Removing Biofilm using Natural Medicines to the Stone Culture Properties. Master's Thesis, The Graduate School Chung-Ang University, Seoul, Korea, 2013; p. 66.
103. Jeong, S.H.; Lee, H.J.; Kim, D.W.; Chung, Y.J. New biocide for eco-friendly biofilm removal on outdoor stone monuments. *Int. Biodeterior. Biodegrad.* **2017**, *131*, 19–28. [CrossRef]

104. Bastian, F.; Jurado, V.; Nováková, A.; Alabouvette, C.; Saiz-Jimenez, C. The microbiology of Lascaux cave. *Microbiology* **2010**, *156*, 644–652. [CrossRef]
105. Ma, Y.; Zhang, H.; Du, Y.; Tian, T.; Xiang, T.; Liu, X.; Wu, F.; An, L.; Wang, W.; Gu, J.-D.; et al. The community distribution of bacteria and fungi on ancient wall paintings of the Mogao Grottoes. *Sci. Rep.* **2015**, *5*, 7752. [CrossRef]
106. Li, Y.; Huang, Z.; Petropoulos, E.; Ma, Y.; Shen, Y. Humidity governs the wall-inhabiting fungal community composition in a 1600-year tomb of emperor Yang. *Sci. Rep.* **2020**, *10*, 1–10. [CrossRef]
107. Bindschedler, S.; Cailleau, G.; Braissant, O.; Millière, L.; Job, D.; Verrecchia, E.P. Unravelling the enigmatic origin of calcitic nanofibres in soils and caves: Purely physicochemical or biogenic processes? *Biogeosciences* **2014**, *11*, 2809–2825. [CrossRef]
108. Gonzalez-Pimentel, J.L.; Jurado, V.; Laiz, L.; Miller, A.Z.; Hermosin, B.; Saiz-Jimenez, C. Biodeterioration of Roman tombs: The role of pigmented actinobacteria. In *Geophysical Research Abstracts, Proceedings of the EGU General Assembly, Vienna, Austria, 7–12 April 2019*; EGU: Munich, Germany, 2019; Volume 21, p. 8949.
109. Onofri, S.; Zucconi, L.; Isola, D.; Selbmann, L. Rock-inhabiting fungi and their role in deterioration of stone monuments in the Mediterranean area. *Plant Biol.* **2014**, *14*, 384–391. [CrossRef]
110. Favero-Longo, S.E.; Gazzano, C.; Girlanda, M.; Castelli, D.; Tretiach, M.; Baiocchi, C.; Piervittori, R. Physical and chemical deterioration of silicate and carbonate rocks by meristematic microcolonial fungi and endolithic lichens (*Chaetothyriomycetidae*). *Geomicrobiol. J.* **2011**, *28*, 732–744. [CrossRef]
111. Vasanthakumar, A.; De Araujo, A.; Mazurek, J.; Schilling, M.; Mitchell, R. Microbiological survey for analysis of the brown spots on the walls of the tomb of King Tutankhamun. *Int. Biodeterior. Biodegrad.* **2013**, *79*, 56–63. [CrossRef]
112. Caneva, G.; Ceschin, S.; De Marco, G. Mapping the risk of damage from tree roots for the conservation of archaeological sites: The case of the Domus Aurea, Rome. *Conserv. Manag. Archaeol. Sites* **2006**, *7*, 163–170. [CrossRef]
113. Caneva, G.; Galotta, G.; Cancellieri, L.; Savo, V. Tree roots and damages in the Jewish catacombs of Villa Torlonia (Roma). *J. Cult. Herit.* **2009**, *10*, 53–62. [CrossRef]
114. Caneva, G.; Salvadori, O.; Ricci, S.; Ceschin, S. Ecological analysis and biodeterioration processes over time at the Hieroglyphic Stairway in the Copán (Honduras) archaeological site. *Plant Biol.* **2005**, *139*, 295–310. [CrossRef]
115. Bruno, L.; Bellezza, S.; De Leo, F.; Urzì, C. A study for monitoring and conservation in the Roman Catacombs of St. Callistus and Domitilla, Rome (Italy). In *The Conservation of Subterranean Cultural Heritage*; Saiz-Jimenez, C., Ed.; CRC Press, Taylor & Francis Group: London, UK, 2014; pp. 37–44.
116. Caneva, G.; De Nuntiis, P.; Fornaciari, M.; Ruga, L.; Valenti, P.; Pasquariello, G. Aerobiology applied to the preventive conservation of cultural heritage. *Aerobiologia* **2019**, *36*, 1–5. [CrossRef]
117. Jurado, V.; Sanchez-Moral, S.; Saiz-Jimenez, C. Entomogenous fungi and the conservation of the cultural heritage: A review. *Int. Biodeterior. Biodegrad.* **2008**, *62*, 325–330. [CrossRef]
118. Kim, D.W.; Jeong, S.H.; Lee, M.Y.; Chung, Y.J. Thermal environment analysis for preserving ancient mural painting in Songsan-ri Tomb No. 6, Gongju, Korea. *J. Conserv. Sci.* **2016**, *32*, 521–534. [CrossRef]
119. Kim, J.M.; Lee, J.H.; Yeo, M.S.; Kim, K.W. Transient analysis of thermal and humidity environment in Korean ancient tumulus. *Archit. Inst. Korea Symp. Proc.* **2016**, *34*, 163–164.
120. Massa, S. Esigenze conservative di un ipogeo interessato da vegetazione sovrastante. In *Ruderi & Vegetazione. Questioni di Restauro*; Mancini, R., Rossi Doria, I., Eds.; Ginevra Bentivoglio Editoria: Rome, Italy, 2017; pp. 265–289.
121. Bastian, F.; Alabouvette, C.; Saiz-Jimenez, C. Bacteria and free-living amoeba in the Lascaux Cave. *Res. Microbiol.* **2009**, *160*, 38–40. [CrossRef] [PubMed]
122. Martin-Sanchez, P.M.; Nováková, A.; Bastian, F.; Alabouvette, C.; Saiz-Jimenez, C. Use of biocides for the control of fungal outbreaks in subterranean environments: The case of the Lascaux Cave in France. *Environ. Sci. Tech.* **2012**, *46*, 3762–3770. [CrossRef] [PubMed]
123. Caneva, G.; Tescari, M. Stone biodeterioration: Treatments and preventive conservation. In Proceedings of the 2017 International Symposium of Stone Conservation, Conservation Technologies for Stone Cultural Heritages: Status and Future Prospects, Seoul, Korea, 21 September 2017; National Research Institute of Cultural Heritage of Korea: Daejeon, Korea, 2017; pp. 95–114.

124. Fidanza, M.R.; Caneva, G. Natural biocides for the conservation of stone cultural heritage: A review. *J. Cult. Herit.* **2019**, *38*, 271–289. [CrossRef]
125. Jeong, S.H.; Lee, H.J.; Lee, M.Y.; Chung, Y.J. Conservation environment for mural tomb in Goa-ri, Goryeong. *J. Conserv. Sci.* **2017**, *33*, 189–201. [CrossRef]
126. Caneva, G.; Fidanza, M.R.; Tonon, C.; Favero-Longo, S.E. Biodeterioration patterns and their interpretation for potential applications to stone conservation: A hypothesis from allelopathic inhibitory effects of lichens on the Caestia Pyramid (Rome). *Sustainability* **2020**, *12*, 1132. [CrossRef]

*applied sciences*

MDPI

*Article*

# *Thymus vulgaris* Essential Oil and Hydro-Alcoholic Solutions to Counteract Wooden Artwork Microbial Colonization

**Silvia Sparacello [1], Giuseppe Gallo [1], Teresa Faddetta [1], Bartolomeo Megna [1], Giovanna Nicotra [2], Beatrice Bruno [2], Belinda Giambra [1] and Franco Palla [1,*]**

[1] Dipartimento di Scienze e Tecnologie Biologiche, Chimiche e Farmaceutiche, Università degli Studi di Palermo, Via Archirafi 38, 90123 Palermo, Italy; silvia.sparacello@you.unipa.it (S.S.); giuseppe.gallo@unipa.it (G.G.); teresa.faddetta@unipa.it (T.F.); bartolomeo.megna@unipa.it (B.M.); info@belindagiambra.it (B.G.)

[2] EPO Srl, Via Stadera 19, 20141 Milano, Italy; gnicotra@eposrl.com (G.N.); bbruno@eposrl.com (B.B.)

* Correspondence: franco.palla@unipa.it; Tel.: +39-91-23891224

**Featured Application: We improve the knowledge on plant extracts application assuming their use as bio-pesticides, safe for both humans and the environment from the point of view of a green strategy.**

**Abstract:** Aromatic plants represent a source of natural products with medicinal properties, and are also utilized in the food and pharmaceutical industries. Recently, the need for eco-compatible and non-toxic products, safe for both the environment and human health, have been proposed for the sustainable conservation of historic–artistic artifacts. In this study, in order to counteract microbial colonization (*Aspergillus* sp., *Streptomyces* sp., *Micrococcus* sp.) on wooden artwork surfaces, *Thymus vulgaris* L. (Lamiaceae) essential oil (EO) and hydro-alcoholic (HA) solutions were applied in a polyphasic approach. The antimicrobial activities of EO and HA solutions were preliminarily assessed by agar disc diffusion (ADD) and well plate diffusion (WPD) in vitro methods, defining the specific concentration useful for bacterial and fungal genera, identified by optical microscopies, in vitro cultures (nutrient or Sabouraud agar), and DNA base molecular biology investigations. Specifically, the microbial patina was directly removed by a hydro-alcoholic solution (while evaluating the potential colorimetric change of the artwork's surface) combined with exposure to EO volatile compounds, performed in a dedicated "clean chamber". This study proposes, for the first time, the combined use of two plant extracts to counteract microbial development on wooden artworks, providing supplementary information on these products as bio-agents.

**Keywords:** biodeterioration; bacteria; fungi; essential oil; hydro-alcoholic extract; plant products; green strategy

**Citation:** Sparacello, S.; Gallo, G.; Faddetta, T.; Megna, B.; Nicotra, G.; Bruno, B.; Giambra, B.; Palla, F. *Thymus vulgaris* Essential Oil and Hydro-Alcoholic Solutions to Counteract Wooden Artwork Microbial Colonization. *Appl. Sci.* **2021**, *11*, 8704. https://doi.org/10.3390/app11188704

Academic Editors: Filomena De Leo, Valme Jurado and Antonio Valero Díaz

Received: 29 August 2021
Accepted: 15 September 2021
Published: 18 September 2021



## 1. Introduction

This study concerned demo-ethno-anthropological assets from a conservative point of view, proposing an integrated approach as much as was possibly safe for the health of the operators and the environment, from a green restoration perspective. The B731 wooden sculpture used in this study refers to a female subject made from a single block of wood, and part of the Sogo Bò collection at the Museo Internazionale delle Marionette "Pasqualino Noto" in Palermo, Italy (Figure 1), which features ancient artworks of unknown ages.

Sogo Bò is a very old theatre linked to local Mali folklore. Traditionally, to build the puppets, a group of young boys, and, only in recent times, young girls (not yet married) come together in cooperative groups called ton. Sogo Bò theatre is intrinsically linked to music, so much so, that no character or show can be identified without the relevant songs. The construction of the puppets and the staging of the shows are strongly linked to religions and to a spiritual path reserved only for actors and initiates [1–3]. In order to respect the intrinsic value of the works, experimental studies were developed to have a more complete awareness of the materials, thus guaranteeing the realization of a coherent intervention, even when using innovative materials.

(a) (b)

**Figure 1.** The Sogo Bò theatre wooden sculpture (B731) from the International Puppet Museum, Palermo, Italy: (**a**) recto; (**b**) verso. The largest central sculpture is 64.5 cm in height and 25.0 cm in width; the two small ones on the sides are 20.0 cm in height and 12.0 cm in width.

The conservation and sustainable fruition of cultural assets must be based on evaluating physical, chemical, and biological factors that accelerate the deterioration process [4–7]. Concerning biological wood, decay, bacteria, fungi, and insects represent a threat to the conservation, acting by bioactive agents (such as enzymes), and hydrolysing the chemical constituents of wood (cellulose, hemicelluloses, and lignin). Moreover, several other biological systems can use wood as a supportive substrate where nutrient molecules (carbohydrates and proteins) can easily be found. Although different degradation steps can be distinguished in relation to woody plant species and preservation strategies, a relevant role is played by temperature, relative humidity, and lighting, particularly in indoor environments, such as museums, libraries, and archives [8–15]. Environmental conditions are closely related to microbial colonization and insect infestation. Insect larvae utilize wood as a source of food and fungi such as *Alternaria*, *Aspergillus*, *Botrytis*, *Cladosporium*, *Fusarium*, and *Penicillium*, have a deteriorative effect due to their metabolic activities. The wood decay process can also be related to bacteria of the genera *Bacillus*, *Micrococcus*, and *Streptomyces*. Moreover, even if bacterial and fungal species are not directly involved in wood deterioration, they can play the role of "first colonizer", facilitating the spreading of wood decay microorganisms (ecological niche succession) [16–20]. To counteract microbial colonization, a biocidal treatment is generally applied.

Although commercial synthetic biocides are easy to apply and are effective against a wide range of microorganisms, they do not have a single biological target, and can induce several negative side effects such as persistent residues in the environment, which are detrimental to human health. The use of commercial biocides has been taking place for many years, but there is still no detailed information on their effect over time on both historical–artistic artifacts, human health, and the environment [21–26].

Plant antimicrobial extracts generally correspond to hydro-alcoholic or essential oil solutions, these are mainly used in food, pharmaceutical and medical industries since they have a marked activity against bacteria and fungi [27–32]. They have recently been utilized in the conservation of cultural heritage [33–37], where bioactive molecules were isolated from marine organisms [38] to replace synthetic biocides.

In our previous studies, *Thymus vulgaris* L. (Lamiaceae family) essential oil was proposed as an alternative to commercial biocides [39–42]. In this work, the application of hydro-alcoholic solutions (HA) was combined with exposure to essential oil (EO) volatile compounds. Specifically, *T. vulgaris* HA and EO solutions were utilized to counteract microbial colonization, spread over the surface of a kapok wood (*Ceiba pentrada* L. Gaertn) sculpture (Sogo Bò theatre, Mali, Africa, artifact B731).

## 2. Materials and Methods

### 2.1. Wooden Artwork

The tradition of puppet theatre in Africa has very ancient origins and an exhaustive historical–geographical path is very difficult to define [1,2]. Originating in Sogo Bò theatre, the wooden sculptures were made of kapok wood (Figure 2), *Ceiba pentrada* L. Gaertn, traditionally used by Mali population. The artifact shows a dark-brown color, attributable to treatments by hot spatulas in order to induce a superficial combustion of the wood, creating a brown, smooth surface, as suggested by Mary Jo Arnoldi's personal communication [3]. Samples have been collected from a fracture in the wooden structure with flexible razor blades.

(a) (b) (c)

**Figure 2.** Microscopy observation of the thin wood sections, three anatomical directions: (**a**) transverse; (**b**) tangential; (**c**) radial. The bars correspond to 100 μm.

### 2.2. Microbial Taxa Identification

A spread microbiological patina was clearly distinguished on the artwork's surface, (Figure 3) where three sampling areas (6 cm × 6 cm) were selected. Sampling was carried out by sterile swabs used to inoculate nutrient or Sabouraud agar plates (OXOID). After incubation at 30 ± 1 °C for 16–48 h, bacterial and fungal colonies were isolated. Fungal mycelia, fruiting structures, and spores were collected by a gentle pressing of small pieces of transparent adhesive tape on isolated colonies. Morphology of conidia and conidiophore structures were observed by optical microscopy (Leica, Wetzlar, Germany), after Lugol's iodine staining (Lugol's reactive = 1 g of iodine and 1 g of potassium iodide in 100 mL of distilled water [43]). Moreover, isolated bacteria were taxonomically characterized on the base of their 16S target sequences [44,45]; universal ITS1-ITS 4 primers were utilized for the identification of isolated fungi [46,47]. PCR products were purified according to the manufacturer's protocol using a GenElute kit (Sigma-Aldrich, Darmstadt, Germany), and were sequenced by Eurofins Genomics' service (https://www.eurofinsgenomics.eu/ Ebersberg, Germany) based on the Sanger PCR-sequencing method. Sequences were analyzed (% of similarity), referring to genomic databases (EMBL-Germany, NIH-USA), with the BLAST platform [48].

**Figure 3.** The Sogo Bò sculpture's surface, whitish spots are recognizable.

### 2.3. *In Vitro Antimicrobial Activity Assays*

Bacterial (*Bacillus* sp., *Streptococcus* sp.) and fungal (*Penicillium* sp., *Aspergillus* sp.) colonies were preliminarily grown as a liquid culture (nutrient or Sabouraud broth), incubating for 16–36 h at 30 ± 1 °C. The antimicrobial activity of the EO and HA solutions was evaluated by agar diffusion disc (ADD) or well plate diffusion (WPD) in vitro methods (Figure 4). Aliquots (10–20 μL) of bacterial or fungal liquid cultures were uniformly spread by a sterile Drigalsky spatula on nutrient or Sabouraud surface media, and the surface was allowed to dry. The assays were performed, in duplicate, using bacterial or fungal liquid cultures, normalized to the concentration of $1 \times 10^6$ CFU/mL and $1 \times 10^4$ conidia/mL, respectively.

In the ADD assay, 10 μL aliquots of *T. vulgare* essential oil (50%, 25%,12.5%) solutions were dropped on a sterile paper disc (6 mm in diameter, Dutscher papier, FR) centred on a nutrient or Sabouraud agar medium (9 cm Petri dishes) surface, previously sown with microbial cultures. Incubating at 30 ± 1 °C, confluent microbial growth was clearly observed except in the inhibition halo (i.h.) area, the diameter (mm) of which was related to the antimicrobial activity of the main components. In control assays, paper discs were soaked with 70% ethanol or 3% (vol/vol) benzalkonium chloride solutions.

In the WPD assay, a hole 4 mm in diameter was drilled aseptically on the agar, and a 10 μL aliquot of *T. vulgare* HA solution was loaded. After incubation at 30 ± 1 °C, the diameters (mm) of the growth inhibition halos were measured.

The antimicrobial activity vs. the microbial colony was outlined as: a sensitive strain = i.h. > 9 mm; a resistant strain = i.h. < 6 mm. As summarized in Table 1, both EO and HA solutions were able to produce inhibition halos of at least 9 mm, showing good antimicrobial action. Reduced halos in diameter were revealed for benzalkonium-Cl (>6 mm) and for Et-OH (almost no activity).

Biocide or biostatic activity was also evaluated, distinguishing minimum inhibitory concentration (MIC) and minimum biocidal concentration (MBC). MIC corresponds to any visible microbial growth at the lowest concentration after incubation at 30 + 1 °C, and MFC was measured by an antimicrobial-free sub-culture, defining the lowest concentration of essential oil able to kill 99.5% of the original inoculum. In 96-well microtiters, 30 microliter aliquots of 12.5%, 25%, 50%, and 100% EO solutions were mixed with an equal volume of microbial cultures and a liquid nutritive medium. Microbial growth after 18–36 h of incubation at 30 + 1 °C, and the absorbance at 500–600 nm, were measured; as the control, the commercial biocide benzalkonium chloride (3%, $v/v$) was utilized [48]. Assays were performed in duplicate.

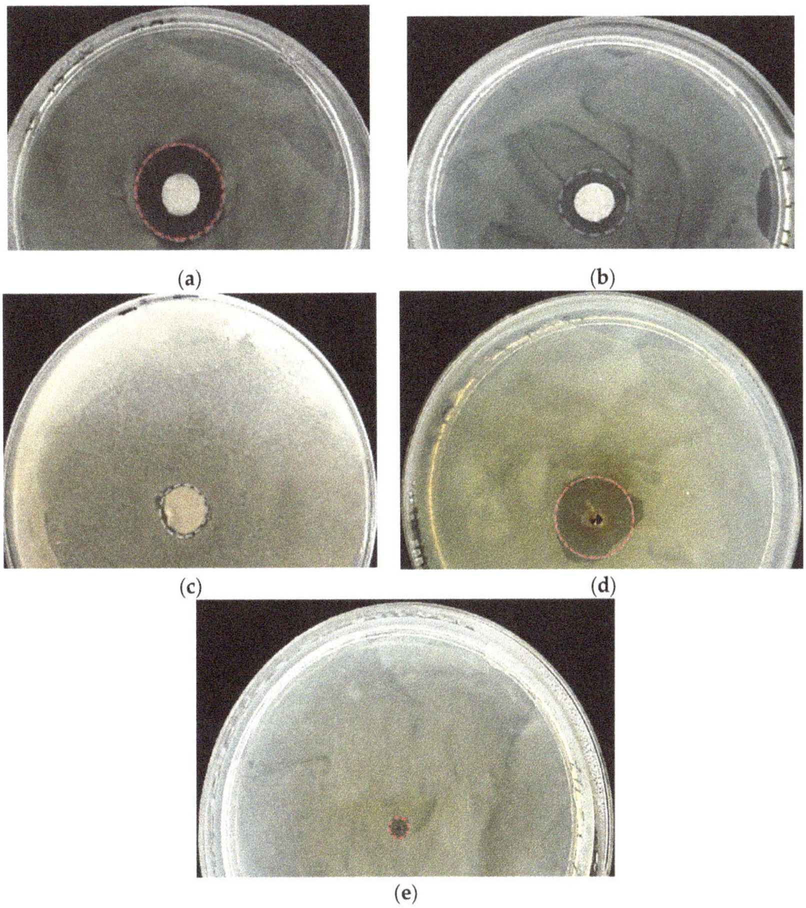


**Figure 4.** Antimicrobials vs. *Streptomyces* sp. colonies and related inhibition halo assayed with the ADD method: (**a**) *T. vulgaris* essential oil (12.5%); (**b**) benzalkonium chloride (3% *v*/*v*); (**c**) Et-OH (70% *v*/*v*). The diameter of >9 mm in (**a**,**b**) confirms antibacterial activity in these solutions. A very small diameter (almost not observable) is seen in (**c**), the 70% Et-OH solution. WPD method: whereas an inhibition halo is clearly recognizable in (**d**) related to the *T. vulgare* hydro-alcoholic solution (100%), the i.h. is barely detectable for (**e**), the 70% Et-OH solution.

**Table 1.** Antimicrobial activity of the *T. vulgaris* and control solutions vs. microbial taxa isolated in this study. Values of inhibition halos (mm) were derived from duplicate assays.

| Biocide Solution | Microbial Taxa | ADD Method Inhibition Halo (mm) |
|---|---|---|
| E.O. 12.5% | *Bacillus* sp. | *21.5–22.0* |
| | *Streptococcus* sp. | *17.4–18.0* |
| | *Aspergillus* sp. | *8.5–9.0* |
| | *Penicillum* sp. | *12.0–12.5* |
| | | WPD method Inhibition halo (mm) |

**Table 1.** *Cont.*

| Biocide Solution | Microbial Taxa | ADD Method Inhibition Halo (mm) |
|---|---|---|
| H.A. 100%. | *Bacillus* sp. | *11.0–11.0* |
| | *Streptococcus* sp. | *9.3–10.0* |
| | *Aspergillus* sp. | *8.7–9.0* |
| | *Penicillum* sp. | *8.1–9.0* |
| **Control solutions** | | *ADD or* WPD method Inhibition halo (mm) |
| Et-OH 70% | All bacterial and fungal isolated colonies | *1.8–2.0* |
| | | *0.4–0.5* |
| Benz. Clor. 3% | | *5.5–6.0* |
| | | *3.8–4.0* |

## 2.4. *T. vulgaris EO and HA Solutions*

The chemical compounds of the *Thymus vulgaris* L. essential oil were identified by GC–MS (retention indices on the HP 5MS column) [40], highlighting the richness in carvacrol (64.96%), thymol (8.25%), and their biogenetic precursor *p*-cymene (11.29%). The hydro-alcoholic solution, supplied by EPO Srl., Milano, had been assessed by gas-chromatography according to the European Council Pharmacopoeia: carvacrol = 0.0010%; thymol = 0.12%; camphor and eucalyptol <0.0010% [49].

## 2.5. *Artwork Surface Analysis*

In order to identify the finishing layers/binders on the artwork's surface, wood samples were analyzed by Fourier-transform infrared spectroscopy (FTIR) and environmental scanning electron microscopy equipped with an X-ray energy dispersive system (ESEM-EDS) [50].

The FTIR was performed in ATR mode, using a Perkin Elmer Spectrum 2 spectroscope, an ATR module with a diamond cell, and a spectral resolution 4 $cm^{-1}$ (measurement was performed four times/sample, range 400–4000 $cm^{-1}$). Furthermore, to evaluate any variation on the artwork's surface after the *T. vulgaris* HA application, FTIR and ESEM-EDS (FEI QUANTA 600 ESEM equipped with an FEG electron gun and an EDAX –EDS microprobe) investigations were carried out.

ESEM-EDS samples were observed in a low-vacuum mode, i.e., 0.4 mbar of water vapor pressure, in order to observe the samples without gilding.

## 2.6. *Microbial Colonies' Direct Removal by HA Solution*

In order to improve the biocidal effects, the microbial colonization on the artwork's surface was preliminarily removed with HA *T. vulgaris* solutions, by cotton swab or a soft brush, in relation to surface conservation conditions (mechanical action was gently carried out for few seconds).

Potential variation in the color, before and after surface treatment, was evaluated by a portable colorimeter, PCE CSM 7.

## 2.7. *Exposure to T. vulgaris EO Volatile Compounds*

As in our previous studies, artifacts were exposed to EO volatile compounds in ad hoc assembled "clean chambers" with gas-barrier thermo-sealed film, in environmental conditions [40,42]. In order to maintain a saturate atmosphere, 0.9 mL/1000 $cm^3$ of *T. vulgaris* EO was dispensed in three small glass containers, placed at equidistant points. The thermo-hygrometric parameters (inside–outside, min–max temperature, and relative humidity) were continuously monitored by an Oregon Scientific datalogger equipped with a microprobe. Treatments were carried out at temp = 22 + 2 °C and U.R. = 56 + 2% values for seven weeks.

### 2.8. Microbial Monitoring

Microbial growth monitoring was performed, assessing the total microbial load in the three areas selected on the artwork's surface. Sampling was performed before, during, and at the end (seven weeks) of treatment, with sterile cotton swabs.

Before combined treatment, the total microbial load was high in all selected.

Keeping one area as the control, identified as *T. vulgaris* EO exposure without surface treatment (orange line), the surface of the other two selected areas was treated with a *T. vulgaris* HA or Et-OH solution.

During exposure, sampling on the selected areas was performed by making small cuts on the gas-barrier film with a sterile scalpel, through which a sterile swab was easily introduced without changing the environmental conditions; immediately after sampling, barrier film cuts were thermo-sealed. Each swab was used to inoculate nutrient and Sabouraud agar plates.

Microbial load was assessed as the total number of bacterial and fungal colonies grown on nutrient and Sabouraud agar plates per sample.

## 3. Results

Thin sections of the sculpture's wood were analyzed with a Leitz Laborlux Pol 12 petrographic microscope (Figure 2), and images were acquired using a Leica MD170HD camera and Leica Application Suite v4.3 software polarizing filters (used in parallel mode). Specifically, vessels were gathered in radial chains, often placed side by side, with decreasing dimensions according to the growth period of the tree. The perforation plates showed a simple organization. Multi-seriate rays, composed of fewer than 10 uniform cells with different heights, were observed. These characteristics, in agreement with M.J. Arnoldi's suggestion, allowed us to consider *Ceiba* sp. as the original plant, *Angiosperm* (hardwood).

On the artwork's surface was a clearly distinguishable whitish biological patina (Figure 3). Microscopy observations, in vitro cultures, and molecular investigations allowed us to identify the bacterial (*Bacillus* sp., *Streptococcus* sp.) and fungal (*Penicillium* sp., *Aspergillus* sp.) colonies. In our previous studies, the *Thymus vulgaris* EO was successful when applied as an antimicrobial agent to control microbial colonization, acting on bacterial and fungal colonies, as well as on complex biofilm [40–42,51].

Several studies showed that compounds of natural origin, such as plant extracts, display antimicrobial and antioxidant activity that differ in relation to several factors: organic/inorganic artefacts; indoor/outdoor environments; bacterial or fungal colonizer; biological activity of the EO and HA solutions and mechanisms of action [51]. The antimicrobial actions of essential oil and hydro-alcoholic solutions have been preliminarily evaluated in vitro, using ADD and WPD methods, against both bacterial- and fungal-isolated colonies.

Shown in Figure 4 is the antimicrobial activity of the 12.5% *T. vulgaris* EO (Figure 4a), the control solution 3% $v/v$ benzalkonium chloride (Figure 4b), and the 70% $v/v$ Et-OH (Figure 4c) assayed vs. *Streptomyces* sp. colonies with the ADD method. Different inhibition halo (i.h.) diameters were distinguishable: (**a**) i.h. > 9 mm; (**b**) i.h. > 6 mm; (**c**) i.h. = almost not observable.

For the *T. vulgare* hydro-alcoholic solution, the antimicrobial activity was evaluated by the WPD method: the 100% HA solution produced an i.h. > 9 mm (Figure 4d); i.h. < 4 mm was obtained with the 70% Et-OH solution (Figure 4e).

In the ADD and WPD assays, 3% $v/v$ benzalkonium chloride and 70% $v/v$ Et-OH solutions represent control solutions; the benzalkonium chloride (phase) solution was diluted in water to achieved a final concentration equal to 3% (vol/vol).

Assays were performed in duplicate.

Results of the ADD and WPD assays performed in this study, as shown vs. the Streptomyces sp. colonies in Figure 4, are summarized in Table 1. Although the EO showed increased antimicrobial activity, the HA solution was able to produce inhibition halos of at least 9 mm. A reduced diameter is shown for the benzalkonium-Cl and Et-OH control solutions, producing halos less than 6 mm in diameter.

The minimum inhibitory concentration (MIC) and minimum bactericidal/fungicidal concentration (MBC/MFC) were assessed with the microdilution method, performed in a 96-well microtiter. The minimum inhibitory concentration (MIC) corresponds to the lowest concentration of the sample at which the tested microorganisms did not demonstrate any visible growth. The minimum bactericidal/fungicidal (MBC/MFC) concentrations were determined as the lowest concentration of the antimicrobial able to kill the 99.5% of the original inoculum in the sub-culture (antimicrobial-free media) [39].

In 96-well microtiters, 30 microliter aliquots of 12.5%, 25%, 50%, and 100% EO solutions were mixed with an equal volume of a microbial culture and a liquid nutritive medium, measuring the absorbance at 500–600 nm, after 18–36 h of incubation at 30 + 1 °C. As the control, the commercial biocide benzalkonium chloride (3%, $v/v$) was utilized. The *T. vulgaris* EO showed a good, but different, inhibitory effect vs both bacterial and fungal colonies; MBC/MFC values measured are closely related to phenolic compounds [48,51,52].

A reduced variance was revealed, which could be related to different factors, such as oil solubility or the different susceptibility of the microbes to the substances, because all microbes present differences in cell wall structure, lipid and protein composition of the cytoplasmic membrane, as well as in specific physiological processes.

The values of inhibition halos (mm) in Table 1 were derived from ADD and WPD assays, performed in duplicate for each antimicrobial and control solution. A variable activity of EO and HA solutions was distinguished, and attributable to the different concentrations of chemical compounds such as carvacrol and thymol. Furthermore, different activity on microbial systems could be related to difference on cell wall structures, lipid/protein composition of cytoplasmic membranes, and to potential influence on specific physiological processes [53–55].

A survey to assess possible changes pre- and post-removal by HA solution application was carried out, performing colorimetric measurement on selected points of the artwork's surface (Figure 5). The ΔE results were lower than 4.0 (3.821 on the lighter area and 2.75 on the darker area, respectively), corresponding to a variation not perceptible by the human eye, as shown in Figure 6. The measurement did not show any significant variation in all three parameters (L*, a*, b*), both on the darkened (surface) and lighter (lacuna) areas.

(a) (b)

**Figure 5.** Selected area of the wood sculpture surface pre- (**a**) and post- (**b**) microbial colony removal by the *T. vulgaris* hydro-alcoholic solution. Notable is the complete removal of whitish spots and any distinguishable change in color in the brownish and light areas (lacuna) respectively.

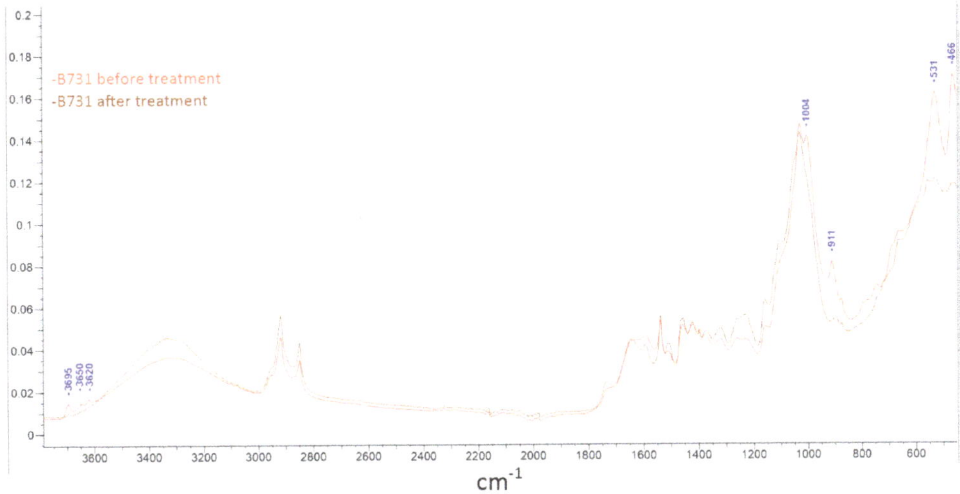


(**a**)

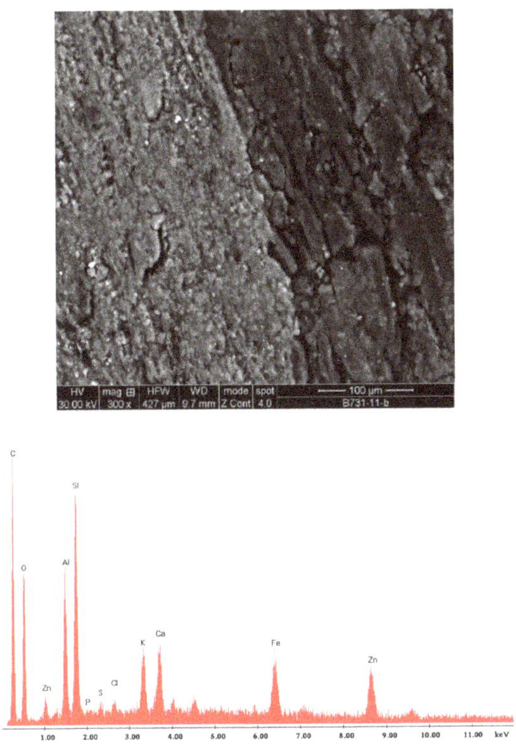


(**b**)

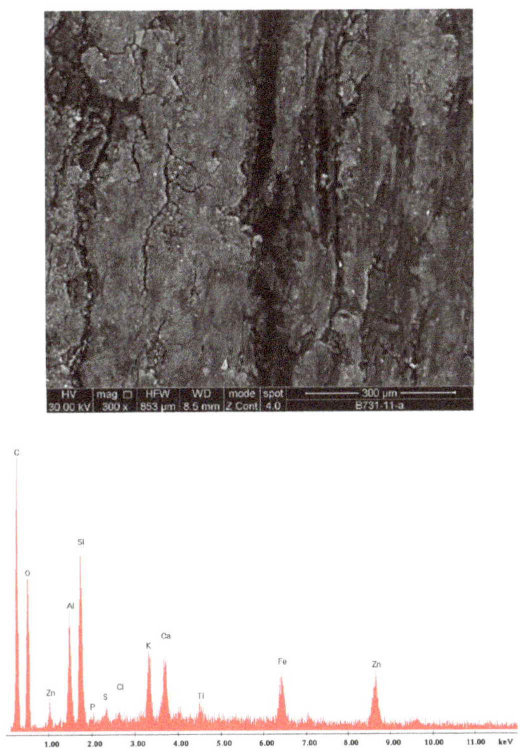


(**c**)

**Figure 6.** Evaluation of potential changes on the artwork's surface pre- and post- treatment, respectively: (**a**) comparison of FTIR spectra; ESEM-EDS analysis pre (**b**) and post treatment (**c**).

Removal action by HA solution, was preceded by FTIR and ESEM-EDS analyses of the wooden sculpture's surface, in order to exclude the presence of binder or finishing layers. As seen in Figure 6a, the FTIR spectrum revealed a silico-aluminous layer classified as a soiled contaminant due to conservation conditions, identified by the highlighted peaks. Furthermore, an E-SEM analysis of the wood surface was performed, pre- and post-treatment by the HA solution, and any changes have been highlighted (Figure 6b,c). Particularly E-SEM images showed no increase in fractures or changes in the elemental composition of the surface.

To test the hypothesis that the antimicrobial efficiency of *T. vulgaris* extracts can be implemented performing a cooperative action of EO exposure and HA solutions, a dedicated "clean chamber" was assembled. Specifically, after the microbial colonies' removal by the HA solution, the wooden artwork was exposed to EO volatile compounds (Figure 7). During the seven weeks of permanence in the "clean chamber", a monitoring of microbial colonization was carried out weekly. Sampling on the selected surface areas was performed with sterile swabs, utilized to inoculate nutrient and Sabouraud agar media in an in vitro microbial culture. The monitoring continued for another ten weeks, after which the artifact was brought back to the museum's exhibition hall.

**Figure 7.** "Clean chamber" assembled for exposure (7 weeks) of the wooden sculpture to *T. vulgaris* EO volatile compounds. Inside/outside temperature (22 + 1 °C) and relative humidity (56 + 2%) values were continuously monitored by an Oregon thermo-hygrometer equipped with a microprobe.

Results of the microbial growth monitoring, performed for total of seventeen weeks, are shown in Figure 8a,b, evaluating the total microbial load (TML). In the first seven weeks: (a) TML initially revealed on artwork's surface (black line), slightly decreased after exposure to EO volatile compounds (orange line). Alternatively, if, before EO volatile compounds exposure, a treatment with 70% Et-OH (green line) or an HA solution (blue line) was carried out, a reduced or very low microbial load was recognizable, and the microbial load reached zero when the cooperative action of *T. vulgaris* solutions (HA–EO) was achieved (blue line).

Moreover, microbial load was monitored for a further ten weeks after the re-exposition of the wooden artifact into the museum's exposure-hall environment. As shown in (b), the best results can still be seen for the combined HA–EO treatment (blue line).

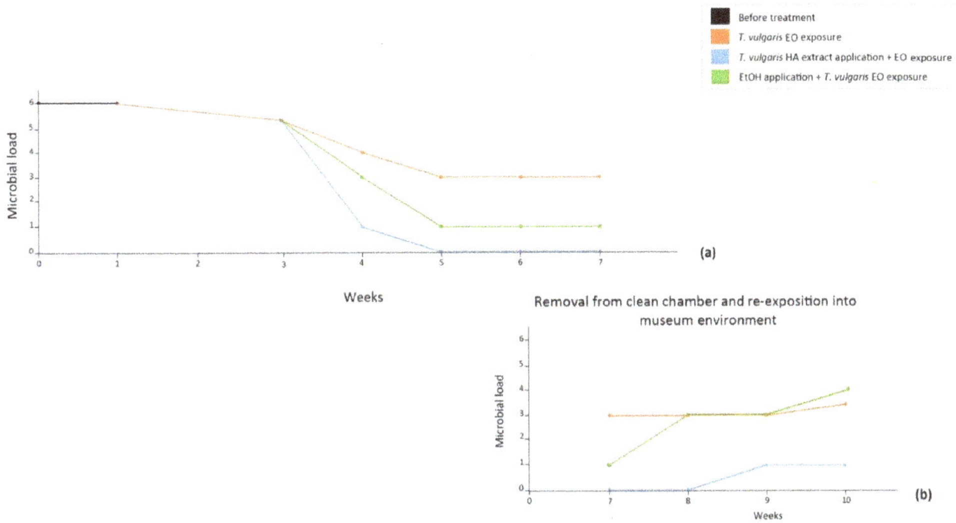


**Figure 8.** Monitoring of the microbial load on selected areas of the artwork's surface, pre- and-post treatment, performed (**a**) for seventeen weeks and (**b**) for ten weeks after re-exposition into museum exposition hall.

## 4. Discussion

Generally, to control and eradicate microbial colonization, synthetic biocides are applied, but several factors, such as environmental pollution and toxicity to humans, need to be evaluated [23–26]. In recent times, the choice of sustainable methods to counteract microbial colonization on cultural assets is focused on plant bioactive compounds, representing valid alternatives to synthetic preservatives [56–61]. Specifically, essential oils can certainly be considered antimicrobial agents, containing several chemical compounds (phenols, quinines, tannins, etc.) acting on microbial systems through different mechanisms and their potential synergic effect [31]. Undoubtedly, the main limit of the use of plant extracts is the method of application on cultural assets, as avoiding direct contact may induce unpredictable modifications on surface layers (varnishing, pigments, or other substances). In this study, *Thymus vulgaris* essential oil and hydro-alcoholic solutions are proposed as a novel protocol to counteract microbial-colonized Sogo Bò wooden sculptures.

Gas-chromatography analysis showed that carvacrol and thymol are the main compounds in both EO and HA solutions; the antimicrobial activity of both is reported in the literature [62,63].

In this combined treatment, the wooden sculpture's surface was preliminarily treated with a *T. vulgaris* HA solution, followed by exposure to EO volatile compounds in a dedicated "clean chamber", under environmental conditions. On the bigger section of the wooden sculpture, three areas were selected to monitor the action of thymus products on microbial colonization. As summarized in Figure 8a, before treatment, a high microbial load was revealed, slightly decreasing after exposure to EO volatile compounds in the ad hoc assembled "clean chamber". Total microbial load was further reduced when combining the exposure to EO volatile compounds with an application of 70% Et-OH, becoming zero when the *T. vulgaris* HA solution was applied.

Monitoring was continued for a further ten weeks after the artifacts were re-exhibited in the museum hall, which highlighted that the microbial load was still very low.

## 5. Conclusions

Historical–artistic assets represent an inestimable heritage whose value is not only related to artistic characteristics, but also to past history, culture, and traditions. These significant treasures need to be protected from damage induced by air pollution, physical and chemical agents, and microbial colonization.

Aiming to set up specific protocols to increase the efficacy of exposure to EO volatile compounds, in our previous study, the exposition of microbial colonized parchment artworks to *Thymus vulgaris* L. and *Crithmum maritimum* L. EOs was performed under vacuum conditions that significantly reduced the exposure time [64].

Evaluating that plant bioactive compounds are extracted as EO or HA solutions, in the present work, for the first time, a synergic effect of *T. vulgaris* extracts (essential oil and hydro-alcoholic solutions) was proposed for a specific wooden sculpture, performing the exposure under environmental conditions.

This strategy turned out to be ideal for the Sogo Bò theatre (Mali, Africa) artifact, the finishing treatment of which is not present on the sculpture's surface.

In our hypothesis, after alcohol evaporation (HA solution), the *T. vulgaris* antimicrobial compound should remain on the artwork's surface, enhancing the antimicrobial activity of EO volatile compounds, applied in an ad hoc assembled "clean chamber".

This is a further applicative protocol, respectful of both operators and the environment, based on green conservation strategy, which allowed us to hypothesize its use in replacing synthetic biocides in a sustainable conservation of cultural assets.

**Author Contributions:** Conceptualization, S.S., G.N. and F.P.; methodology, S.S., T.F., G.G., B.M. and B.G.; software, S.S. and F.P.; investigation and data curation, S.S., B.M., B.B. and G.G.; writing—original draft preparation, S.S., B.B. and F.P.; writing—review and editing, S.S., G.N. and F.P. All authors have read and agreed to the published version of the manuscript.

**Funding:** This research received no external funding. Publication fee was supported by EPO Srl, Milano, Italy.

**Institutional Review Board Statement:** The study did not involve humans or animals.

**Acknowledgments:** The authors especially thank the Director and the staff of "Museo Internazionale delle Marionette A. Pasqualino" Palermo, Italy, for yielded collaboration, and M.J. Arnoldi for the precious suggestions on wooden species. Results on application of *T. vulgaris* essential oil and hydro-alcoholic solution are part of S. Sparacello's final examination of the Master's degree in Conservation and Restoration of Cultural Heritage at the University of Palermo (Italy), graduated cum laude and qualified as an Italian Restorer of Cultural Heritage (MiC, Italian Ministry of Culture).

**Conflicts of Interest:** The authors declare no conflict of interest.

## References

1. Arnoldi, M.J. Playing the Puppets: Innovation and Rivalry in Bamana Youth Theatre of Mali. *Drama Rev.* **1988**, *32*, 2. [CrossRef]
2. Linking, W.W. Marionnettes du Mali, Paris. 1980. Available online: http://www.gourcuff-gradenigo.com/mali.html (accessed on 24 September 2020).
3. Arnoldi, M.J.; Curator Emerita, Africa Department of Anthropology, Smithsonian Institution, Washington, DC, USA. Personal communication, 2021.
4. Pournou, A. Biology of wood deteriogens. In *Biodeterioration of Wooden Cultural Heritage*; Springer Nature: Basingstoke, UK, 2020; Chapter 3; p. 99.
5. Allsopp, D.; Seal, K.J.; Gaylarde, C.C. *Introduction to Biodeterioration*; Cambridge University Press: Cambridge, UK, 2004.
6. Dakal, T.C.; Cameotra, S.S. Microbially induced deterioration of architectural heritages: Routes and mechanisms involved. *Environ. Sci. Eur.* **2012**, *24*, 1–13. [CrossRef]
7. Ranalli, G.; Zanardini, E.; Sorlini, C. Biodeterioration—Including Cultural Heritage, in Encyclopedia of Microbiology. In *Encyclopedia of Microbiology*, 4th ed.; Schmidt, T.M., Ed.; Academic Press: Cambridge, MA, USA, 2019.
8. Di Carlo, E.; Barresi, G.; Palla., F. Biodeterioration. In *Biotechnology and Conservation of Cultural Heritage*; Palla, F., Barresi, G., Eds.; Springer International Publishing: Basel, Switzerland, 2017; pp. 1–30.
9. Blanchette, R.A.; Nilsson, T.; Daniel, G.; Abad, A. Biological degradation of wood. In *Archaeological Wood*; Rowell, R.M., Barbour, R.J., Eds.; American Chemical Society: Washington, DC, USA, 1990; Volume 74, pp. 141–174.
10. Reinprect, L. *Wood Deterioration, Protection and Maintenance*; John Wiley & Sons, Ltd.: Chichester, UK, 2016.
11. Ljaljević Grbić, M.; Stupar., M.; Vukojević, J.; Maričić, I.; Bungur, N. Molds in Museum Environments: Biodeterioration of Art Photographs and Wooden Sculptures. *Arch. Biol. Sci.* **2013**, *65*, 955–962. [CrossRef]

12. Valentin, N. Microbial contamination in museum collections: Organic materials. In *Molecular Biology and Cultural Heritage*; Saiz-Jimenez, C., Ed.; Swets & Zeitlinger: Lisse, The Netherlands, 2003; pp. 85–91.
13. Palla, F. Biological risks for the conservation and exhibition of historical artistic artefacts in confined spaces. In *Risk Management in the Cultural Heritage Sector: Museums, Libraries and Archives*; Lorusso, S., Natali, A., Palla, F., Eds.; MIMESIS Edizioni: Milano, Italy, 2015; pp. 95–108.
14. Biological Deterioration & Damage to Furniture & Wooden Objects. Available online: https://www.si.edu/mci/english/learn_more/taking_care/biodetwood.html (accessed on 29 March 2018).
15. Tornari, V.; Basset, T.; Andrianakis, M.; Kosma, K. Impact of Relative Humidity on Wood Sample: A Climate Chamber Experimental Simulation Monitored by Digital Holographic Speckle Pattern Interferometry. *J. Imaging* **2019**, *5*, 65. [CrossRef] [PubMed]
16. Clausen, A.C. Bacterial association with decaying wood: A review. *Int. Biodeterior. Biodegrad.* **1996**, *37*, 101–107. [CrossRef]
17. Blanchette, R.A. A review of microbial deterioration found in archaeological wood from different environments. *Int. Biodeterior. Biodegrad.* **2000**, *46*, 189–204. [CrossRef]
18. Pangallo, D.; Šimonovičová, A.; Chovanová, K.; Ferianc, P. Wooden art objects and the museum environment: Identification and biodegradative characteristics of isolated microflora. *Lett. Appl. Microbiol.* **2007**, *45*, 87–94. [CrossRef]
19. Romero, S.; Giudicessi, S.L.; Vitale, R.G. Is the fungus Aspergillus a threat to cultural heritage? *J. Cult. Her.* **2021**, *51*, 107–124. [CrossRef]
20. Hyde, K.D.; Jones, E.B.G. Introduction to fungal succession. *Fungal Divers.* **2002**, *10*, 241–253.
21. Chojnackah, M.K. Biocides. In *Encyclopedia of Toxicology*, 3rd ed.; Biomedical Sciences, Elsevier: Amsterdam, The Netherlands, 2014; pp. 461–463.
22. Young, M.E.; Alakomi, H.L.; Fortune, A.A.; Goburshina, A.A.; Krumbein, W.E.; Maxwell, I.; McCullagh, C.; Robertson, P.; Saarela, M.; Valero, J.; et al. Development of a biocidal treatment regime to inhibit biological growths on cultural heritage: BIODAM. *Environ. Geol.* **2008**, *56*, 631–641. [CrossRef]
23. Marcotte, S.; Estel, L.; Leboucher, S.; Minchin, S. Occurrence of organic biocides in the air and dust at the Natural History Museum of Rouen, France. *J. Cult. Herit.* **2014**, *15*, 68–72. [CrossRef]
24. Arias-Estévez, M.; López-Periago, E.; Martínez-Carballo, E.; Simal-Gándara, J.; Juan-Mejuto, J.C.; García-Río, L. The mobility and degradation of pesticides in soils and the pollution of groundwater resources. *Agric. Ecosyst. Environ.* **2008**, *123*, 247–260. [CrossRef]
25. Hernández, A.F.; Parron, T.; Tsatsakis, M.A.; Requena, M.; Alarcón, R.; López-Guarnido, O. Toxic effects of pesticide mixtures at molecular level: Their relevance to human health. *Toxicology* **2013**, *307*, 136–145. [CrossRef]
26. Hernández, A.F.; Gil, F.; Lacasagña, M. Toxicological interactions of pesticide mixtures: An update. *Arch. Toxicol.* **2017**, *91*, 3211–3223. [CrossRef]
27. Petrovska, B.B. Historical review of medicinal plants usage. *Pharmacogn. Rev.* **2012**, *6*, 1–5. [CrossRef]
28. Xie, Y.; Wang, Z.; Huang, Q.; Zhang, D. Antifungal activity of several essential oils and major components against wood-rot fungi. *Ind. Crop. Prod.* **2017**, *108*, 278–285. [CrossRef]
29. Fernandez-Lopes, J.; Viuda-Martos, M. Application of essential oils in food systems. *Foods* **2018**, *7*, 1–4.
30. Abu-Shanab, B.; Adwan, G.; Abu-Safiya, D.; Jarrar, N.; Adwan, K. Antibacterial Activities of Some Plant Extracts Utilized in Popular Medicine in Palestine. *Turk. J. Biol.* **2004**, *28*, 99–102.
31. Bakkali, F.; Averbeck, S.; Averback, D.; Idaomar, M. Biological effects of essential oils—A review. *Food Chem. Toxicol.* **2008**, *46*, 446–475. [CrossRef]
32. Reichling, J.; Schnitzler, P.; Suschke, U.; Saller, R. Essential oils of aromatic plants with antibacterial, antifungal, antiviral, and cytotoxic properties—An overview. *Complement. Med. Res.* **2009**, *16*, 79–90. [CrossRef]
33. Afifi, H.A.M. Comparative efficacy of some plant extracts against fungal deterioration of stucco ornaments in the Mihrab of Mostafa Pasha, Ribat, Cairo, Egypt. *Am. J. Biochem. Mol. Biol.* **2012**, 2, 40–47. [CrossRef]
34. Gatti, L.; Troiano, F.; Vacchini, V.; Cappitelli, F.; Balloi, A. An In Vitro Evaluation of the Biocidal Effect of Oregano and Cloves' Volatile Compounds against Microorganisms Colonizing an Oil Painting—A Pioneer Study. *Appl. Sci.* **2021**, *11*, 78. [CrossRef]
35. Fierascu, I.; Ion, R.M.; Radu, M.; Bunghez, I.R.; Avramescu, S.M.; Fierascu, R.C. Comparative study of antifungal effect of natural extracts and essential oils of *Ocimum basilicum* on selected artefacts. *Rev. Roum. Chim.* **2014**, *59*, 207–211.
36. Liu, Z.; Zhang, Y.; Zhang, F.; Hu, C.; Liu, G.; Pan, J. Microbial community analyses of the deteriorated storeroom objects in the Tianjin Museum using culture-independent and culture-dependent approaches. *Front. Microbiol.* **2018**, *9*, 802. [CrossRef] [PubMed]
37. Rakotonirainy, M.S.; Lavédrine, B. Screening for antifungal activity of essential oils and related compounds to control the biocontamination in libraries and archives storage areas. *Int. Biodeter. Biodegrad.* **2005**, *55*, 141–147. [CrossRef]
38. Barresi, G.; Di Carlo, E.; Trapani, M.R.; Parisi, M.G.; Chille, C.; Mulè, M.F.; Cammarata, M.; Palla, F. Marine organisms as source of bioactive molecules applied in restoration projects. *Herit. Sci.* **2015**, *3*, 17. [CrossRef]
39. Rotolo, V.; Barresi, G.; Di Carlo, E.; Giordano, A.; Lombardo, G.; Crimi, E.; Costa, E.; Bruno, M.; Palla, F. Plant extracts as green potential strategies to control the biodeterioration of cultural heritage. *Int. J. Conserv. Sci.* **2016**, *7*, 839–846.
40. Palla, F.; Bruno, M.; Mercurio, F.; Tantillo, A.; Rotolo, V. Essential oil as natural biocides in conservation of cultural heritage. *Molecules* **2020**, *25*, 730. [CrossRef] [PubMed]

41. Ebadollahi, A.; Ziaee, M.; Palla, F. Essential oils extracted from different species of the Lamiaceae plant family as prospective bioagents against several detrimental pests. *Molecules* **2020**, *25*, 1556. [CrossRef]
42. Palla, F.; Caruana, E.; Di Carlo, E.; Rotolo, V. Plant essential oils in controlling fungal colonization on wooden substrate. *Borziana* **2021**, *2*, 5–14.
43. Di Carlo, E.; Chisesi, R.; Barresi, G.; Barbaro, S.; Lombardo, G.; Rotolo, V.; Sebastianelli, M.; Travagliato, G.; Palla, F. Fungi and bacteria in indoor cultural heritage environments: Microbial related risks for artworks and human health. *Environ. Ecol. Res.* **2016**, *4*, 257–264. [CrossRef]
44. Frank, J.A.; Reich, C.I.; Sharma, S.; Weisbaum, J.S.; Wilson, B.A.; Olsen, G.J. Critical evaluation of two primers commonly used for amplification of bacterial 16S rRNA genes. *Appl. Environ. Microbiol.* **2008**, *74*, 2461–2470. [CrossRef]
45. Cardinale, M.; Brusetti, L.; Quatrini, P.; Borin, S.; Puglia, A.M.; Rizzi, A.; Zanardini, E.; Sorlini, C.; Corselli, C.; Daffonchio, D. Comparison of different primer sets for use in automated ribosomal intergenic spacer analysis of complex bacterial communities. *Appl. Environ. Microbiol.* **2004**, *70*, 6147–6156. [CrossRef] [PubMed]
46. Gardes, M.; Bruns, T.D. ITS primers with enhanced specificity for basidiomycetes-application to the identification of mycorrhizae and rusts. *Mol. Ecol.* **1993**, *2*, 113–118. [CrossRef] [PubMed]
47. Martin, K.J.; Rygiewicz, P.T. Fungal-specific PCR primers developed for analysis of the ITS region of environmental DNA extracts. *BMC Microbiol.* **2005**, *5*, 28. [CrossRef] [PubMed]
48. Barresi, G.; Cammarata, M.; Palla, F. Biocide. In *Biotechnology and Conservation of Cultural Heritage*; Palla, F., Barresi, G., Eds.; Springer International Publishing: Basel, Switzerland, 2017; pp. 49–66.
49. European Pharmacopoeia 10th Edition, 10.2 n.0865, July 2020. Available online: www.edqm.eu (accessed on 24 September 2020).
50. Timofeeff, M.N.; Lowestein, T.K.; Blackburn, W.H. ESEM-EDS: An improved technique for major element chemical analysis of fluid inclusions. *Chem. Geol.* **2000**, *164*, 171–181. [CrossRef]
51. Rotolo, V.; De Caro, M.L.; Giordano, A.; Palla, F. Solunto archaeological park in Sicily: Life under tesserae. *Flora Medit* **2018**, *28*, 233–245.
52. Barry, A.L. *The Antimicrobic Susceptibility Test: Principles and Practices*; Lea and Febiger: Philadelphia, PA, USA, 1976.
53. Reichling, J.; Suschke, U.; Schneele, J.; Geiss, H.K. Antibacterial activity and irritation potential of selected essential oil components structure activity relationship. *Nat. Prod. Commun.* **2006**, *1*, 1003–1012. [CrossRef]
54. Popovici, R.A.; Vaduva, D.A.; Pinzaru, I.; Dehelean, C.A.; Farcas, C.G.; Coricovac, D.; Danciu, C.; Popescu, I.; Alexa, E.; Lazureanu, V.; et al. A comparative study on the biological activity of essential oil and total hydro-alcoholic extract of *Satureja hortensis* L. *Exp. Ther. Med.* **2019**, *18*, 932–942. [CrossRef]
55. Saad, N.; Muller, C.D.; Lobstein, A. Major bioactivities and mechanism of action of essential oil and their components. *Fragance Flavours J.* **2013**, *28*, 269–279. [CrossRef]
56. Cappitelli, F.; Cattò, C.; Villa, F. The control of cultural heritage microbial deterioration. *Microorganisms* **2020**, *8*, 1542. [CrossRef] [PubMed]
57. Fidanza, M.R.; Caneva, G. Natural biocides for the conservation of stone cultural heritage: A review. *J. Cult. Herit.* **2019**, *38*, 271–286. [CrossRef]
58. Lavin, P.; Gómez de Saravia, S.; Guiamet, P. *Scopulariopsis* sp. and *Fusarium* sp. in the Documentary Heritage: Evaluation of their biodeterioration ability and antifungal effect of two essential oils. *Microb. Ecol.* **2016**, *71*, 628–633. [CrossRef] [PubMed]
59. Nuță, D.C.; Limban, C.; Ioniță, C.C.; Nicolau, I.; Zarafu, I. The Use of Essential Oils as a Strategy to Combat Microbial Biofilms. A Review. *Processes* **2021**, *9*, 537. [CrossRef]
60. Borrego, S.; Valdés, O.; Vivar, I.; Lavin, P.; Guiamet, P.; Battistoni, P.; Gómez de Saravia, S.G.; Borges, P. Essential oils of plants as biocides against microorganisms isolated from Cuban and Argentine documentary heritage. *ISRN Microbiol.* **2012**, *2012*, 826786. [CrossRef]
61. Sakr, A.A.; Ghaly, M.F.; El-Sayed, F.; Abdel-Haliem, M. The efficacy of specific essential oils on yeasts isolated from the royal tomb paintings at Tanis, Egypt. *Int. J. Conserv. Sci.* **2012**, *3*, 87–92.
62. Gavaric, N.; Mozina, S.S.; Kladar, N.; Bozin, B. Chemical profile, antioxidant and antibacterial activity of Thyme and Oregano essential oils, Thymol and Carvacrol and their possible synergism. *J. Essent. Oil Bear. Plants* **2015**, *18*, 1013–1021. [CrossRef]
63. Klarić, M.Š.; Kosaleć, I.; Mastelić, J.; Piecková, E.; Pepeljnak, S. Antifungal activity of thyme (*Thymus vulgaris* L.) essential oil and thymol against moulds from damp dwellings. *Lett. Appl. Microbiol.* **2007**, *44*, 36–42. [CrossRef]
64. D'Agostino, G.; Giambra, B.; Palla, F.; Bruno, M.; Badalamenti, N. The Application of the Essential Oils of *Thymus vulgaris* L. and *Crithmum maritimum* L. as Biocidal on Two Tholu Bommalu Indian Leather Puppets. *Plants* **2021**, *10*, 1508. [CrossRef]

*applied sciences*

MDPI

*Article*

# An In Vitro Evaluation of the Biocidal Effect of Oregano and Cloves' Volatile Compounds against Microorganisms Colonizing an Oil Painting—A Pioneer Study

**Lucrezia Gatti [1], Federica Troiano [1], Violetta Vacchini [2], Francesca Cappitelli [1] and Annalisa Balloi [2,3,*]**

1 Department of Food, Environmental and Nutritional Sciences, Università degli Studi di Milano, Via Celoria 2, 20133 Milano, Italy; lucrezia.gatti2@unibo.it (L.G.); federica.troiano@unimi.it (F.T.); francesca.cappitelli@unimi.it (F.C.)
2 Micro4yoU s.r.l., Via Morozzo della Rocca 6, 20123 Milano, Italy; violetta.vacchini@unimi.it
3 Accademia delle Belle Arti di Brera, Via Brera 28, 20121 Milano, Italy
* Correspondence: annalisa.balloi@polimi.it; Tel.: +39-3479041042

**Featured Application: We hypothesize an effective and potentially color respectful method of essential oils (EOs) application in a cleaning procedure for biodeteriorated oil paintings. The procedure consists of flowing a thin film with the EOs onto an evaporating surface and then placing it parallel to the painting by using some supports, so that the vapors of the EOs may homogeneously reach the painting surface, thereby avoiding a direct contact of the EOs with pigments. Future work should be conducted to verify the feasibility of this methodology in a real case study.**

**Abstract:** In this study, the biocidal activity of two plant derivatives (oregano and cloves' essential oils—EOs) was evaluated, as a potential innovative and eco-friendly cleaning method for canvas paintings. The object of the study was the oil painting on canvas entitled "Studio di nudo" (Giovanni Maria Mossa, 1921), showing stains caused by microorganisms. The research focused on: (1) isolation and identification of microorganisms associated with discolorations on the obverse and reverse sides of the canvas; (2) evaluation of biocidal activity of selected EOs against fungal and bacterial collections. The phylogenetic identification was conducted with both cultivation and molecular methods. The canvas was mainly colonized by *Penicillium*, *Aspergillus*, and *Cephaloteca* fungal genera and by bacteria of the *Bacillus* genus. To evaluate the biocidal effect of the EOs' volatile components only, an antibiogram assay (agar disc diffusion method) and a customized assay (named the contactless test) were conducted. Tested EOs showed antimicrobial activity on fungi and bacteria. However, compared to cloves, oregano EO exhibited a better inhibition activity both in contact and contactless tests. The work is pioneering for the use of EOs' volatile compounds against oil painting biodeteriogens, and gives insights into possible extended, innovative and eco-friendly cleaning methods for painting control procedures.

**Keywords:** antimicrobial activity; canvas painting biodeterioration; cleaning procedure; contactless test; cultural heritage; control; plant essential oils; volatile components

**Citation:** Gatti, L.; Troiano, F.; Vacchini, V.; Cappitelli, F.; Balloi, A. An In Vitro Evaluation of the Biocidal Effect of Oregano and Cloves' Volatile Compounds against Microorganisms Colonizing an Oil Painting—A Pioneer Study. *Appl. Sci.* **2021**, *11*, 78. https://dx.doi.org/10.3390/app11010078

Received: 9 November 2020
Accepted: 21 December 2020
Published: 24 December 2020



## 1. Introduction

In current conservation practices, the use of biocides is still the most popular method in cleaning procedures for different biodeteriorated artefacts. However, the strong negative impact of these chemicals on human health, object surfaces and ecosystems are pushing researchers and conservators to find alternative solutions that are selective against biodeteriogens and at the same time environmentally friendly and safe for humans. Moreover, the choice of ecofriendly biocides is led by the EU regulation U Directive 98/8/EC, which recommends the withdrawal from the market of biocides harmful to humans and non-target organisms. Alternative products involve the use of natural molecules, such as, for example, plant derivates [1,2].

In particular the essential oils (EOs) obtained by distillation or pressing of aromatic plants are well known for their bactericidal and fungicidal properties [3]. They are mixtures of volatile compounds, insoluble in water but soluble in organic substances, characterized by low molecular weights and strong odors [4]. Their antimicrobial activity is due to their bioactive compounds such as phenols, quinines, and tannins, acting through various mechanisms, such as modification of the membrane structure or alteration of the enzymatic activity [5].

Today EOs are widely used in the food processing, phyto-sanitary, pharmaceutical, and cosmetic sectors [6–8]. Moreover, in recent years, a large number of studies have investigated the use of EOs for the control of microbial colonization on cultural heritage surfaces [3].

Oregano (*Origanum vulgare*) proved to be effective against fungi isolated on wood and stone artefacts [9,10]. Clove (*Syzygium aromaticum*) and garlic (*Allium sativum*) oils proved potent antimicrobials against different fungal species including *Aspergillus niger* [11].

Methodologies of application of EOs to the surfaces vary, and in almost all cases they imply a direct contact between oils and artefact surfaces, such as the application by brush, with packaging or padding methods or in a thickening solution (e.g., cellulose and/or sepiolite) [12].

EOs' roles as antimicrobial agents have been tested on paper documents [13–15], historical textiles [16], wooden and stone artefacts [17,18], and also on objects of large dimensions [19]. Elsayed and Shabana [20] evaluated the effect of some EOs on fungal infestation on simulated painting models. However, despite the increasing interest, EOs have never been applied on real oil paintings. This is not surprising, as by coming into contact with pigments, EOs could act as a solvent and cause irreversible damage on painted surfaces. Citrus essential oils are in use in some "green" cleaning formulations and are known to have powerful solubilization abilities [21,22]. Moreover, paintings possess a multi-material nature (sometimes with unknown composition due to the artists having prepared their own blends), and their reaction with EOs might be unpredictable.

We hypothesize the possibility to use the volatile organic compounds (VOCs) of the essential oils for the cleaning procedure used on oil paintings, in order to avoid direct contact between the EOs and the painting surface. In the present work, the isolation and identification of the fungal and bacterial taxa from stained areas on the obverse and reverse surfaces of an oil painting on canvas were conducted, and the antimicrobial activity of oregano and cloves EOs were assessed against the isolates. Moreover, oregano EO was also tested using its volatile components only. Even if this is a preliminary Proof of Concept work, aimed to verify if indeed the VOCs of the EOs could have a significant inhibiting action on biodeteriogens of oil paintings, it can open up new ways to preserve oil paintings.

## 2. Materials and Methods

### 2.1. Artwork Description and Conservation State

The painting which is the object of the study is an oil on canvas, entitled "Studio di Nudo", painted in 1921 by the Italian painter Giovanni Maria Mossa (1896, Tempio Pausania, Sardinia—1982, Pianico, Lombardy). With this portrait, the young artist won the Hayez Prize that allowed him to attend the "Accademia delle Belle Arti" di Brera in Milan. The painting represents a dancer, sitting frontally but with her face turned in profile. The woman is wearing a dark tutu that covers the lower part of her body. The figure stands out against a dark background, without a spatial perspective. Stylistically, even if the work belongs to the author's youthful period, the characteristics of the painter are recognizable, but it is possible to find uncertainties in the drawing of the body and in the disproportion of the neck and arms compared to the rest of the body.

According to the claims by the conservators working with this painting, the canvas showed a microbial attack resulting in numerous discolorations both on the reverse, where brown dot-like areas are clearly visible, and on the obverse sides, visible especially in the areas of the body skin, due to the chromatic contrast with the pigment. The canvas surface also shows signs of laceration, dabbed during previous restorations with glue.

*2.2. Sampling Method*

Samples were collected by using a sterile nitrocellulose membrane (Sartorius AG, Gottingen, Germany—17.34 $cm^2$): 18 sampling points were identified from the obverse side of the canvas, in correspondence with visible biological attack (Figure 1a): four on the background (sampling points 8F, 9F, 23F, 24F), 2 on hair (20F, 21F), 2 on the face (18F, 22F), 4 on rosy parts (10F, 15F, 16F, 17F), 2 on drapery (13F, 14F), and 4 on glue (11F, 12F, 19F, 25F). Seven sampling points were collected from the reverse side as well (Figure 1b): four samples were taken in correspondence with the lower-external edges of the canvas (1R, 2R, 4R, 5R) a central one (3R), and two in correspondence with the glue used to dab the tears in the canvas (6R, 7R).

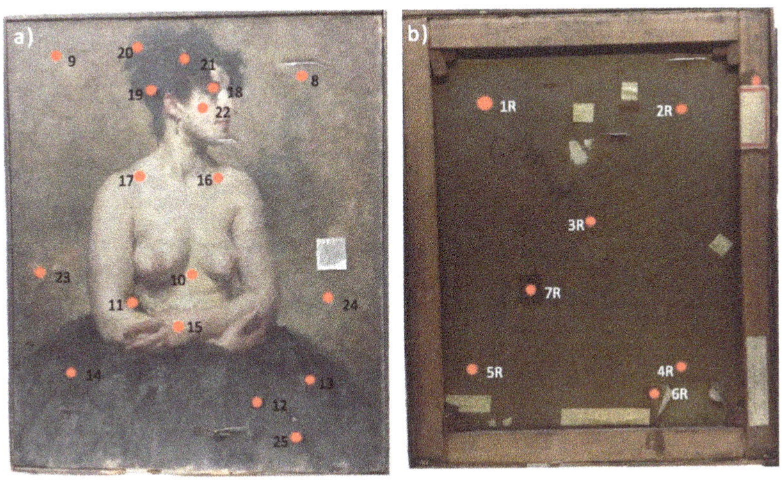


**Figure 1.** The artwork "Studio di nudo" and position of sampling points from observe (**a**) and reverse (**b**) sides of the canvas.

The membranes were placed on the surface to be sampled, giving a homogeneous pressure for thirty seconds, with a sterile cotton swab [23]. Each area was sampled twice, and each nitrocellulose membrane was transferred into petri dishes with either of two different media: Potato Dextrose Agar (PDA; 4 g/L potato extract, 20 g/L glucose, 15 g/L agar) for fungi cultivation; Nutrient Broth Agar (NB; 10 g/L peptone, 5 g/L sodium chloride, 10 g/L beef extract, 15 g/L agar) for bacteria cultivation, with the addition of cycloheximide (50 $\mu g\ mL^{-1}$) to prevent fungal growth.

*2.3. Isolation and Identification of Microbial Community (Fungi and Bacteria)*

Petri dishes inoculated with nitrocellulose membranes were incubated under sterile conditions at 30 °C: bacterial and fungal growth was followed up to 7 and 20 days, respectively. For each sampling point, results are expressed as CFU/$dm^2$. On the basis of macroscopic features, bacterial and fungal colonies were selected and transferred into new single petri dishes with a specific growth medium and incubated at 30 °C.

After a visual examination, bacterial and fungal isolates were clustered according to their morphological characteristics and an evaluation of the relative abundance of each morphological group on the painting was possible. DNA was extracted from all isolates of each cluster according to Troiano et al. [24]. Once extracted, DNA was amplified through the Polymerase Chain Reaction. 16S region was amplified for bacteria, by using 16F (5′-AGAGTTTGATCCTGGCTCAG-3′) and 16R (5′-CTACGGCTACCTTGTTACGA-3′) primers, and a chemical protocol according to Rizzi et al. [25]. The thermal protocol provides for one cycle at 94 °C for 4 min, 35 cycles at 94 °C for 45 s, at 55 °C for 1 min and at 72 °C for 2 min, a second cycle at 72 °C for 10 min and a last cycle at 12 °C constant until samples removal. The end volume of the reaction was 50 µL, 48 µL of mix solution and 2 µL of DNA from each sample. ITS fungal region was amplified using ITS1F (5′-CTTGGTCATTTAGAGGAAGTAA-3′) and ITS4 (5′-TCCTCCGCTTATTGATATGC-3′) primers and chemical protocol as follows: buffer 1X, magnesium chloride ($MgCl2$) 1.5 mM, normal deoxynucleotides triphosphate (dNTP) mix 0.12 mM, 16F 0.30 µM, 16R 0.30 µM, thermostable DNA polymerase (Taq) 1 U. The thermic protocol was conducted according to Manter et al. [26]. The end volume of reaction was 1 µL of DNA in 49 µL of Master Mix buffer. PCR products were sequenced by Macrogen Inc. (Seoul, Korea) and then analyzed through BLASTn software [27] and Classifier Ribosomal Database Project [28] for bacteria, while for fungi Mycobank software [29] was used (accessed in May 2019). The 16S rRNA genes and fungal ITS sequences were deposited in the European Nucleotide Archive (ENA) under the unique accession numbers ERZ1667197 and ERZ1667410, respectively, and were registered under the study PRJEB40902.

### 2.4. Essential Oils

The essential oils used in this study were oregano (*Origanum vulgare* L., 1753, Fitomedical) and clove (*Syzygium aromaticum* (L.) Merr. et L.M. Perry, Fitomedical, Binasco, Italy). EOs were used diluted with ethanol 70% in the ratio 2:1 and ethanol 70% (EtOH) was used as a negative control as suggested in the research by Borrego et al. 2012 [15].

#### 2.4.1. Contact Test with Essential Oils

One bacterial or fungal strain for each morphological cluster was inoculated in Petri dishes containing NB and PDA respectively. Three filter paper discs (47 mm) were placed on each plate after being soaked with 10 µL of oregano EO or clove EO or 70% ethanol, the latter used as a negative control. The test was performed in duplicate. Plates were incubated at 30 °C until the growth was visible. After growth, the inhibition halo around each bacterial colony was measured and the standard deviation was calculated. Since for fungi an inhibition halo radius was difficult to measure, it was observed by naked eye and recorded using the following scale: [+] for medium inhibition (approximately 0.2–1 cm), [++] for high inhibition (approximately 1–2 cm), [+++] for total inhibition and [–] for no inhibition.

#### 2.4.2. Contactless Test with Essential Oils

In order to evaluate the inhibition properties exclusively of the EOs volatile components, a contactless experiment was conducted. One fungal or bacterial strain for each morphological cluster was plated on PDA and NBA respectively. The test was performed in duplicate. Plates were overturned and 100 µL of oregano, 100 µL of cloves, and 100 µL of 70% ethanol, used as a negative control, were placed on the lids of petri dishes separately. The plates were incubated in reverse at 30 °C. After growth an inhibition halo was measured for bacterial strains, whereas for fungal strains the inhibited area was evaluated by naked eye using the following scale: [+] for medium inhibition (approximately one third of the petri dish), [++] for high inhibition (approximately the half of the petri dish), and [+++] for total inhibition. Regarding the inhibition scale degrees, see Table S1.

## 3. Results and Discussion

Modern oil paintings have a complex composition, with a mixture of inorganic and organic materials [30]. For these reasons, the choice of the best method for controlling painting biodeterioration is critical for successfully treating contaminated artworks. Biocidal treatments are usually employed for controlling microbial growth and their efficacy as well as their drawbacks have been widely discussed in literature [3]. One of the well-known drawbacks is the difficulty in applying antimicrobial compounds without damaging the pictorial layers [31,32]. However, natural products, such as essential oils and plant derivatives, due to their biocidal activity, represent a useful tool in the control of biodeterioration of cultural heritage, without negative environmental and human impacts.

Previous studies reported that the major components of oregano EO are the terpenes thymol and carvacrol [33], whereas the principal component of clove EO is eugenol [34]. The antimicrobial activity of oregano is mainly due to the presence of thymol and carvacrol. Actually, this activity is based on the molecular hydrophobicity of terpenes which promotes the partition of the EOs in the lipids of the cell membrane, leading to membrane permeabilization and leakage of cytoplasmic content [35]. Furthermore, thymol acts on fungi, modifying the morphology of the hyphae and causing their aggregation, with consequences on the diameter of the hyphae themselves and on the breakdown of the cell barrier [36]. Eugenol is an amphipathic hydroxyphenyl propene, active against fungi and a wide range of Gram-negative and Gram-positive bacteria [37]. The mechanisms of action are different and include changes in the morphology and disruption of the cytoplasmatic membrane, the production of Reactive Oxygen Species (ROS) and the inhibition of some essential enzymes, such as proteases, histidine carboxylases, amylases, and ATPases [37].

In this framework, the aim of the study was to determine whether oregano and clove EOs could be applied in cleaning procedures for oil canvas paintings attacked by microorganisms. For the study, an oil canvas with a visible microbial attack was selected. Once the fungi and bacteria colonizing the canvas were identified, the isolates were exposed directly and indirectly to the selected EOs.

The results showed that the biocidal effect of EOs' volatile components were effective on the isolates, suggesting their potential for possible application in a real cleaning procedure.

### *3.1. Isolation and Morphological Characterization of Microbial Community (Fungi and Bacteria)*

By using nitrocellulose membranes as a sampling method, coupled with cultural analysis, 62 fungi and 20 bacteria were isolated from the stained areas of the oil painting. A quantitative microbial risk assessment was evaluated via a direct count of colonies on Petri dishes. The results of the bacterial count were of 0.7 CFU/dm$^2$ on the reverse side and 1.7 CFU/dm$^2$ on the obverse side. The results showed a higher concentration of colonies on the obverse side of the painting, especially on sample points 15F, 17F, 19F, and 25F. As regards fungal counts, 0.14 CFU/dm$^2$ on the reverse and 2.9 CFU/dm$^2$ on the obverse sides were recorded. In general the microbial count on both obverse and reverse sides of the canvas case study was not particularly high [23,38]. This can be explained by the use of a non-invasive sampling technique using the nitrocellulose membrane and the possibility that not all the microbial strains might have been cultured due to the limits of the cultivation methods [39].

According to the results obtained by the cultural analysis, a greater level of bacterial and fungal contamination was observed on the obverse side of the canvas painting. This is in contrast with what is reported in the literature, where a greater microbial contamination is usually found on the reverse of the canvas [38,40]. The more consistent colonization of the obverse rather than the reverse sides could be due to the way in which the canvas has been stored over the years, probably stacked with other paintings.

In order to get an insight into the biodiversity corresponding to the stained areas of the painting, bacterial and fungal isolates were clustered according to their morphology (Table S2). Among the 13 different bacterial clusters, Clusters I and VII were the most numerous (Figure S1a). Twenty-four different fungal clusters were identified as well; among these, 12 were represented by only one isolate, while the most abundant was Cluster IV, counting 17 colonies collected on the obverse side of the painting (Figure S1b).

*3.2. Molecular Characterization of Microbial Community (Fungi and Bacteria)*

Molecular characterization was conducted via a culture-dependent approach in order to isolate and carry out further investigation on the microbial community present on the artwork and in particular the potential biodeteriorative microorganisms harbored.

DNA was extracted from all bacterial and fungal strains of all clusters isolated from the oil on canvas painting; they were identified by 16S rRNA gene and ITS sequencing respectively. Results are shown in Tables S3 and S4.

*Bacillus* was the predominant bacterial genus (clusters I, VII, and XIII). Cluster I was the most abundant, with 20% of bacterial isolates. One member of cluster I, isolated from two sampling points on the reverse side of the painting, showed 96% similarity with *Bacillus* genus. Another member of cluster I, isolated in correspondence with sample 6R from the reverse side of the canvas was similar to *Bacillus subtilis* with 100%. *Bacillus* spp. are Gram-positive spore-forming bacteria, ubiquitous in the environment (soil and water) and dominant over artefact surfaces [31], such as deteriorated historical paper [41,42]. *Bacillus* is also responsible for structural changes of carboxymethyl cellulose during biodeterioration processes [43] and it is able to produce amylase, cellulase, protease, and acids, which are known to contribute to archaeological manuscript biodeterioration [44]. Three sampling points (21F, 18F, and 22F) chosen from the obverse side of the canvas allowed 98–99% identification of isolates with *Bacillus thuringiensis* (cluster I), *B. simplex* (cluster XIII), and *B. luteus* (cluster VII). Cluster VII is the second most abundant cluster, with 13.80% of bacterial isolates. Bacterial stains belonging to cluster VII were isolated from sampling points 11F, 13F, 16F, 22F, and 24F. Other identified bacteria on the obverse side belong to *Cellulosimicrobium*, *Paenibacillus*, *Pseudomonas*, *Stenotrophomonas*, and *Micrococcus* genera, while isolates affiliated to the Xanthomonadaceae family and *Streptomyces* were collected from the reverse side of the canvas. *Streptomyces*, isolated from stained areas of the paint layer of oil paintings, have been responsible of bio-pigment production [45]. *Pseudomonas* sp. can cause different types of surface deterioration, such as pigmentation, efflorescence and patinas [46]. *Paenibacillus* genus was isolated from human and environmental samples [47]. *Xanthomonas* shows cellulose structure degradation activity [48]. *Stenotrophomonas* was isolated and identified from an oil painting on canvas, which showed visible signs of biodeterioration [49]. Gram-positive *Micrococcus luteus* is known to attack cellulosic materials by lytic enzymes and pigmented components [50]. *Pichia occidentalis* can be used for biological detoxification of lignocellulosic hydrolysate, because it can degrade volatile fatty acids [51]. A Phaeosphaeriaceae species was isolated and identified on aged oil sludge-contaminated soil [52].

Among the fungi, sequence analysis for ITS fragments revealed that the isolates exhibited 99–100% similarity to three main genera, among which the *Penicillium* genus was the most abundant.

Sequences retrieved from sampling points on the obverse side of the canvas showed 99–100% similarity with *Penicillium* sp. According to the literature, *Penicillium* is associated with poorly ventilated, moist environments, representing a risk for human health and cultural heritage [39]. Several studies reported the enzymatic activity in biodeterioration of archaeological documents, and cultural heritage in general [53]. *Penicillium chrysogenum*, a member of the most abundant cluster II, was identified from sampling points on both obverse and reverse sides and it is known for its ability to deteriorate cellulose and lignin [54]. The majority of the isolates showed affiliation to *Cephaloteca foveolata* and to *Aspergillus versicolor*, *Aspergillus insuetus*, and other *Aspergillus* spp. and to *Cladosporium* sp., *Alternaria* sp., *Pichia occidentalis*, and *Phaeosphaeriaceae* sp. *Cephalotheca foveolata* is known as a human pathogen, responsible for skin infection [55]. According to literature, *Aspergillus* is accounted as one of the most commonly occurring fungal genera recorded on canvas paintings [39]. Together with *Alternaria alternata*, *A. niger* is reported as the most common fungal species detected on oil paintings and artworks, often the two species are isolated on the same objects [15,39,56,57]. *Cladosporium*, *Chaetomium* and *Alternaria* are cellulose degraders, commonly present on biodeteriorated oil paintings and cellulosic materials, as are some members of the Trichocomaceae family [23,46,56], while *Pichia* is a yeast genus able to ferment sugar and assimilate nitrates [58]. Sequencing did not give reliable results for some bacterial and fungal clusters: although they were not identified, one member of each non-identified cluster was tested in contact and contactless tests.

### 3.3. Assessment for Antifungal and Antibacterial Activity of EOs in Contact Tests

The results of the antimicrobial activity of oregano and clove extracts in contact tests revealed that the two oils had an inhibitory effect against the growth of all bacteria and almost all fungal clusters (23 out of 24) (Table 1).

**Table 1.** Inhibition halo of oregano and clove essential oils on bacteria and fungi in contact tests. Tests were performed in duplicate and standard deviation was calculated. NI = not identified strain. Ethanol 70% was used as a negative control as it did not show any inhibitory effect against the isolates.

| CONTACT TEST | | | | |
|---|---|---|---|---|
| | ISOLATES | TAXA | OREGANO | CLOVE |
| BACTERIA | $I_c$ | *Bacillus subtilis* subsp. *subtilis* | 1.0 ± 0.2 cm | 0.4 ± 0.1 cm |
| | $II_a$ | Xanthomonadaceae | 1.1 ± 0.2 cm | 0.4 ± 0.1 cm |
| | $III_a$ | *Streptomyces* sp. | 1.0 ± 0.1 cm | 0.5 ± 0.1 cm |
| | $IV_a$ | NI | 1 ± 0 cm | 0.3 ± 0.2 cm |
| | $V_a$ | NI | 1.3 ± 0.2 cm | 0.4 ± 0.1 cm |
| | $VI_a$ | *Stenotrophomonas* | 0.9 ± 0.1 cm | 0.4 ± 0.1 cm |
| | $VII_a$ | *Pseudomonas psychrotolerans* | 0.3 ± 0.1 cm | 0.2 ± 0 cm |
| | $VIII_a$ | Xanthomonadaceae | 1 ± 0 cm | 0.3 ± 0.2cm |
| | $IX_a$ | *Cellulosimicrobium cellulans* | 0.9 ± 0.1 cm | 0.5 ± 0 cm |
| | $X_a$ | Penibacillaceae | 1.4 ± 0.1 cm | 0.4 ± 0.2 cm |
| | $XI_a$ | NI | 1.3 ± 0.2 cm | 0.7 ± 0.1 cm |
| | $XII_a$ | *Paenibacillus* sp. | 1.2 ± 0 cm | 0.9 ± 0.2 cm |
| | $XIII_a$ | *Bacillus simplex* | 2 ± 0 cm | 0.6 ± 0.2 cm |
| FUNGI | $I_a$ | *Penicillium chrysogenum* | ++ | + |
| | $II_c$ | NI | + | + |
| | $III_a$ | *Penicillium chrysogenum* | ++ | ++ |
| | $IV_b$ | *Cephalotheca foveolata* | ++ | ++ |
| | $V_a$ | *Aspergillus* sp. | + | + |
| | $VI_a$ | *Cephalotheca foveolata* | ++ | ++ |
| | $VII_a$ | *Cladosporium parahalotolerans* | + | + |
| | $VIII_a$ | NI | ++ | + |
| | $IX_a$ | *Cephalotheca foveolata* | ++ | ++ |
| | $X_a$ | *Aspergillus versicolor* | ++ | ++ |
| | $XI_a$ | NI | ++ | ++ |
| | $XII_a$ | NI | - | - |
| | $XIII_a$ | *Penicillium chrysogenum* | ++ | ++ |
| | $XIV_a$ | Trichocomaceae | + | + |
| | $XV_a$ | Chaetomiaceae | ++ | ++ |
| | $XVI_a$ | *Penicillium chrysogenum* | ++ | ++ |
| | $XVII_a$ | *Cephalotheca foveolata* | ++ | ++ |
| | $XVIII_a$ | NI | + | + |
| | $XIX_a$ | Phaeosphaeriaceae | + | + |
| | $XX_a$ | *Penicillium* sp. | + | + |
| | $XXI_b$ | NI | + | + |
| | $XXII_b$ | NI | + | + |
| | $XXIII_b$ | NI | ++ | + |
| | $XXIV_a$ | NI | ++ | + |

In particular, fungi belonging to *Penicillium chrysogenum*, *Cephaloteca foveolata*, *Cephaloteca* sp., *Aspergillus versicolor*, and Chaetomiaceae were highly inhibited in the presence of oregano EO (indicated with "++"), while a mild effect (indicated with "+") was observed for *Aspergillus* sp., *Cladosporium parahalotolerans*, *Penicillium* sp., and strains of the families Phaeosphaeriaceae and Trichocomaceae. Clove EO exhibited a high inhibitory activity on *Penicillium chrysogenum, Cephaloteca foveolata, Cephaloteca* sp., *Aspergillus versicolor,* and Chaetomiaceae and a medium inhibition on the other clusters (Table 1). A previous work reported the antifungal activity exhibited by clove and garlic oils against different fungal species including *A. niger* [59]. Camphor and clove EO showed antifungal activity against *A. niger* and *A. alternata* in agar plate tests as well as in simulated canvas painting models [20]. Previous studies showed the antifungal properties of different EOs, among which *O. vulgare* was active against several fungal species including *A. niger* and *A. ochraceus* by means of micro-, macro-dilution, and micro-atmosphere methods [18,55]. All the bacterial strains were inhibited by both EOs (Table 1). The minimum inhibition halo (0.3 $\pm$ 0.1 cm and 0.2 cm for oregano and clove, respectively) was recorded for bacteria identified as *Pseudomonas psychrotolerans* (Cluster VII). On the other hand, the maximum inhibition halo in the presence of oregano was measured for isolates grouped in Cluster XIII (2 cm), which showed similarity with *B. simplex* and in Cluster XII, affiliated with *Paenibacillus* sp. in the presence of clove (0.9 $\pm$ 0.2 cm). These data are in line with previous works which highlighted the inhibitory effect of EOs, and the concentration of EOs required for growth inhibition [60].

### 3.4. *Assessment for Antifungal and Antibacterial Activity in the Contactless Test*

In order to develop an application method able to prove the remote effect of EOs on microorganisms, avoiding direct contact with the pigments of paintings, a preliminary in vitro contactless test was designed in this work. Since the oregano EO gave the most promising results with the contact test, it was chosen for the contactless test. In this test microorganisms, grown on agar plates, come in contact solely with the volatile components of the EO. In detail, once the microorganisms were plated on the growth medium, a drop of the essence was placed on the lid of the plate, and the plate was turned over and incubated under sterile conditions. The result of the test is shown in Table 2.

In the test, a total inhibition effect was exhibited by oregano EO on bacterial clusters II (Xanthomonadaceae), IX (*Cellulosimicrobium cellulans*), and XII (*Paenibacillus* sp.). Moreover, if compared with the inhibition halo obtained with the contact test, oregano showed a higher effect in contactless tests both on fungi and bacteria (Tables 1 and 2). Even if a comparison between the two tests cannot be done because of the use of different EOs' quantities (10 µL in the contact test and 100 µL in the contactless test), the results obtained provide clues for the sole utilization of the volatile components in control practices. Previous work confirmed the antimicrobial efficacy of EOs' vapor phase for disinfection of textiles, reporting no changes in terms of structural parameters of the object [12,16]. However, although these data provided support to the plethora of promising uses and properties exhibited by EOs, they did not avoid the procedure of direct contact of the oils with the surface of the object under study.

**Table 2.** Inhibition halo of oregano and cloves essential oils on bacteria and fungi in contactless tests. The experiment was performed in duplicate. NI = not identified strain. Ethanol 70% was used as a negative control as it did not show any inhibitory effect against the isolates.

| | CONTACTLESS TEST | | |
|---|---|---|---|
| | ISOLATES | TAXA | OREGANO |
| BACTERIA | $I_c$ | *Bacillus subtilis* subsp. *subtilis* | 1.7 ± 0.3 cm |
| | $II_a$ | Xanthomonadaceae | Total inhibition |
| | $III_a$ | *Streptomyces* | 1.6 ± 0.4 cm |
| | $IV_a$ | NI | 1.6 ± 0.4 cm |
| | $V_a$ | NI | 1.8 ± 0.2 cm |
| | $VI_a$ | *Stenotrophomonas* | 1.6 ± 0.2 cm |
| | $VII_a$ | *Pseudomonas psychrotolerans* | 1.5 ± 0.5 cm |
| | $VIII_a$ | Xanthomonadaceae | 1.1 ± 0.4 cm |
| | $IX_a$ | *Cellulosimicrobium cellulans* | Total inhibition |
| | $X_a$ | Penibacillaceae | 1.9 ± 0.1 cm |
| | $XI_a$ | NI | 1.6 ± 0.2 cm |
| | $XII_a$ | *Paenibacillus* sp. | Total inhibition |
| | $XIII_a$ | *Bacillus simplex* | 2 ± 0 cm |
| FUNGI | $I_a$ | *Penicillium chrysogenum* | ++ |
| | $II_c$ | NI | + |
| | $III_a$ | *Penicillium chrysogenum* | +++ |
| | $IV_b$ | *Cephalotheca foveolata* | ++ |
| | $V_a$ | *Aspergillus* sp. | +++ |
| | $VI_a$ | *Cephalotheca foveolata* | ++ |
| | $VII_a$ | *Cladosporium parahalotolerans* | + |
| | $VIII_a$ | NI | +++ |
| | $IX_a$ | *Cephalotheca foveolata* | +++ |
| | $X_a$ | *Aspergillus versicolor* | +++ |
| | $XI_a$ | NI | ++ |
| | $XII_a$ | NI | + |
| | $XIII_a$ | *Penicillium chrysogenum* | +++ |
| | $XIV_a$ | Trichocomaceae | + |
| | $XV_a$ | Chaetomiaceae | ++ |
| | $XVI_a$ | *Penicillium chrysogenum* | ++ |
| | $XVII_a$ | *Cephalotheca foveolata* | +++ |
| | $XVIII_a$ | NI | ++ |
| | $XIX_a$ | Phaeosphaeriaceae sp. | + |
| | $XX_a$ | *Penicillium* sp. | + |
| | $XXI_b$ | NI | +++ |
| | $XXII_b$ | NI | + |
| | $XXIII_b$ | NI | + |
| | $XXIV_a$ | NI | +++ |

## 4. Conclusions

The research gave insight into the possibility of extending the use of EOs in the conservation of oil paintings. Indeed, the trial presented in this work was planned not with the purpose of the proper selection of suitable EOs related to the microorganisms isolated from an oil painting, but to lay the groundwork for the development of new control practices which would be suited for this kind of artwork.

In conservation, the main limit in the use of plant derivates is the methodology of application, which usually implies the direct contact of EOs with the artwork surfaces. This may cause problems due to the unpredictable reaction with pigments (and other substances used in making paintings) and EOs, such as the potential solvent effect. For this reason, here we tested oregano and cloves EOs, already reported in the literature for their inhibitory activities, focusing the attention solely on the volatile components of EOs, avoiding direct contact with pigments.

The results obtained in the present work proved that the oregano and clove EOs, and in particular the volatile components of oregano EO, are able to inhibit the growth of potential oil paintings' biodeteriogens. We hypothesize that an effective method for EOs application could be to flow a thin film of EOs onto an evaporating surface and to place it close to the painting using some supports, so that the vapors of the EOs may homogeneously reach the painting surface, therefore avoiding direct contact of EOs with the pigments. With the aim of maximizing the antimicrobial effect of the EOs, we suggest treating the painting in a confined area, such as a display case, in order to have a saturating effect. Future studies will be devoted to verifying if this method is effectively color respectful and to establish the role of other parameters such as the minimum inhibitory concentration, exposure to light, temperature, the treatment period and modality in a real case study.

**Supplementary Materials:** The following are available online at https://www.mdpi.com/2076-3417/11/1/78/s1, Table S1: Inhibition scale degree; Table S2: Clusterization of bacterial isolates according to their morphology; Table S3: Clusterization of fungi isolates according to their morphology; Table S4: Clustering of bacterial strains isolated from canvas painting and their phylogenetic identification by 16S rRNA gene sequencing; Table S5: Clusterization of fungal strains isolated from canvas painting and their phylogenetic identification by ITS gene sequencing; Figure S1: Clusterization and relative abundance of bacterial (a) and fungal (b) isolates.

**Author Contributions:** Conceptualization: A.B.; funding acquisition: A.B.; investigation: L.G.; methodology: V.V.; project administration: A.B.; supervision: A.B. and V.V.; validation: F.C.; visualization: F.T.; writing—original draft: L.G. and V.V.; writing—review and editing: F.T. All authors have read and agreed to the published version of the manuscript.

**Funding:** The APC was funded by Università degli Studi di Milano, PIANO DI SOSTEGNO ALLA RICERCA 2015–2017—TRANSITION GRANT LINEA 1A PROGETTO "UNIMI PARTENARIATI H2020", coordinator Francesca Cappitelli.

**Institutional Review Board Statement:** Not applicable.

**Informed Consent Statement:** Not applicable.

**Data Availability Statement:** Data available in a publicly accessible repository.

**Acknowledgments:** We thank Francesca Mapelli for the assistance in the submission of phylogenetic sequences. The authors also thank Chiara Nenci for her help during the sampling campaigns and for making available the painting history report.

**Conflicts of Interest:** The authors declare no conflict of interest.

## References

1. Casiglia, S.; Bruno, M.; Senatore, F. Activity against Microorganisms Affecting Cellulosic Objects of the Volatile Constituents of Leonotis nepetaefolia from Nicaragua. *Nat. Prod. Commun.* **2014**, *9*, 1637–1639. [CrossRef]
2. Barresi, G.; Di Carlo, E.; Trapani, M.R.; Parisi, M.G.; Chille, C.; Mule, M.F.; Cammarata, M.; Palla, F. Marine organisms as source of bioactive molecules applied in restoration projects. *Herit. Sci.* **2015**, *3*, 15–18. [CrossRef]
3. Fidanza, M.R.; Caneva, G. Natural biocides for the conservation of stone cultural heritage: A review. *J. Cult. Herit.* **2019**, *38*, 271–286. [CrossRef]
4. Raut, J.S.; Karuppayil, S.M. A status review on the medicinal properties of essential oils. *Ind. Crop. Prod.* **2014**, *62*, 250–264. [CrossRef]
5. Zoubi, Y.E.; Farah, A.; Zaroual, H.; El Ouali, A. Antimicrobial activity of Lavandula stoechas phenolic extracts against pathogenic bacteria isolated from a hospital in Morocco. *Vegetos Int. J. Plant Res.* **2020**, *33*, 703–711. [CrossRef]
6. Rios, J.L.; Recio, M.C. Medicinal plants and antimicrobial activity. *J. Ethnopharmacol.* **2005**, *100*, 80–84. [CrossRef]
7. Chouhan, S.; Sharma, K.; Guleria, S. Antimicrobial Activity of Some Essential Oils-Present Status and Future Perspectives. *Medicines* **2017**, *4*, 58. [CrossRef]
8. Predoi, D.; Iconaru, S.L.; Buton, N.; Badea, M.L.; Măruţescu, L. Antimicrobial Activity of New Materials Based on Lavender and Basil Essential Oils and Hydroxyapatite. *Nanomaterials* **2018**, *8*, 291. [CrossRef]
9. Xie, Y.; Wang, Z.; Huang, Q.; Zhang, D. Antifungal activity of several essential oils and major components against wood-rot fungi. *Ind. Crop. Prod.* **2017**, *108*, 278–285. [CrossRef]
10. Romano, I.; Granata, G.; Poli, A.; Finore, I.; Napoli, E.; Geraci, C. Inhibition of bacterial growth on marble stone of 18th century by treatment of nanoencapsulated essential oils. *Int. Biodeterior. Biodegrad.* **2020**, *148*, 104909. [CrossRef]

11. Rotolo, V.; Barresi, G.; Di Carlo, E.; Giordano, A.; Lombardo, G.; Crimi, E.; Costa, E.; Bruno, M.; Palla, F. Plant extracts as green potential strategies to control the biodeterioration of cultural heritage. *Int. J. Conserv. Sci.* **2016**, *7*, 839–846.
12. Jugreet, B.S.; Suroowan, S.; Rengasamy, R.K.; Mahomoodally, M.F. Chemistry, bioactivities, mode of action and industrial applications of essential oils. *Trends Food Sci. Technol.* **2020**, *101*, 89–105. [CrossRef]
13. Campanella, L.; Angeloni, R.; Cibin, F.; Dell'Aglio, E.; Grimaldi, F.; Reale, R.; Vitali, M. Capsulated essential oil in gel spheres for the protection of cellulosic cultural heritage. *Nat. Prod. Res.* **2019**, 1–8. [CrossRef]
14. Čabalová, I.; Češek, B.; Mikala, O.; Gojný, J.; Kačík, F.; Tribulová, T. The influence of selected efficient compounds of essential oils for paper protection. *J. Cult. Herit.* **2019**, *37*, 148–154. [CrossRef]
15. Borrego, S.; Valdés, O.; Vivar, I.; Lavin, P.; Guiamet, P.; Battistoni, P.; De Saravia, S.G.; Borges, P. Essential Oils of Plants as Biocides against Microorganisms Isolated from Cuban and Argentine Documentary Heritage. *ISRN Microbiol.* **2012**, *2012*, 1–7. [CrossRef] [PubMed]
16. Matusiak, K.; MacHnowski, W.; Wrzosek, H.; Polak, J.; Rajkowska, K.; Śmigielski, K.; Kunicka-Styczyńska, A.; Gutarowska, B. Application of Cinnamomum zeylanicum essential oil in vapour phase for heritage textiles disinfection. *Int. Biodeterior. Biodegrad.* **2018**, *131*, 88–96. [CrossRef]
17. Geweely, N.; Afifi, H.; Ibrahim, D.; Soliman, M. Inhibitory Effect of Essential Oils on Growth and Physiological Activity of Deteriorated Fungal Species Isolated from Three Archeological Objects, Saqqara excavation, Egypt. *Geomicrobiol. J.* **2020**, *37*, 520–533. [CrossRef]
18. Stupar, M.; Grbić, M.L.; Džamić, A.; Unković, N.; Ristić, M.; Jelikić, A.; Vukojević, J. Antifungal activity of selected essential oils and biocide benzalkonium chloride against the fungi isolated from cultural heritage objects. *S. Afr. J. Bot.* **2014**, *93*, 118–124. [CrossRef]
19. Di Vito, M.; Bellardi, M.G.; Violi, M. Gli oli essenziali entrano nei musei. *Natural* **2018**, *1*, 69–73.
20. Elsayed, Y.; Shabana, Y. The effect of some essential oils on Aspergillus niger and Alternaria alternata infestation in archaeological oil paintings. *Mediterr. Archaeol. Archaeom.* **2018**, *18*, 71–87.
21. Garti, N.; Yaghmur, A.; Leser, M.E.; Clement, V.; Watzke, H.J. Improved Oil Solubilization in Oil/Water Food Grade Microemulsions in the Presence of Polyols and Ethanol. *J. Agric. Food Chem.* **2001**, *49*, 2552–2562. [CrossRef]
22. Tokuoka, Y.; Uchiyama, H.; Abe, M.; Christian, S.D. Solubilization of Some Synthetic Perfumes by Anionic-Nonionic Mixed Surfactant Systems. 1. *Langmuir* **1995**, *11*, 725–729. [CrossRef]
23. Okpalanozie, O.E.; Adebusoye, S.A.; Troiano, F.; Polo, A.; Cappitelli, F.; Ilori, M. O Evaluating the microbiological risk to a contemporary Nigerian painting: Molecular and biodegradative studies. *Int. Biodeter. Biodegr.* **2016**, *114*, 184–192. [CrossRef]
24. Troiano, F.; Polo, A.; Villa, F.; Cappitelli, F. Assessing the microbiological risk to stored sixteenth century parchment manuscripts: A holistic approach based on molecular and environmental studies. *Biofouling* **2014**, *30*, 299–311. [CrossRef] [PubMed]
25. Rizzi, E.; Lari, M.; Gigli, E.; De Bellis, G.; Caramelli, D. Ancient DNA studies: New perspectives on old samples. *Genet. Sel. Evol.* **2012**, *44*, 21. [CrossRef]
26. Manter, D.K.; Vivanco, J.M. Use of the ITS primers, ITS1F and ITS4, to characterize fungal abundance and diversity in mixed-template samples by qPCR and length heterogeneity analysis. *J. Microbiol. Methods* **2007**, *71*, 7–14. [CrossRef]
27. Available online: https://blast.ncbi.nlm.nih.gov/Blast.cgi (accessed on 9 November 2020).
28. Available online: http://rdp.cme.msu.edu/classifier/classifier.jsp (accessed on 9 November 2020).
29. MycoBank. Available online: http://www.mycobank.org (accessed on 9 November 2020).
30. Bonaduce, I.; Duce, C.; Lluveras-Tenorio, A.; Lee, J.; Ormsby, B.; Burnstock, A.; Berg, K.J.V.D. Conservation Issues of Modern Oil Paintings: A Molecular Model on Paint Curing. *Accounts Chem. Res.* **2019**, *52*, 3397–3406. [CrossRef]
31. Poyatos, F.; Morales, F.; Nicholson, A.W.; Giordano, A. Physiology of biodeterioration on canvas paintings. *J. Cell. Physiol.* **2018**, *233*, 2741–2751. [CrossRef]
32. Kakakhel, M.A.; Wu, F.; Gu, J.-D.; Feng, H.; Shah, K.; Wang, W. Controlling biodeterioration of cultural heritage objects with biocides: A review. *Int. Biodeterior. Biodegrad.* **2019**, *143*, 104721. [CrossRef]
33. Leyva-López, N.; Gutiérrez-Grijalva, E.P.; Vazquez-Olivo, G.; Heredia, J.B. Essential Oils of Oregano: Biological Activity beyond Their Antimicrobial Properties. *Molecules* **2017**, *22*, 989. [CrossRef]
34. Pramod, K.; Ansari, S.H.; Ali, J. Eugenol: A Natural Compound with Versatile Pharmacological Actions. *Nat. Prod. Commun.* **2010**, *5*, 1999–2006. [CrossRef] [PubMed]
35. Mauriello, E.; Ferrari, G.; Donsì, F. Effect of formulation on properties, stability, carvacrol release and antimicrobial activity of carvacrol emulsions. *Colloids Surf. B Biointerfaces* **2021**, *197*, 111424. [CrossRef] [PubMed]
36. Klarić, M. Šegvić; Kosalec, I.; Mastelić, J.; Piecková, E.; Pepeljnak, S. Antifungal activity of thyme (*Thymus vulgaris* L.) essential oil and thymol against moulds from damp dwellings. *Lett. Appl. Microbiol.* **2007**, *44*, 36–42. [CrossRef]
37. Marchese, A.; Barbieri, R.; Coppo, E.; Orhan, I.E.; Daglia, M.; Nabavi, S.M.; Izadi, M.; Abdollahi, M.; Ajami, M. Antimicrobial activity of eugenol and essential oils containing eugenol: A mechanistic viewpoint. *Crit. Rev. Microbiol.* **2017**, *43*, 668–689. [CrossRef] [PubMed]
38. Capodicasa, S.; Fedi, S.; Porcelli, A.M.; Zannoni, D. The microbial community dwelling on a biodeteriorated 16th century painting. *Int. Biodeterior. Biodegrad.* **2010**, *64*, 727–733. [CrossRef]
39. Pasquarella, C.; Balocco, C.; Saccani, E.; Capobianco, E.; Viani, I.; Veronesi, L.; Pavani, F.; Pasquariello, G.; Rotolo, V.; Palla, F.; et al. Biological and microclimatic monitoring for conservation of cultural heritage: A case study at the De Rossi room of the Palatina library in Parma. *Aerobiologia* **2019**, *36*, 105–111. [CrossRef]

40. Montanari, M.; Iotti, M.; Innocenti, G. Isolamento, identificazione e attività cellulosolitica di microrganismi fungini associati al biodeterioramento di un dipinto su tela del XIXesimo secolo. *Micol. Ital.* **2009**, *3*, 19–24.
41. Lech, T. Evaluation of a Parchment Document, the 13th Century Incorporation Charter for the City of Krakow, Poland, for Microbial Hazards. *Appl. Environ. Microbiol.* **2016**, *82*, 2620–2631. [CrossRef]
42. Kumar, S.; Priyanka; Kumar, U. *Microbial Biotechnology Approaches to Monuments of Cultural Heritage*; Springer Science and Business Media LLC: Singapore, 2020; pp. 1–12.
43. Orehek, J.; Dogsa, I.; Tomšič, M.; Jamnik, A.; Kočar, D.; Stopar, D. Structural investigation of carboxymethyl cellulose biodeterioration by Bacillus subtilis subsp. subtilis NCIB 3610. *Int. Biodeterior. Biodegrad.* **2013**, *77*, 10–17. [CrossRef]
44. Fouda, A.; Abdel-Maksoud, G.; Abdel-Rahman, M.A.; Salem, S.S.; Hassan, S.E.-D.; El-Sadany, M.A.-H. Eco-friendly approach utilizing green synthesized nanoparticles for paper conservation against microbes involved in biodeterioration of archaeological manuscript. *Int. Biodeterior. Biodegrad.* **2019**, *142*, 160–169. [CrossRef]
45. Sakr, A.A.; Ghaly, M.F.; Edwards, H.G.M.; Ali, M.F.; Abdel-Haliem, M.E.F. Involvement of Streptomyces in the Deterioration of Cultural Heritage Materials Through Biomineralization and Bio-Pigment Production Pathways: A Review. *Geomicrobiol. J.* **2020**, *37*, 653–662. [CrossRef]
46. Mazzoli, R.; Giuffrida, M.G.; Pessione, E. Back to the past: Find the guilty bug—Microorganisms involved in the biodeterioration of archeological and historical artifacts. *Appl. Microbiol. Biotechnol.* **2018**, *102*, 6393–6407. [CrossRef] [PubMed]
47. Sáez-Nieto, J.; Medina-Pascual, M.; Carrasco, G.; Garrido, N.; Fernandez-Torres, M.; Villalón, P.; Valdezate, S. Paenibacillus spp. isolated from human and environmental samples in Spain: Detection of 11 new species. *New Microbes New Infect.* **2017**, *19*, 19–27. [CrossRef] [PubMed]
48. Cao, J.; Chu, C.; Zhang, M.; He, L.; Qin, L.; Li, X.; Yuan, M. Different Cell Wall-Degradation Ability Leads to Tissue-Specificity between Xanthomonas oryzae pv. oryzae and Xanthomonas oryzae pv. oryzicola. *Pathogens* **2020**, *9*, 187. [CrossRef] [PubMed]
49. Zhgun, A.A.; Avdanina, D.; Shumikhin, K.; Simonenko, N.; Lyubavskaya, E.; Volkov, I.; Ivanov, V. Detection of potential biodeterioration risks for tempera painting in 16th century exhibits from State Tretyakov Gallery. *PLoS ONE* **2020**, *15*, e0230591. [CrossRef] [PubMed]
50. Su, X.; Zhang, S.; Mei, R.; Zhang, Y.; Hashmi, M.Z.; Liu, J.; Lin, H.; Ding, L.; Sun, F. Resuscitation of viable but non-culturable bacteria to enhance the cellulose-degrading capability of bacterial community in composting. *Microb. Biotechnol.* **2018**, *11*, 527–536. [CrossRef] [PubMed]
51. Marjakangas, J.M.; Lakaniemi, A.-M.; Koskinen, P.E.; Chang, J.-S.; Puhakka, J.A. Lipid production by eukaryotic microorganisms isolated from palm oil mill effluent. *Biochem. Eng. J.* **2015**, *99*, 48–54. [CrossRef]
52. Wang, H.; Kuang, S.; Lang, Q.; Yu, W. Effects of Aged Oil Sludge on Soil Physicochemical Properties and Fungal Diversity Revealed by High-Throughput Sequencing Analysis. *Archaea* **2018**, *2018*, 1–8. [CrossRef]
53. Polo, A.; Cappitelli, F.; Villa, F.; Pinzari, F. Biological invasion in the indoor environment: The spread of *Eurotium halophilicum* on library materials. *Int. Biodeterior. Biodegrad.* **2017**, *118*, 34–44. [CrossRef]
54. Yang, Y.; Yang, J.; Liu, J.; Wang, R.; Liu, L.; Wang, F.; Yuan, H. The composition of accessory enzymes of *Penicillium chrysogenum* P33 revealed by secretome and synergistic effects with commercial cellulase on lignocellulose hydrolysis. *Bioresour. Technol.* **2018**, *257*, 54–61. [CrossRef]
55. Tsang, C.-C.; Tang, J.Y.M.; Chan, K.-F.; Lee, C.-Y.; Chan, J.; Ngan, A.H.Y.; Cheung, M.; Lau, E.C.L.; Li, X.; Ng, R.H.Y.; et al. Diversity of phenotypically non-dermatophyte, non-Aspergillus filamentous fungi causing nail infections: Importance of accurate identification and antifungal susceptibility testing. *Emerg. Microbes Infect.* **2019**, *8*, 531–541. [CrossRef] [PubMed]
56. Kurowski, G.; Vogt, O.; Ogonowski, J. Paint-degrading microorganisms. *Czas. Tech.* **2017**, *2017*, 81–92. [CrossRef]
57. Veneranda, M.; Blanco-Zubiaguirre, L.; Roselli, G.; Di Girolami, G.; Castro, K.; Madariaga, J.M. Evaluating the exploitability of several essential oils constituents as a novel biological treatment against cultural heritage biocolonization. *Microchem. J.* **2018**, *138*, 1–6. [CrossRef]
58. Navarrete, C.; Martínez, J.L. Non-conventional yeasts as superior production platforms for sustainable fermentation based bio-manufacturing processes. *AIMS Environ. Sci.* **2020**, *7*, 289–305. [CrossRef]
59. Borrego, S.; Gómez de Saravia, S.; Valdés, O.; Vivar, I.; Battistoni, P.; Guiamet, P. Biocidal activity of two essential oils on fungi that cause biodeterioration of paper documents. *Int. J. Conserv. Sci.* **2016**, *7*, 369–380.
60. Paster, N.; Juven, B.J.; Shaaya, E.; Menasherov, M.; Nitzan, R.; Weisslowicz, H.; Ravid, U. Inhibitory effect of oregano and thyme essential oils on moulds and foodborne bacteria. *Lett. Appl. Microbiol.* **1990**, *11*, 33–37. [CrossRef]

*applied sciences*

*Article*

# Biocidal Activity of Phyto-Derivative Products Used on Phototrophic Biofilms Growing on Stone Surfaces of the Domus Aurea in Rome (Italy)

**Lorenza Rugnini [1], Giada Migliore [2], Flavia Tasso [2], Neil Thomas William Ellwood [3], Anna Rosa Sprocati [2] and Laura Bruno [1,*]**

1 LBA-Laboratory of Biology of Algae, Department of Biology, University of Rome "Tor Vergata", via Cracovia 1, 00133 Rome, Italy; rgnlnz01@uniroma2.it
2 ENEA, Territorial and Production Systems Sustainability Department, via Anguillarese 301, 00123 Rome, Italy; giada.migliore@enea.it (G.M.); flavia.tasso@enea.it (F.T.); annarosa.sprocati@enea.it (A.R.S.)
3 Department of Science, University of Roma Tre, Viale G. Marconi 446, 00146 Rome, Italy; ellwood@uniroma3.it
* Correspondence: laura.bruno@uniroma2.it

Received: 23 July 2020; Accepted: 16 September 2020; Published: 21 September 2020

**Abstract:** Hypogean or enclosed monuments are important cultural heritage sites that can suffer biodegradation. Many of the stone walls of the prestigious Domus Aurea are overwhelmed by dense biofilms and so need intervention. Room 93 was chosen as a study site with the aim to test the efficacy of phyto-derivatives as new biocides. Laboratory studies were performed comparing the effects of liquorice leaf extract (*Glycyrrhiza glabra* L.), lavender essential oil (*Lavandula angustifolia* Mill.) and a combination of both. In situ studies were also performed to test the effect of liquorice. The results were compared with those of the commonly used synthetic biocide benzalkonium chloride. The effects on the biofilms were assessed by microscopy along with chlorophyll fluorescence analysis. The phototrophs in the biofilms were identified morphologically, while the heterotrophs were identified with culture analysis and 16S gene sequencing. Results showed that the mixed solution liquorice/lavender was the most effective in inhibiting the photosynthetic activities of biofilms in the laboratory tests; while, in situ, the effect of liquorice was particularly encouraging as an efficient and low-invasive biocide. The results demonstrate a high potential biocidal efficacy of the phyto-derivatives, but also highlight the need to develop an efficient application regime.

**Keywords:** cultural heritage; biodeterioration; biofilms; cyanobacteria; biocides; phyto-derivative; liquorice; lavender; PAM

## 1. Introduction

Conservation of stone surfaces, in hypogea or in confined monuments, is threatened by the addition of artificial lighting that is essential for visibility; however, when it is combined with humidity, nutrient availability and particular physical substrate characteristics, light facilitates the colonisation by biodeteriorative biofilms [1–3]. These biofilms are formed by diverse microbial communities of cyanobacteria, bacteria and, to a lesser extent, by algae and fungi [4,5]. Different studies have demonstrated their deleterious effects (aesthetical, chemical and physical) on diverse substrates (stone and painted surfaces [6,7], as well as other valuable cultural heritage materials [8–10]). While there is the desire to rid surfaces from these deteriogenic biofilms, the general confinement of these sites hinders the use of chemical biocides, even if they are effective, as they cause major health issues for operators. These biocides also tend to be polluting substances that can persist in the natural environment [11]. While biocides regularly used in the restoration field (e.g., benzalkonium chloride, nystatin and sodium fluoride, among others) have known bacterial, fungal and insecticidal activities [12], their effectiveness on algae and cyanobacteria is still not fully understood.

The challenge is now to develop more safe and sustainable methods in the restoration and conservation of lithic cultural heritage using products and practices that take environmental, social and economic aspects into account [13–17]. This challenge will involve players from multiple disciplines, i.e., not only restorers and art historians, but also scientific researchers, politicians, entrepreneurs and economists. The most notably harmful products in use are being eliminated from the market and others should follow, and the search for new bio-based and sustainable products, as replacements, could open a new potential market in the bioeconomy chain. Substitution with safer alternatives and greener technologies is strongly driven by regulation and contributes to the overarching EU objectives for a non-toxic environment and a circular economy [18].

Among the alternative treatments to be investigated, plant extracts are showing promise as biocides by having many advantages over the traditional synthetic options. They are environmentally friendly (biodegradable) and have a functional complexity, which is thought to evade the possibility of resistance, one of the main negative characteristics of synthetic biocides. Plant-based biocides are generally alcohol extracts or essential oils that have been used in medical, food and pharmaceutical industries. Their effectiveness against common human pathogens including bacteria, fungi and yeasts is well-known, but more recently tests on their antimicrobial activity against organisms associated with biodeterioration of archives, libraries, museums and stone monuments have been carried out [12,13,19]. In the present study, liquorice leaf extract and lavender essential oil were chosen for two reasons: the biofilms to be treated are complex communities, and the ineffectiveness of the synthetic biocides previously used, which required ever more frequent treatments (data not published).

While the antimicrobial activity (against bacteria and fungi) of liquorice (*Glycyrrhiza glabra* L.) root extracts has been well established [20–22], less is known about the full biocidal potential of liquorice leaf extracts. Two antimicrobial flavonones were isolated from the leaves of *G. glabra*: Pinocembrina and lycoflavanone [23]. A comparative study on the antimicrobial effects of leaf and root ethanolic extracts carried out on human pathogenic microorganisms demonstrated that there was overlapping activity and that the leaf extract was more effective against some strains [24,25]. Leaf extract has also been used as a fungicide for agricultural purposes [26]. The biocidal effect of the leaf extract was shown to be effective against 20 bacterial strains and 10 fungal strains isolated from hypogean environments (held in the ENEA collection), frescoes, wall and canvas paintings [27]. Furthermore, the use of an agri-food waste, such as liquorice leaves, fully meets the circular economy criterium. The use of a mixture of essential oils, *Lavandula angustifolia* Mill. (10%) and *Thymus vulgaris* L. (1%) (1:1 *v:v* mixture) has also been used on hypogean biofilms with positive biocidal effect on photosynthetic biofilms [13].

The aim of the present study is to determine the biocidal activity of liquorice leaf extract and lavender essential oils on biofilms growing on enclosed illuminated stone surfaces. The site chosen for the study is the Domus Aurea, which is a highly valuable Roman cultural heritage site, originally commissioned by Emperor Nero as a new residence after the devastating fire of 64 AD. The palace, famous for its sumptuous decoration, consisted of a series of buildings separated by gardens, woods, vineyards and an artificial lake (Figure 1). After Nero's death, successive emperors decided to erase all traces of Nero and his palace and the luxurious chambers were deprived of any valuables and filled up with soil to be used as foundations for other buildings and they remained undiscovered until the Renaissance (15th and 16th centuries). Unbeknownst to Nero's successors, the infilling of the Domus Aurea's rooms resulted in their conservation throughout the centuries.

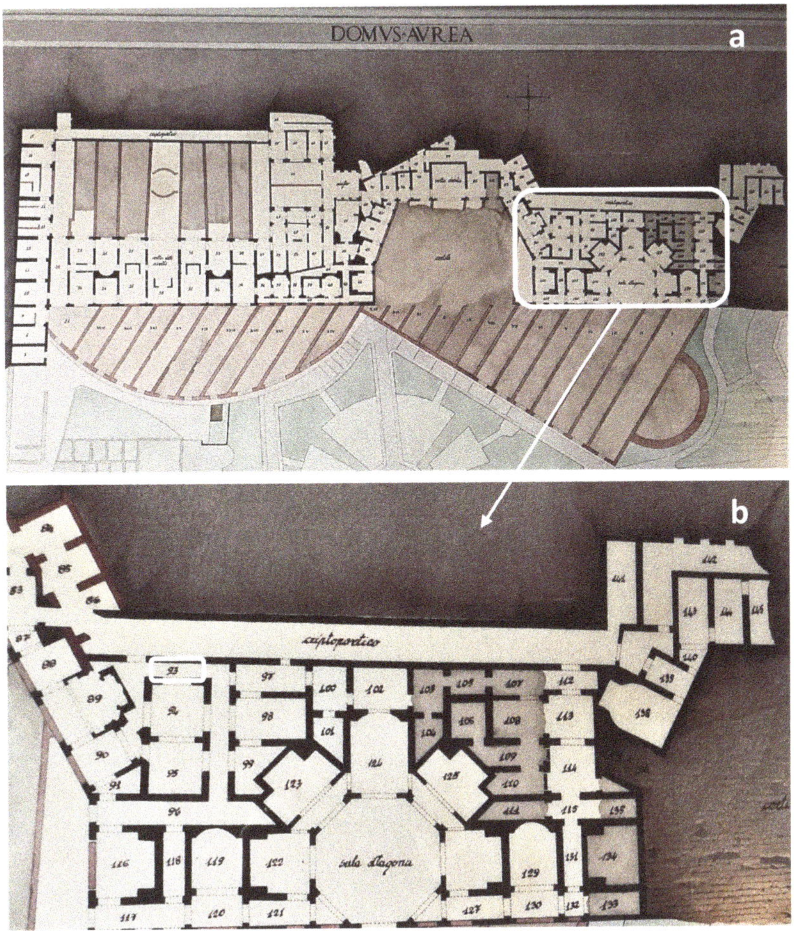


**Figure 1.** (**a**) Map of the remains of the Domus Aurea which consists of a series of buildings separated by gardens, woods and vineyards; (**b**) a detail of the area where the study site, Room 93, is located.

The Domus Aurea was opened to the public in 1999, but six years later was closed due to detachments and security problems, to open again in 2007 with restricted visits. However, due to artificial lighting, high humidity levels and residual soil heaps, many illuminated surfaces have been heavily infested by both phototrophic and heterotrophic biofilms. In the present study, phyto-derivatives were initially tested for their biocidal activity in a set of well plate experiments on the biofilms collected from surfaces in the Domus Aurea. Tests were then carried out on the natural biofilms occurring on a selected wall of Room 93 to show the efficacy of the treatment in situ. It was expected that the biocidal activity of the phytoextracts would be significant, and the tests should help develop safe, yet effective, protocols for the control of biofilm growth in enclosed, lithic environments.

## 2. Materials and Methods

### 2.1. Sampling Site

The biofilms used for the tests reported in this article were collected at the Domus Aurea in the Palatine Hill, Rome, Italy (Figure 1). A wall in Room 93 was selected for the sampling due to the evident biodeterioration of the surface (white, green and blue-green patinas). The wall in *opus latericium* is covered with soil which was not completely removed during the excavations. The biofilm was consistently and evenly distributed (Figure 2). Room 93 is adjacent to the Criptoportico and, although not part of the tourist route, it is illuminated by fluorescent lamps. In 2012, the wall was treated with Preventol® RI80, which is a solution of quaternary ammonium salts with a broad spectrum of activity (fungi, bacteria and algae). Previous intervention by the restorers consisted in spraying the biocidal product (diluted to 3% with de-ionised water) on the surface and this was effective only after multiple applications with prolonged laying time. However, the recolonization was very fast (data not published).

(**a**) (**b**)

**Figure 2.** Two different views of the sampling site in Room 93 with the evident phototrophic biofilms causing discolouration of the walls. (**a**) front view of the corridor present in Room 93; (**b**) detailed image of the wall object of the investigation with the evident presence of the green biofilm; scale bar = 2 m.

Environmental parameters in Room 93 were measured on each sampling trip and during each visit during the in-situ experimental period (about 75 days). Temperature (T °C) and relative humidity (RH%) were measured using a digital humidity and temperature meter (CEM, model DT-625). Surface temperatures were made using a DT-8833 infrared thermometer (Thermosense, UK) and cross-checked using a Type K thermocouple resistor (Thermosense, model HK-400). Irradiance was measured using a radiometer (model LI-185B; LI-COR Inc., Lincoln, NE, USA) equipped with a quantum sensor (LI-190SB) to give PPFD in μmol photons $m^{-2}$ $s^{-1}$.

*2.2. Sampling of Biofilms and Microscopy Observations*

Ten samples of biofilm were collected using the method of adhesive tape strips (MAT; Fungitape Did, Milan, Italy) as a non-destructive sampling method [28]. Fresh samples were mounted on glass sides and were observed with a Zeiss AxioScope light microscope with a 40× objective; images were acquired with a digital camera (Canon EOS 600D—Canon S.P.A., Milan, Italy). The adhesive tape samples were also observed using a Confocal Laser Scanning Microscope (CLSM) FV1000 (Olympus Corp., Tokyo, Japan), with a 60× objective using the autofluorescence channels for chlorophyll *a* and phycobiliproteins (excitation 488, 543, 635 nm, emission 520, 572 and 688 nm). Three-dimensional images were constructed from a series of 2D cross-sectional images (x-y plane) that were captured at 0.5-μm intervals along the z-axis using IMARIS 6.2.0 software (Bitplane AG Zurich, Switzerland). The identification of the cyanobacteria present in the samples was made according to the recent taxonomy of cyanobacteria [29,30].

### 2.3. Heterotrophic Microorganisms

The biofilm was fully characterised by culture-dependent and independent techniques. The cultivable portion of the heterotrophic community was first studied, to isolate the microorganisms and use them for subsequent in vitro investigation for individual sensitivity to the tested biocides. To characterise the cultivable portion of the heterotrophic community inhabiting the biofilm, 20 g of biomass was sampled from random points in a selected area of 2 × 2 m by gently scraping off the biofilm with sterile scalpels; 5 g of the mixed sampled biofilms were then suspended in a 50 mL solution of sodium pyrophosphate and Tween 80 (0.05% and 0.001% *v*/*v* respectively). The suspension was then kept in an orbital shaking incubator (New Brunswick™ Innova 4430—New Brunswick Scientific Co Inc, Edison, NJ, USA) at 28 °C and 150 rpm for 1 h. The number of cultivable heterotrophic microorganisms was obtained by plating 100 μL of suspension and following serial dilutions in three different agar media: Tryptic Soy Agar (TSA, Conda, Spain) and Mineral Medium (MM, Schmidt and Schlegel, 1989) added with 0.5% (*w*/*v*) sodium gluconate and 1.5% agar (Agar, Conda, Spain) for bacteria and Potato Dextrose Agar (PDA, Conda, Spain) for fungi. The plates were then incubated at room temperature until the complete growth of colonies (up to one month).

The isolation of bacterial and fungal strains was based on colony morphologies. Colonies were streaked three times on the same agarized medium from which they were isolated, in order to obtain pure cultures. The plates were then incubated at room temperature until there was complete colony development (24–72 h). Bacterial strains were identified by 16S r-DNA sequencing. Genomic DNA was extracted by resuspending two fresh colonies of each isolate in 20 μL of lysis buffer (NaOH 0.5 M and SDS 0.5% *w*/*v*), heating at 95 °C for 15 min and then transferring to an ice bath for 15 min. The 16S r-DNA complete gene was PCR amplified by using 9 bfm as a forward and 1512 uR as a reverse primer [31]. The reaction mixture contained 1× PCR Master Mix (Thermo Scientific, Waltham, MA, USA), 2 μL of template DNA and 20 pmol of each primer in the total reaction volume of 25 μL. The PCR products were sequenced by a commercial facility. Sequence identities were analysed using the National Center for Biotechnology Information (NCBI) BLAST program (https://blast.ncbi.nlm.nih.gov/Blast.cgi) and the GenBank database. Each isolate was compared against known taxa present in the database. A preliminary identification of the fungal isolates was done according to typical colonial and conidial morphology using a stereo microscope.

### 2.4. Laboratory Treatments with Phyto-Derivatives

Two phyto-derivatives were tested for their biocide activity: liquorice alcoholic leaf extract (*Glycyrrhiza glabra* L.) (LIQ) and lavender essential oil (*Lavandula angustifolia* Mill.) (LAV). The liquorice alcoholic leaf extract at 50% (*w*/*v*) was provided by Trifolio-M GmbH (Lahnau, Germany), while the lavender essential oil was supplied by Sarandrea Marco & CO. S.r.l. (Collepardo, Italy).

In the first set of experiments the collected biofilms (5 g) were homogenised prior to inoculation on to agarized BG11 growth medium. Once extensive growth of the biofilm was observed, they were treated with LAV (5% *v*/*v*), LIQ (10 and 30% *v*/*v*) and sterile saline solution (NaCl 0.9% *w*/*v*) as control (CTRL). Each treatment was tested in triplicate. The extracts were prepared by diluting with distilled water to the desired concentration. In the preparation of the LAV 5%, the solution was emulsified using 1% TWEEN 20. Two applications of 100 μL of each treatment were performed at time 0 and 5 days.

Photosynthesis, the most fundamental and intricate physiological process in phototrophic organisms, is highly sensitive to the majority of stresses (biotic or abiotic) [32]. Therefore, a measure of chlorophyll fluorescence could be considered as a good indicator of the effect of plant extracts on the vitality of the biofilm phototrophic component. A mini-PAM portable fluorometer was used to measure the maximum quantum yield immediately before each treatment (days 1 and 5) and then at 5, 30 and 90 min, 1 and 5 days after each treatment (10 days in total).

The second test involved the remaining 15 g of biofilm sampled from the selected area on the wall, which was then homogenized and transferred into 24 well plates. In each well, 250 mg of biofilm was placed on a sterile nitrocellulose filter and treated with either LIQ (10%), LAV (5%) or a mixture of

LIQ10% and LAV5% (LIQLAV). As a positive control the chemical biocide benzalkonium chloride 50% (Antichità Belsito, Rome, Italy) was applied at 0.6% (v/v) (BENZ). As negative control, the biofilm was treated with sterile saline solution (CTRL). All the tests were done in triplicate. Measures of maximum quantum yield were conducted immediately before the application of each treatment (100 μL) and then after 5 days. In order to avoid the cross contamination of treatments, one well plate was used for each substance.

To evaluate the biocide effect on the heterotrophic microorganisms, the nitrocellulose filters supporting the biofilm were weighed and suspended 1:10 w/v in a solution of sodium pyrophosphate (0.05%) and Tween 80 (0.001%). The suspensions were incubated in an orbital shaker at 28 °C and 150 rpm for 1 h. The number of cultivable heterotrophic microorganisms was obtained by plating 100 μL of suspension and following decimal dilutions in TSA, MM and PDA agar media. The plates were then incubated at room temperature until the complete growth of colonies. Counts of viable microorganisms were referred to as colony forming units per gram of wet biofilm (CFU/g wet biofilm).

### 2.5. *In Situ Application of Phyto-Derivatives*

Following the well plate experiments, the phyto-derivatives were tested in situ. Unfortunately, according to the safety limitation for the application of substances in the Domus Aurea's archaeological restoration site, it was not possible to test in situ the lavender essential oil.

Three areas of Room 93 (15 × 6 cm, each one) were selected that had a homogenous biofilm covering. Each area was treated by spraying separate treatments directly onto the surface using a stencil protector to restrict the application to the treatment area only; LIQ (10%), BENZ (0.6% v/v) and CTRL (sterile saline solution) were employed. Two applications were performed, at time 0 and after 14 days. Effective quantum yield was measured immediately before each treatment application and then 5 min afterwards; follow-up measurements were made on days 7, 14, 35 and 76. Coinciding with the yield measurements, in situ micrographs (200× max) were taken before and after treatments on biofilms using a Handheld Digital Microscope (USB) Dino-Lite Premier 1.3MP (AnMo Electronics Corporation, New Taipei City, Taiwan). Observations of biofilm samples collected using adhesive tape strips were also conducted at the light microscope (40×) and at the stereo microscope (1.6×) Zeiss, Stemi 508 (Carl Zeiss Microscopy GmbH, Jena, Germany) equipped with a digital camera Axiocam ERc 55 (Carl Zeiss Microscopy GmbH, Jena, Germany).

### 2.6. *Photosynthetic Parameters*

To evaluate the biocidal effects of the applied phyto-derivative products, photosynthetic parameters (chlorophyll fluorescence) of the biofilms were measured using the pulse amplitude modulated fluorometer Mini-PAM of Heinz Walz (Effeltrich, Germany) both on-site and in the laboratory. In situ measurements were made using a holder that maintained the fibre-optic probe at 60° to the surface (to not block the natural light levels), and 6 mm from the biofilm to allow comparisons to be made among sites (in the laboratory, the probe was held perpendicular to the samples). In the laboratory, the maximum potential quantum yields ($F_v/F_m$) were assured by performing the measurements made in quasi-darkness (<1 μmol photons $m^{-2}$ $s^{-1}$), whilst the effective quantum yield ($F_{v'}/F_{m'}$) was measured in situ [33]. Correlations, T-tests, ANOVA and post hoc tests were carried out using SPSS (v. 23 IBM).

## 3. Results and Discussions

### 3.1. *Macro and Micro-Environmental Conditions*

As expected, the environmental conditions were relatively stable in Room 93 with temperatures of 17.0 ± 0.5 °C, humidity 88 ± 2% and light irradiance of 1.0 ± 0.2 μmol photons $m^{-2}$ $s^{-1}$. Surface temperature, where the MAT method was employed for biofilm collection, was around 8.0 °C. These environmental conditions are typical of hypogean and confined environments [3,5] and remained stable due to the lower number of visitors in this area of the site.

### 3.2. Biofilm Characterization

Observations using light and confocal microscopy (Figure 3) showed that the biofilm was mainly composed of filamentous cyanobacteria typical of low light environments [4,34]. In particular, the cyanobacterium *Scytonema julianum* was shown to be the dominant species in the biofilm growing in Room 93 of the Domus Aurea, along with *Leptolyngbya* sp. and *Symphyonemopsis* sp. As previously observed [13], growth of these heterocytous-nitrogen fixing species is normally favoured in these nutrient poor environments. The dominance by *S. julianum* was particularly worrisome as it is considered highly biodeteriogenic in that it causes extensive calcium carbonate precipitates by dissolution of minerals from the substrata [11,13].

**Figure 3.** Biofilm sample taken from the wall of Room 93 and observed at the Confocal Laser Scanning Microscope (CLSM); the biofilm is dominated by *S. julianum* (thick filaments) and *Leptolyngbya* sp. (thin filaments). Scale bar = 40 µm.

### 3.3. Associated Heterotrophic Microorganisms

The microbial load composed of heterotrophic bacteria in the collected biofilm was 5.55 ± 0.6 ($\times 10^7$ CFU/g of wet biofilm), 5.4 ± 0.8 ($10^7$ CFU/g of wet biofilm) and 3.6 ± 0.3 ($10^5$ CFU/g of wet biofilm) on TSA, MM and PDA mediums, respectively. Based on colony morphologies, ten bacterial strains were isolated and identified (Table 1), belonging to three main phyla: Proteobacteria (class Alpha and Gamma), Actinobacteria and Bacteroidetes.

**Table 1.** Phylogenetic affiliation by 16S r-DNA sequencing of the bacterial strains co-occurring in the sampled biofilm. The strains have been named with the abbreviation DA followed by a progressive number.

| Strain Name | Phylum | Species | ID % |
|---|---|---|---|
| DA1 | Bacteroidetes | *Chriseobacterium plytrichastri* | 98 |
| DA2 | Proteobacteria | *Ensifer adhaerens* | 98 |
| DA3 | Proteobacteria | *Inquilinus ginsengisoli* | 96 |
| DA4 | Proteobacteria | *Pseudomonas asplenii* | 97 |
| DA5 | Proteobacteria | *Pseudomonas glareae* | 97 |
| DA6 | Proteobacteria | *Pseudomonas helmanticensis* | 99 |
| DA7 | Proteobacteria | *Pseudomonas vancouverensis* | 99 |
| DA8 | Actinobacteria | *Rhodococcus jostii* | 98 |
| DA9 | Actinobacteria | *Rhodocossus koreensis* | 98 |
| DA10 | Actinobacteria | *Streptomyces sproraveus* | 99 |

Only a few colonies of fungal strains were grown on PDA plates, but not yet identified.

The presence of both heterotrophic and phototrophic microorganisms within the biofilm suggests that they may be acting in synergy and could actually be causing the enhancement of hypogean biodeterioration processes. Bacteria and fungi can proliferate using the organic matter released by phototrophs and in turn secreting acidic organic compounds that can solubilize the minerals of the substratum, which are in turn utilised by the algal component [35]. The strain *Ensifer adhaerens* DA2 participate as a nitrogen-fixing agent.

In Roman Catacombs Actinobacteria (especially *Streptomyces*) and filamentous fungi (e.g., *Sporotrichum, Aspergillus, Cladosporium, Penicillium*, etc.) are commonly detected together with photosynthetic microbes [36]. They are widely distributed on stone monuments because of their filamentous growth and their ability to use a large range of nitrogen and carbon sources [37]. Moreover, in this study we have isolated some slime-forming bacteria, belonging to the genus *Pseudomonas*, already described in surface samples of the Roman Catacombs of St. Callixtus and Domitilla [37,38]. Slime-forming bacteria play an essential role in biofilm establishment, producing extracellular complex carbohydrates which ensure adhesion of microbial cells to surfaces, provide protection from environmental antimicrobials, serve as reservoirs for nutrients and create distinct architectures to facilitate further microbial adherence [39,40]. Members of Alpha Proteobacteria and Bacteroidetes were frequently found as heterotrophic components of biofilms affecting stone monuments [41,42].

### 3.4. Effects of Phyto-Derivatives Treatments

Preliminary tests on biofilms grown on agarised BG11 medium were carried out to compare the effects of lavender and liquorice extracts on the vitality of phototrophs. As shown in Figure 4, among the yield values of the controls there were no significant differences ($p > 0.05$); however, those of all the treated biofilms were significantly lower than the controls ($p < 0.001$).

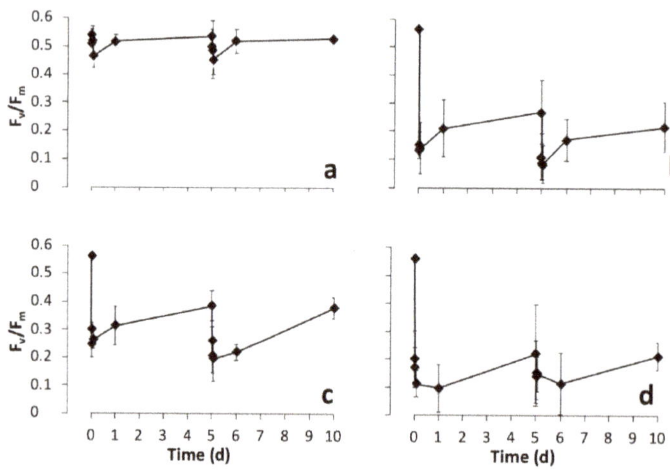


**Figure 4.** Changes over time of photosynthetic yield ($F_v/F_m$) of cultured biofilms treated at days 0 and 5 with diverse phyto-derivatives, (**a**) control (CTRL), (**b**) lavender essential oil (*Lavandula angustifolia* Mill.) (LAV), (**c**) liquorice alcoholic leaf extract (*Glycyrrhiza glabra* L.) (LIQ) at 10% and (**d**) 30%. After each treatment measurements were made immediately, then at 5, 30 and 90 min, 1 and 5 days (10 days in total).

The immediate reduction in yield of the controls and, to some extent, in the treatments, may be due to the partial submergence of the biofilm by addition of the treatment solution, which may have reduced gas exchange to the cell [43]. However, even after removal of this effect from the treatment yields (i.e., $F_v/F_m$ treated biofilm—$F_v/F_m$ control), there was still a 50% (LAV), 30% (LIQ10) and 60%

(LIQ30) reduction in $F_v/F_m$. There was a significant reduction in yields of the lavender and both the liquorice treated biofilms after five minutes ($p < 0.001$) and it remained so for the full 10 days. Therefore, it seems yield is susceptible to the two plant extracts, although the mode of action is not easy to determine. The minimal fluorescence ($F_o$) values either remained relatively unchanged (LAV) or decreased (LIQ), actually indicating no photodamage [44], while a larger decrease in $F_m$ was observed that could represent enhanced nonradiative energy loss (heat dissipation) that may act in photo-protection or -repair [45]. In all treatments, yield did recover to some extent, rapidly within the first day, then slower until day five: Recovery from the maximum reduction was 20% for both treatments of LAV, 10% and 6% for LIQ10, 22% and 25% for LIQ30. This recovery is not thought to be a result of new growth; as observations of the biofilms did not indicate this, more likely, it was the occurrence of cellular repair of the photosynthetic apparatus. Yield responded similarly to the second treatment (day 5)—an initial decrease, followed by a slow partial recovery. This time the recovery resulted in yield values that were not significantly different to the initial (day 5) value ($p > 0.05$). However, yields did remain significantly lower than the original levels (day 0) during the experiment ($p < 0.01$). In a study of the recovery of structure and function of freshwater biofilms after being treated with short pulses of herbicide (Diuron) or bactericide (Triclosan), it was established that short biocide pulses can efficiently cause transient effects on biofilms, while longer pulses would result in more persistent effects with chronic ones that could be lethal to the biofilm [46]. In the scenario here using phytoextracts, given their volatile nature, they may behave in a similar manner to the through-flow pulse method and more frequent applications would be needed to determine if there is a chronic effect on biofilm status. A second test has been carried out inside well plates containing a piece of biofilm (250 mg each) evaluating the effect of two applications of LAV, LIQ and a mixed solution of both (LIQLAV) on photosynthetic and heterotrophic microorganisms. As shown in Figure 5 there were no significant differences between the yields before treatments and those of the control biofilms ($p > 0.05$).

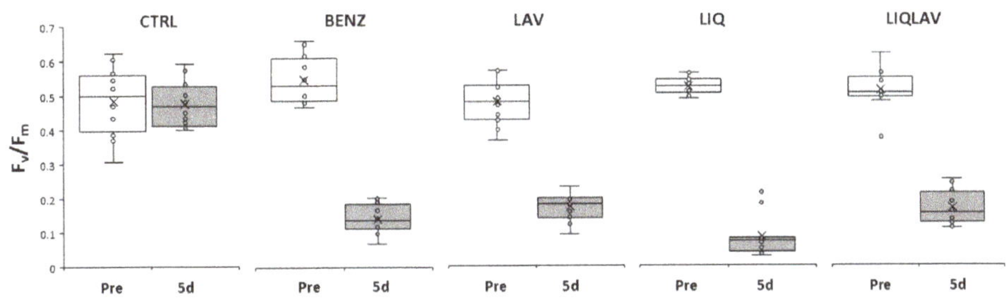


**Figure 5.** Changes in photosynthetic yield before each treatment (Pre) and after 5 days (5d) (n = 12). Control (CTRL), benzalkonium (BENZ, positive control), lavender at 5% (LAV), liquorice at 10% (LIQ) and a mixture of lavender 5% and liquorice 10% (LIQLAV).

The effects of the phyto-derivatives on the photosynthetic yield were comparable to the synthetic BENZ ($p > 0.05$) and resulted in significant lowering of yield ($p < 0.01$–$0.001$). One interesting aspect is the similar $F_v/F_m$ values for all treatments; the level indicating the presence of no viable photosynthesis was not determined. However, $F_v/F_m$ values lower than 0.1 have been proposed for other biofilms to indicate a total loss of photosynthetic activity [47,48], which is similar to the values found in this study and may explain a great deal of the similarity in the response of the biofilms to the treatments. Concerning the effect of the treatments on the heterotrophic bacteria, in Table 2 are reported the microbial load and the number of bacterial strains found in the biofilms after the first and second application of the different phyto-derivatives in the well plates.

**Table 2.** Colony forming units (CFUs) of cultivable heterotrophic bacteria in the different treatments after first and second application. CFU as referred to gram of wet biofilm. In the last column are reported the number of bacterial strains isolated after the second application.

| Treatments | I (CFU/g Wet Biofilm) | II (CFU/g Wet Biofilm) | No. Bacterial Strains after II Application |
|---|---|---|---|
| CTRL | $(2.50 \pm 0.3) \times 10^8$ | $(1.00 \pm 0.5) \times 10^8$ | 6 |
| LIQ | $(1.20 \pm 0.2) \times 10^9$ | $(8.50 \pm 0.3) \times 10^8$ | 1 |
| LAV | $(6.20 \pm 0.5) \times 10^9$ | $(4.40 \pm 0.2) \times 10^8$ | 5 |
| LIQLAV | $(1.10 \pm 0.2) \times 10^9$ | $(3.50 \pm 0.4) \times 10^8$ | 3 |
| BENZ | $(2.60 \pm 0.3) \times 10^8$ | $(2.00 \pm 0.5) \times 10^8$ | 6 |

Based on colony morphologies bacterial strains were isolated and identified. In Table 3 the bacterial strains isolated from the sampled biofilms and from the different treatments after the second application are reported. The DNA sequences have been submitted and we are awaiting the GenBank accession number.

**Table 3.** Phylogenetic identification by 16S r-DNA sequencing of the bacterial strains from the sampled biofilm (in situ) and from the different treatments after the second application. The X indicates the occurrence of each strain in the different treatments.

| Strain nr | ID | Sampled Biofilm | CTRL | LIQ | LIQLAV | LAV | BENZ | Phylum |
|---|---|---|---|---|---|---|---|---|
| DA17 | *Brevundimonas alba* | | | | | | X | Proteobacteria |
| DA1 | *Chriseobacterium plytrichastri* | X | | | | | | Bacteroidetes |
| DA18 | *Dyadobacter sp.* | | | | | | X | Proteobacteria |
| DA2 | *Ensifer adhaerens* | X | X | | X | X | | Proteobacteria |
| DA15 | *Exiguobacterium mexicanum* | | | | | X | | Firmicutes |
| DA3 | *Inquilinus gingengisoli* | X | | | | | | Proteobacteria |
| DA19 | *Lysovacter sp.* | | | | | | X | Proteobacteria |
| DA11 | *Mesorhizobium olivaresii* | | X | | | | | Proteobacteria |
| DA16 | *Microbacterium hydrocarbonoxydans* | | | | | X | | Actinobacteria |
| DA20 | *Phyllobacterium catacumbae* | | | | | | X | Proteobacteria |
| DA4 | *Pseudomonas asplenii* | X | | X | | | | Proteobacteria |
| DA5 | *Pseudomonas glaerae* | X | X | | X | X | X | Proteobacteria |
| DA6 | *Pseudomonas helmanticensis* | X | | | | | X | Proteobacteria |
| DA7 | *Pseudomonas vancouverensis* | X | X | | | | | Proteobacteria |
| DA12 | *Pseudoxanthomonas dokdonensis* | | X | | | | | Proteobacteria |
| DA8 | *Rhodococcus jostii* | X | | | | | | Actinobacteria |
| DA9 | *Rhodococcus korensis* | X | | | | | | Actinobacteria |
| DA13 | *Sphingomonas desiccabilis* | | X | | | | | Proteobacteria |
| DA14 | *Stenotrophomonas rhizophila* | | | | X | X | | Proteobacteria |
| DA10 | *Streptomyces spororaveus* | X | | | | | | Actinobacteria |

Phyto-derivative treatments, as well as benzalkonium chloride, seemed to have little to no effect on the vitality of the heterotrophic bacteria in terms of total microbial load (Table 2), but did have an effect on the composition (Table 3). The diversity of the heterotrophic culturable community of the biofilms was lower in all the treatments, including the control, if compared to the sampled biofilm. However, this may be explained somewhat by the stress induced by the change of culture conditions when moving the biofilm from the original surface to the well plates; in particular, Actinobacteria and Bacteroidetes strains seemed particularly susceptible while the Proteobacteria predominated. This was especially true in the benzalkonium-treated biofilm. Proteobacteria strains, in particular *Pseudomonas* spp., are metabolically active Gram-negative able to use a wide variety of macromolecules, including the organic cell debris originating by the death of biocide-sensitive components of the microflora [11]. In spite of this, the treatment with leaf liquorice extract dramatically reduced species richness to only one isolated strain, *Pseudomonas asplenii*, demonstrating the antibacterial effect already observed on bacteria isolated in clinical trials [24] and more recently against strains isolated from different artwork materials [18]. This effect was still evident in the combined liquorice–lavender treatment, where the number of isolated strains was reduced to three, all belonging to Proteobacteria. The treatment with lavender alone, although it did not reduce biodiversity greatly, did induce a species replacement with the appearance of strains belonging to Firmicutes and Actinobacteria.

### 3.5. In Situ Treatments

Following the well plate experiments, three areas of Room 93 covered by phototrophic biofilms were treated in situ by spraying with liquorice and benzalkonium. Data obtained comparing the effects of the two biocides are reported in Figures 6 and 7.

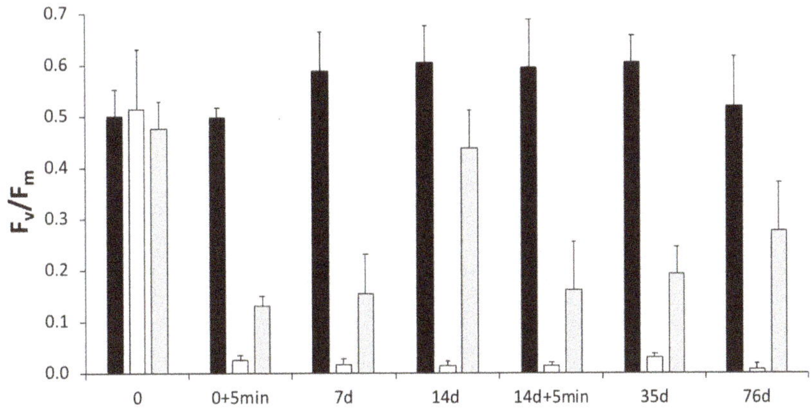


**Figure 6.** In situ treatment of biofilms growing in Room 93 of the Domus Aurea, Rome. Three treatments (CTRL (black), BENZ (white), LIQ (grey)); two applications, the first on day 0 and the second on day 14. PAM fluorometer measurements of yield were made immediately before both treatment applications (0 and 14 day), 5 min after (+5 min), then on days 7 and 14 after the first application, and then on days 35 and 76 (n = 3).

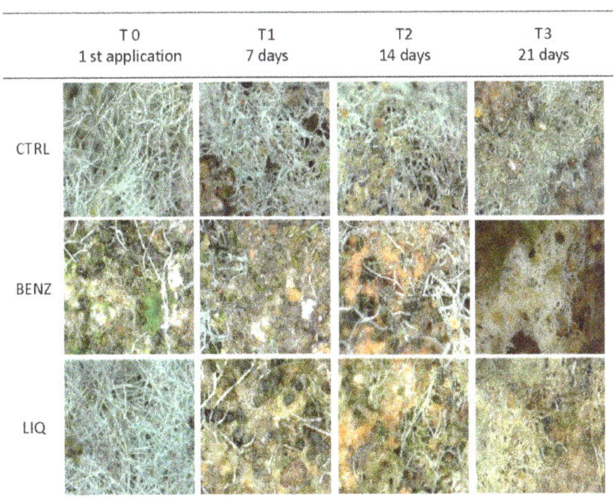


**Figure 7.** In situ Digital Microscope pictures of the treated surfaces at different time intervals acquired with Handheld Digital Microscope (USB) Dino-Lite Premier 1.3 MP (AnMo Electronics Corporation, Taiwan). The effect of LIQ and BENZ on the discolouration of the biofilms is evident.

The photosynthetic yields (Figure 6), measured 5 min after the application of benzalkonium, were significantly lower than the control and the liquorice treated biofilms ($p < 0.05$). However, yield recovery in the biofilms treated with the liquorice extract was almost complete by day 14, when compared to the control ($p < 0.05$), but after the second treatment yield remained 57% lower than the control by day 76. This suggests a possible chronic effect of the treatments or that a higher concentration of the active ingredients is needed to have a lasting effect.

Images of the treated areas taken weekly until 21 days are shown in Figure 7.

The biofilms collected 14 days after the first application were transferred to the lab and observed using light and stereo microscopes (Figure 8). As already observed with PAM measurements, both in situ and lab images showed that BENZ had a stronger effect on the photosynthetic cells, particularly evident in the filaments of *S. julianum* that turned from green to white. The liquorice extract also had a visible biocidal effect, albeit at a lower level, where photosynthetic cells turned from green to light yellow, although this was possibly due to the staining by the liquorice extract itself on the carbonaceous sheath of *S. julianum*. While it was not possible to continuously follow the biofilms, it would be intriguing to know if the presence of some green filaments in the liquorice-treated area were a sign of recovery by some resilient cells that could support the PAM fluorimetry data.

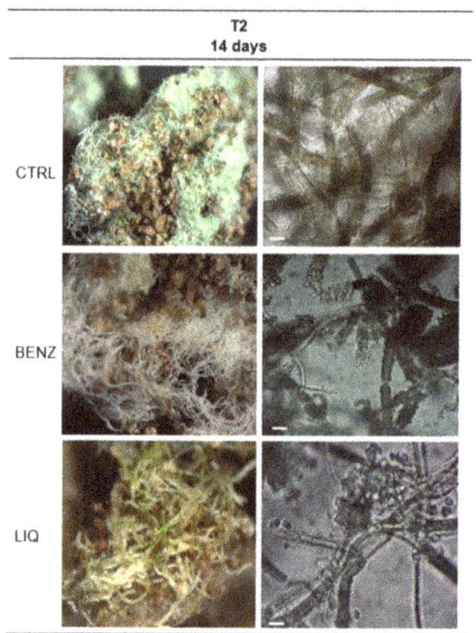


**Figure 8.** Images at stereo microscope (left column) and light microscope (right column) of in situ biofilms growing in Room 93 after 14 days and two applications of biocides. Bar = 10 μm.

To date, there are still no in situ studies containing detailed biofilm monitoring after treatments on a daily basis. This remains an intriguing challenge for future studies that should have a great impact on effective protocol development for biocide application for conservation purposes.

## 4. Conclusions

These studies performed in the prestigious sites of the Domus Aurea, Rome, took into consideration the majority of heterotrophic and phototrophic organisms forming subaerial biofilms, and the effects of biocide treatments on their function and composition. Liquorice alcoholic leaf extract (*Glycyrrhiza glabra* L.) was highly efficient in deterring phototrophic biofilm function and growth and when used in combination with lavender (*Lavandula angustifolia* Mill.) there was a further increase in the biocidal effect. This mixture (liquorice and lavender) also reduced the number of the cultivable heterotrophic bacteria strains in the biofilm, highlighting the widescale effect of these phyto-derivatives and showing the promise that these substances have when used, both alone and in combination. The application of the liquorice extract on the walls of Room 93 revealed how important it is to perform in situ trials to optimize the application of biocides for developing effective protocols, including to define the number of applications and the intervals between treatments.

Overall, the results obtained in this study showed that phyto-derivative products are promising as new non-invasive biocides to be employed for restoration and conservation actions of valuable sites. However, more on-site studies are needed with detailed follow-up monitoring to further understand their mechanisms.

**Author Contributions:** Conceptualization, G.M., F.T., A.R.S., L.B.; methodology, L.R., G.M., F.T., N.T.W.E., A.R.S., L.B.; validation L.R., G.M., F.T., N.T.W.E., L.B.; investigation, L.R., G.M., F.T., N.T.W.E., L.B.; data curation, L.R., G.M., F.T., N.T.W.E., L.B.; writing—original draft preparation, L.R., G.M., F.T., N.T.W.E., L.B.; writing—review and editing, L.R., G.M., F.T., N.T.W.E., A.R.S., L.B.; supervision, A.R.S., L.B. All authors have read and agreed to the published version of the manuscript.

**Funding:** This research received no external funding.

**Acknowledgments:** The authors grateful thank Alessandro D'Alessio e Maria Bartoli from the "Parco Archeologico del Colosseo" for the kind cooperation, the permission to work in the Domus Aurea, the precious opportunity given to us to put scientific research at the service of restoration and conservation, and the constant willingness to evaluate innovative solutions to ancient restoration problems. Isolated heterotrophic strains are included in the microbial collection ENEA-MIRRI-IT—MIcrobial Resource Research Infrastructure—Italian Joint Research Unit http://www.mirri-it.it/index.php/associated/enea/, member of the European MIRRI collection. The authors are thankful to A. Canini for giving us the lavender essential oil to test and the Trifolio-M GmbH for providing us with liquorice leaf extract.

**Conflicts of Interest:** The authors declare no conflict of interest.

## References

1. Albertano, P.; Moscone, D.; Palleschi, G.; Hermosín, B.; Saiz-Jimenez, C.; Sanchez-Moral, S.; Hernández-Mariné, M.; Urzì, C.; Groth, I.; Schroeckh, V.; et al. Cyanobacteria attack rocks (CATS): Control and preventive strategies to avoid damage caused by cyanobacteriaand associated microorganisms in Roman hypogean monuments. In *Molecular Biology and Cultural Heritage*; Saiz-Jimenez, C., Ed.; Balkema: Amsterdam, The Netherlands, 2003; pp. 151–162.
2. Bruno, L.; Valle, V. Effect of white and monochromatic lights on cyanobacteria and biofilms from Roman Catacombs. *Int. Biodeterior. Biodegrad.* **2017**, *123*, 286–295. [CrossRef]
3. Mulec, J.; Kosi, G. Lampenflora algae and methods of growth control. *J. Cave Karst Stud.* **2009**, *71*, 109–115.
4. Albertano, P. Cyanobacterial biofilms in monuments and caves. In *Ecology of Cyanobacteria II: Their Diversity in Space and Time*; Whitton, B.A., Ed.; Springer: Amsterdam, The Netherlands, 2012; pp. 317–343.
5. Urzì, C.; Bruno, L.; De Leo, F. Biodeterioration of paintings in caves, catacombs and other hypogean sites. In *Art, Archaeology and Architecture*; Mitchell, R., Clifford, J., Eds.; Archetype Publications Ltd.: London, UK, 2018; pp. 114–129.
6. Gaylarde, C.; Baptista-Neto, J.A.; Ogawa, A.; Kowalski, M.; Celikkol-Aydin, S.; Beech, I. Epilithic and endolithic microorganisms and deterioration on stone church facades subject to urban pollution in a sub-tropical climate. *Biofouling* **2017**, *33*, 113–127. [CrossRef] [PubMed]
7. Nugari, M.P.; Pietrini, A.M.; Caneva, G.; Imperi, F.; Visca, P. Biodeterioration of mural paintings in a rocky habitat: The Crypt of the Original Sin. *Int. Biodeterior. Biodegrad.* **2009**, *63*, 705–711. [CrossRef]
8. Koziróg, A.; Rajkowska, K.; Otlewska, A.; Piotrowska, M.; Kunicka-Styczy, A.; Brycki, B.; Nowicka-Krawczyk, P.; Ko´scielniak, M.; Gutarowska, B. Colonising organisms as a biodegradation factor affecting historical wood materials at the former concentration camp of Auschwitz II e Birkenau. *Int. J. Mol. Sci.* **2016**, *17*, 1364. [CrossRef] [PubMed]
9. Marano, F.; Di Rita, F.; Palombo, M.R.; Ellwood, N.T.W.; Bruno, L. A first report of biodeterioration caused by cyanobacterial biofilms of exposed fossil bones: A case study of the Middle Pleistocene site of La Polledrara di Cecanibbio (Rome, Italy). *Int. Biodeterior. Biodegrad.* **2016**, *106*, 67–74. [CrossRef]
10. Slimen, A.; Barboux, R.; Mihajlovski, A. High diversity of fungi associated with altered wood materials in the hunting lodge of "La Muette", Saint-Germain-en-Laye, France. *Mycol. Prog.* **2020**, *19*, 139–146. [CrossRef]
11. Urzì, C.; De Leo, F.; Krakòva, L.; Pangallo, D.; Bruno, L. Effects of biocide treatments on the biofilm community in Domitilla's catacombs in Rome. *Sci. Total Environ.* **2016**, *572*, 252–262. [CrossRef]

12. Barresi, G.; Cammarata, M.; Palla, F. Biocide. In *Biotechnology and Conservation of Cultural Heritage*; Palla, F., Barresi, G., Eds.; Springer International Publishing: Cham, Switzerland, 2017; pp. 49–65.
13. Bruno, L.; Rugnini, L.; Spizzichino, V.; Caneve, L.; Canini, A.; Ellwood, N.T.W. Biodeterioration of Roman hypogea: The case study of the Catacombs of SS. Marcellino and Pietro (Rome, Italy). *Ann. Microbiol.* **2019**, *69*, 1023–1032. [CrossRef]
14. Genova, C.; Fuentes, E.; Sanmartín, P.; Favero, G.; Prieto, B. Phytochemical Compounds as Cleaning Agents on Granite Colonized by Phototrophic Subaerial Biofilms. *Coatings* **2020**, *10*, 295. [CrossRef]
15. Ruffolo, S.A.; De Leo, F.; Ricca, M.; Arcudi, A.; Silvestri, C.; Bruno, L.; Urzì, C.; La Russa, M.F. Medium-term in situ experiment by using organic biocides and titanium dioxide for the mitigation of microbial colonization on stone surfaces. *Int. Biodeterior. Biodegrad.* **2017**, *123*, 17–26. [CrossRef]
16. Hsieh, P.; Pedersen, J.Z.; Bruno, L. Photoinhibition of cyanobacteria and its application in cultural heritage conservation. *Photochem. Photobiol.* **2014**, *90*, 533–543. [CrossRef] [PubMed]
17. Sanmartín, P.; Villa, F.; Polo, A.; Silva, B.; Prieto, B.; Cappitelli, F. Rapid evaluation of three biocide treatments against the cyanobacterium *Nostoc* sp. PCC 9104 by color changes. *Ann. Microbiol.* **2015**, *65*, 1153–1158. [CrossRef]
18. European Chemicals Agency. *Impacts of REACH Restriction and Authorisation on Substitution in the EU*; ECHA-20-R-09-EN978-92-9481-618-4; European Chemicals Agency: Helsinki, Finland, 2020. [CrossRef]
19. Genova, C.; Zoppis, E.; Grottoli, A.; Cencetti, C.; Matricardi, P.; Favero, G. An integrated approach to the recovery of travertine biodegradation by combining phyto-cleaning with genomic characterization. *Microchem. J.* **2020**, 156. [CrossRef]
20. Abbas, A.; Zubair, M.; Rasool, N.; Rizwan, K. Antimicrobial Potential of *Glycyrrhiza glabra*. *J. Drug Des. Med. Chem.* **2015**, *1*, 17–20.
21. Gupta, V.K.; Fatima, A.; Faridi, U. Antimicrobial potential of *Glycyrrhiza glabra* roots. *J. Ethnopharmacol.* **2008**, *116*, 377–380. [CrossRef]
22. Fatima, A.; Gupta, V.K.; Luqman, S.S.; Negi, A.S.; Kumar, J.K.; Shanker, K.; Saikia, D.; Srivastava, S.; Darokar, M.P.; Suman, P.S. Antifungal activity of *Glycyrrhiza glabra* extracts and its active constituent glabridin. *Phytother. Res.* **2009**, *23*, 1190–1193. [CrossRef]
23. Fukui, H.; Katsumi, G.; Tabata, M. Two antimicrobial flavanones from the leaves of *Glycyrrhiza glabra*. *Chem. Pharm. Bull.* **2008**, *36*, 4174–4176. [CrossRef]
24. Irani, M.; Sarmadi, M.; Bernard, F.; Ebrahimi, G.H.; Ebrahimi, G.H.; Bazarnov, H.S. Leaves antimicrobial activity of *Glycyrrhiza glabra* L. *Iran. J. Pharm. Res.* **2010**, *9*, 425–428.
25. Bassyouni, R.H.; Kamel, Z.; Megahid, A.; Samir, E. Antimicrobial potential of licorice: Leaves versus roots. *Afr. J. Microbiol. Res.* **2012**, *6*, 7485–7493.
26. Scherf, A.; Treutwein, J.; Kleeberg, H.; Schmitt, A. Efficacy of leaf extract fractions of *Glycyrrhiza glabra* L. against downy mildew of cucumber (*Pseudoperonospora cubensis*). *Eur. J. Plant Pathol.* **2010**, *134*, 755–762. [CrossRef]
27. Sprocati, A.R.; Alisi, C.; Migliore, G.; Marconi, P.; Tasso, F. Sustainable restoration through biotechnological processes: A proof of concept. In *Roles of Microorganisms in Heritage Degradation and Preservation*; Springer: Cham, Switzerland, 2020. (in press)
28. Urzì, C.; De Leo, F. Sampling with adhesive tape strips: An easy and rapid method to monitor microbial colonization on monument surfaces. *J. Microbiol. Methods* **2001**, *44*, 1–11. [CrossRef]
29. Komárek, J. Cyanoprokaryota: 3rd part: Heterocytous genera. In *Süswasserflora von Mitteleuropa 19/3*; Büdel, B., Gärtner, G., Krienitz, L., Schagerl, M., Eds.; Springer Spektrum: Heidelberg/Berlin, Germany, 2013; p. 1130.
30. Komárek, J.; Anagnostidis, K. Cyanoprokaryota. 2nd part: Oscillatoriales. In *Süsswasserflora von Mitteleuropa 19/2*; Büdel, B., Krienitz, L., Gärtner, G., Schagerl, M., Eds.; Elsevier Spektrum: Heidelberg, Germany, 2005; p. 759.
31. Mühling, M.; Woolven-Allen, J.; Murrell, J.C.; Joint, I. Improved group-specific PCR primers for denaturing gradient gel electrophoresis analysis of the genetic diversity of complex microbial communities. *ISME J.* **2008**, *2*, 379–392. [CrossRef] [PubMed]
32. Dayan, F.E.; Zaccaro, M.L.M. Chlorophyll fluorescence as a marker for herbicide mechanisms of action. *Pestic. Biochem. Phys.* **2012**, *102*, 189–197. [CrossRef]

33. Schreiber, U.; Neubauer, C.; Schliwa, U. PAM fluorometer based on medium-frequency pulsed X-flash measuring light: A highly sensitive new tool in basic and applied photosynthesis research. *Photosynth. Res.* **1993**, *36*, 65–72. [CrossRef] [PubMed]
34. Bruno, L.; Bellezza, S.; De Leo, F.; Urzì, C. A study for monitoring and conservation in the Roman Catacombs of St. Callistus and Domitilla, Rome (Italy). In *The Conservation of Subterranean Cultural Heritage*; Saiz-Jimenez, C., Ed.; CRC Press, Taylor & Francis Group: London, UK, 2014; pp. 37–44.
35. Eckhardt, F.E.W. Mechanisms of the microbial degradation of minerals in sandstone monuments, medieval frescoes and plasters. In Proceedings of the 5th International Congress on Deterioration and Conservation of Stone, Lausanne, Switzerland, 25–27 September 1985; Felix, G., Ed.; Presses Polytechniques Romandes: Lausanne, Switzerland, 1985; Volume 2, pp. 643–665.
36. Alberano, P.; Urzì, C. Structural interactions among epilithic cyanobacteria and heterotrophic microorganisms in Roman hypogea. *Microbiol. Ecol.* **1999**, *38*, 244–252. [CrossRef]
37. Saarela, M.; Alakomi, H.L.; Suihko, M.L.; Maunuksela, L.; Raaska, L.; Mattila-Sandholm, T. Heterotrophic microorganisms in air and biofilm samples from Roman catacombs with a special emphasis on actinobacteria and fungi. *Int. Biodeterior. Biodegrad.* **2004**, *1*, 27–37. [CrossRef]
38. Krakova, L.; De Leo, F.; Bruno, L.; Pangallo, D.; Urzì, C. Complex bacterial diversity in the white biofilms of St. Callixtus Catacombs in Rome evidenced by different investigation strategies. *Environ. Microbiol.* **2015**, *17*, 1738–1752. [CrossRef]
39. Di Martino, S. Bacterial adherence: Much more than a bond. *AIMS Microbiol.* **2020**, *4*, 563–566. [CrossRef]
40. Limoli, D.H.; Jones, C.J.; Wozniak, D.J. Bacterial extracellular polysaccharides in biofilm formation and function. In *Microbial Biofilms*, 2nd ed.; Ghannoum, M., Parsek, M., Whiteley, M., Mukherjee, P.K., Eds.; American Society of Microbiology: Washington, DC, USA, 2015. [CrossRef]
41. Miller, A.Z.; Laiz, L.; Gonzalez, J.M.; Dionísio, A.; Macedo, M.F.; Saiz-Jimenez, C. Reproducing stone monument photosynthetic-based colonization under laboratory conditions. *Sci. Total Environ.* **2008**, 278–285. [CrossRef]
42. Mihajlovski, A.; Gabarre, A.; Seyer, D.; Boust, F.; Di Martino, P. Bacterial diversity on rock surface of the ruined part of a French historic monument: The Chaalis abbey. *Int. Biodeterior. Biodegrad.* **2017**, *120*, 161–169. [CrossRef]
43. Blom, C.W.P.M. Adaptations to flooding stress: From plant community to molecule. *Plant Biol.* **1999**, *1*, 263–273. [CrossRef]
44. Murchie, E.H.; Lawson, T. Chlorophyll fluorescence analysis: A guide to good practice and understanding some new applications. *J. Exp. Bot.* **2013**, *64*, 3983–3998. [CrossRef] [PubMed]
45. Kusama, Y.; Inoue, S.; Jimbo, H.; Takaichi, S.; Sonoike, K.; Hihara, Y. Zeaxanthin and echinenone protect the repair of photosystem II from inhibition by singlet oxygen in *Synechocystis* sp. PCC 6803. *Plant Cell Physiol.* **2015**, *56*, 906–916. [CrossRef]
46. Proia, L.; Morin, S.; Peipoch, M.; Romaní, A.M.; Sabater, S. Resistance and recovery of river biofilms receiving short pulses of Triclosan and Diuron. *Sci. Total Environ.* **2011**, *409*, 3129–3137. [CrossRef]
47. Lüttge, U.; Büdel, B. Resurrection kinetics of photosynthesis in desiccation-tolerant terrestrial green algae (Chlorophyta) on tree bark. *Plant Biol.* **2010**, *12*, 437–444. [CrossRef]
48. Gray, D.W.; Lewis, L.A.; Cardon, Z.G. Photosynthetic recovery following desiccation of desert green algae (Chlorophyta) and their aquatic relatives. *Plant Cell Environ.* **2007**, *30*, 1240–1255. [CrossRef]

*applied sciences*

MDPI

*Article*

# High Dimension Granite Pavement Bio-Desalination Practical Implementation

**Pilar Bosch-Roig [1,*], Lourdes Pérez-Castro [2], Ángeles Fernández-Santiago [3] and Ignacio Bosch [4]**

1 Instituto de Restauración del Patrimonio, Universitat Politècnica de València, 46022 Valencia, Spain
2 Consorcio de la Ciudad de Santiago, Rúa do Vilar, 59, 15702 Santiago de Compostela, Spain; lourdespc@consorciodesantiago.org
3 Alfa ArteRestauración S.L., Sobrada, Torrón 23, 36739 Tomiño, Spain; anxos.santiago@gmail.com
4 Instituto de Tecnologías y Aplicaciones Multimedia, Universitat Politècnica de València, 46022 Valencia, Spain; igbosroi@dcom.upv.es
* Correspondence: mabosroi@upvnet.upv.es

**Featured Application: This work has developed an innovative and optimized in situ high dimension bio-desalination application protocol that has been applied for the bio-desalination of the 233 m$^2$ Conxo Chapel granite pavement. This optimized protocol has the potential to be transferable to other large scale in situ biocleaning strategies.**

**Abstract:** Biocleaning technology is based on the use of safe environmental microorganisms for green cultural heritage (CH) restoration. Compared with traditional cleaning products, this biological technique is very specific, effective, and nontoxic. This innovative biotechnological application has been used for recovering diverse monuments and artworks. Most CH in situ surfaces that are treated with microorganisms are small areas; however, some important pathologies, such as salt contamination, can affect high dimension artistic surfaces. The purpose of this study is to analyze and overcome the problems and limitations of scaling up the bio-desalination protocol for in situ applications. Three water-based gel delivery systems and three heating systems were tested in situ and evaluated in terms of performance difficulty, efficacy, and costs. The tests were carried out on the salt contaminated granite pavement of Cristo Chapel of Sta Ma de Conxo in Santiago de Compostela (Spain). Ground agar 2% and a heating electric mat were selected as the best performing systems. The implemented protocol was applied for the bio-desalination of the 233 m$^2$ Chapel pavement. Conductivity, nitrate–nitrite measurements, biological monitoring, and digital image analysis were performed to determine the efficacy of the treatment. This research allowed for the development of an innovative and optimized in situ, high dimension bio-desalination application protocol transferable to other large scale, in situ biocleaning strategies.

**Keywords:** bio-desalination; biocleaning; nitrate; bacteria; cultural heritage; granite pavement



**Citation:** Bosch-Roig, P.; Pérez-Castro, L.; Fernández-Santiago, Á.; Bosch, I. High Dimension Granite Pavement Bio-Desalination Practical Implementation. *Appl. Sci.* **2021**, *11*, 6458. https://doi.org/10.3390/app11146458

Academic Editors: Filomena De Leo and Valme Jurado

Received: 27 May 2021
Accepted: 11 July 2021
Published: 13 July 2021



## 1. Introduction

Cultural heritage (CH) is our past generation's legacy that must be conserved for future generations. Different circumstances, such as pollution, inadequate old restorations, salt contamination, and climate change are damaging these artworks and monuments. In order to properly conserve them, they need to be restored.

Among the many "diseases" present on CH monuments, alterations, such as black crust, salts, organic residues, and graffiti, are similar in that they are all difficult to clean with traditional methodologies. Therefore, there is a need to develop and standardize a more effective and safe technology. The majority of the current cleaning techniques are physical and/or chemical methods. These traditional techniques involve risks for the applicators, general public, artworks, and environment. These risks include health risks (linked, among other things, to the inhalation and touching of toxic cleaning chemicals), environmental contamination (most of the time, chemicals used for cleaning simply flow into the sewage system), and CH alteration (due to the lack of specificity of these traditional techniques).

Due to the risks associated with traditional CH cleaning techniques, an important demand for less risky cleaning products currently exists. The CH sector is asking for more "green" sustainable products as valid alternatives and even to substitute for the chemical methods. An existing demand for less time-consuming cleaning techniques also exists in order to augment the efficiency of the interventions and to reduce personnel costs.

In order to solve these problems, innovative, fast, and safe biological cleaning techniques have been developed in the last two decades. These alternative microbial biotechnologies use natural living organisms instead of toxic chemical substances [1] to avoid health and environmental problems [2]. In addition, these techniques are very specific, allowing proper preservation of the treated artwork while avoiding CH alterations due to the restoration process. They are also time-saving techniques that reduce personnel related costs. Therefore, biotechnology applied to CH restoration has become a successful, novel alternative to traditional conservation and preservation approaches [3].

This biorestoration technology, in the last two decades, has been effectively tested and validated experimentally at the lab scale and on in situ CH materials (such as artworks and monuments). Many authors have reported the use of different appropriately selected microorganisms for removing black crust (mostly sulfates), salts (bio-desalination), organic residues (mostly animal glue), and graffiti, among others, from organic and inorganic artistic materials. Most of the reported biocleaning methods are used as a cleaning alternative alone or in combination with other treatments when traditional techniques are ineffective, aggressive, or dangerous [4–6]. However, some authors show that they can be even the first option. Giuventù et al. (2011) carried out a comparative study showing the efficient removal of black crust on the colored lithotypes of Florence Cathedral by using *Desulfovibrio vulgaris subsp. vulgaris* compared with chemical and laser treatment [7]. Our previous studies have shown that a *Pseudomonas stutzeri* bio-desalination treatment was the most effective method compared to the most frequently used desalination methods [8].

The dimensions of CH artifacts can vary significantly, from a few millimeters (small artworks) to many square meters (large monuments). Also, CH pathologies can affect small or huge areas. One of the pathologies that affects larger surfaces is salt contamination of artistic stone, which frequently affects historical monuments. According to the literature, most of the CH in situ stone surfaces treated with microorganisms are small areas whose dimensions are usually not indicated (n.i.), and of those studies that do report the area, dimensions typically vary from 1 $cm^2$ to 95 $m^2$ [4]. Artistic treated items include sculptures such as: *Pietà Rondanini* by Michelangelo; the marble column and statue (2.2 m high and 60 cm deep) of the writer Neera by Lina Arpesani in the Cemetery of Milan; the Demetra and Cronos limestone sculpture from the courtyard of the Buonconsiglio Castle in Trento created by Jacob Eberle (n.i.); the marble polychrome artifact in Galleria Regionale Sicilia in Palermo (n.i.) [4,9–12], mural paintings such as: the indoor vault fresco lunette (25 $m^2$) in the Santos Juanes church in Valencia; three detached frescoes at Camposanto Monumentale in Pisa (27 $m^2$, 84 $m^2$, 95 $m^2$, respectively); external vault tempera mural paintings in Chiesa S. Apollinare in Trento (n.i.); a mural painting in loggia di Cassina Farnese in Rome (n.i.); a mural painting in Queen Teodolinda Chapel in Monza Cathedral (n.i.); a mural painting in Cappella S. Pietro in Chiesa S. Prudenza in Rome (n.i.); a mural painting by Orazio Riminaldi in cupula del Duomo di Pisa (n.i.); an oil mural painting from Mussei Vaticani in Rome (n.i.) [13–21], and monuments such as: a 5 $m^2$ area of the external sandstone walls at Matera Cathedral; three areas (15 × 15 $cm^2$) of external colored marble lithotypes at Florence Cathedral; a granite pavement slab (60 cm × 40 cm × 15 cm) at the Santa Maria del Conxo chapel in Santiago de Compostela [7,8,22,23]. The frescoes at Camposanto Monumentale of Pisa have been the highest dimension artworks treated with biocleaning technology. These three 14th Century detached frescoes have been completely biocleaned: "Conversione Saint Efisio e battaglia" by S. Aretino (3.5 m × 7.8 m); "Stories of the Holy Fathers" (6.10 m × 15.65 m), and "Trionfo della Morte" by B. Buffalmarco

(5.6 m × 15.0 m) using *Pseudomonas stutzeri* A29. The authors developed an optimized biocleaning protocol for ex situ fresco restoration based on cotton wool as the delivery system and a central heating system for temperature control because the restoration was carried out in a laboratory [14,21,24]. However, to the best of our knowledge, no high dimension biocleaning for real CH in situ applications has yet been reported.

The challenge of this research was to analyze and overcome the problems and limitations of scaling up the protocol for CH in situ application. Therefore, the objective was to implement an innovative and optimized in situ biocleaning protocol that is applicable to high dimension surfaces. Bacterial biomass production, temperature conditioning, delivery systems, performance difficulty, cleaning efficiency, and economic costs were all factors that were considered.

In particular, we focused on the bio-desalination of the high dimension granite pavement of the Cristo Chapel of Santa María de Conxo Monastery. The pavement is affected by high nitrate salt contamination that affects its visual appearance [8,25]. The relative humidity (RH) inside the Chapel is high (75–80% mean values) because the city, proximal to the ocean, also has a high air humidity level most of the year (80% mean values). This high humidity alters the chapel pavement by darkening the granite slabs. In order to reduce water entering the Chapel, in 2013, the Santiago Consortium promoted work to prevent the free fall of rainwater on the walls (gutters and downspouts were placed). The exterior grouting of the chapel's walls was improved, gaps (through which the water entered) were eliminated, and perimeter drainage was executed in order to separate the construction from the cemetery and prevent the entry of water from there. In 2015, a hygrothermal study revealed that the ground was wet under the slabs, and that this moisture was abnormally retained, both in the ground and in the stone, due to the presence of hygroscopic salts [25]. Previous studies have analyzed the pavement using chemical analysis. Fourier Transform Infrared (FTIR) analysis showed that the chemical composition of the pavement was quartz, feldspar, calcite, and kaolinite group minerals correlating with granite stone. Photometric, semi-quantitative test strips and FTIR analyzes of soluble salt content found mostly nitrate [8,25].

Salt contamination is higher on the surface of the slabs than on the lower part. This has a double effect: on the one hand, it retains the humidity that comes from the ground, but on the other, it attracts and condenses water vapor from the air. These two mechanisms prevent the drying of the pavement and, therefore, lead to the darkening of it [25]. To solve the problem of saline contamination, there was a need to proceed with pavement desalination.

On the basis of previous works, we selected *Pseudomonas stutzeri* due to its great capability in completing nitrate elimination, removing 3 g/L (as $KNO_3$) in just 48 h, and its effectiveness on artworks when nitrate biocleaning strategies are applied [8,16,22,26–28]. *Pseudomonas stutzeri* is an aerobic, non-sporulating, gram-negative soil bacterium widely distributed in the environment. It is classified as a risk group 1 bacterium (according to the CECT classification and based on INSHT(National Institute of Safety and Hygiene at Work (Instituto Nacional de Seguridad e Higiene en el Trabajo)) technical guide), which means that there is a low individual and community risk and is, therefore, unlikely to cause disease. Even though there was a low risk, and it has rarely been reported, by itself, in relation to affecting patients or causing human disease (only in combination with other microorganisms and on patients with serious underlying disease) [29], it was important that restorers put in place extreme precautions using a safe handling procedure. When applying the biocleaning technologies we took into account that we were working with living microorganisms, and paid special attention when handling the bacteria and with the correct management of the bio-treatment residuals. The safe handling procedure included the use of personal protection equipment such as working gloves and a lab coat [2].

Once the biocleaning bacteria was selected and the stone pathology characterized, the application protocol was optimized according to two important factors required for the microorganisms to act: temperature and water. Temperature is one of the most important factors conditioning the growth of microorganisms as well as the efficiency of the denitrifying process needed for the bio-desalination to occur [28]. So, in situ high dimension heating systems are needed for the treatment and the optimal and constant temperature must be established while also taking into account cost-efficient features. Water is essential for the viability of the microorganisms and, therefore, for a correct bio-desalination process, but it can also produce material alterations [16,30]. Previous biocleaning studies have shown that minimal water has been enough for the process to succeed so, therefore, the use of a water-based gel appeared to be the best option [31,32].

According to the abovementioned factors, this work analyzes in situ three water-based gel delivery systems and three heating systems and evaluates them in terms of in situ performance difficulty, efficacy, and costs, allowing us to implement an innovative and optimized in situ high scale bio-desalination application protocol.

## 2. Materials and Methods

### 2.1. Site Description

The pavement object of this study is in the Cristo Chapel of the Santa María de Conxo Church, located in Santiago de Compostela, Spain. It is a Baroque construction (17th century) with original granite ashlar walls and vaults. The flooring is composed of 233 $m^2$ of approximately 15 cm thick granite slabs that rest directly on the ground and whose joins are mostly open. Some of the slabs constitute burial covers. The Chapel also has a cemetery on its north side, with niches attached to the facade of this orientation.

### 2.2. Pseudomonas Stutzeri Optimal Denitrification Assessment

*Pseudomonas stutzeri*, strain 930T (CECT), nitrate-reducing bacterium was selected due to its complete denitrification capacity and its wide use in biocleaning strategies [8,16,22,26–28]. The complete denitrification process consisted of a stepwise process of nitrate ($NO_3^-$) reduction to nitrite ($NO_2^-$) and then to nitrogen ($N_2$) gas [33].

In order to determine the optimal time and temperature for its complete denitrification activity, in vitro analysis was carried out. One colony of *Pseudomonas stutzeri 930T* (CECT) was inoculated in 10 mL Nitrate Broth cultivation media (Scharlau SL) tubes and incubated at the three temperatures tested: 4 °C, 20 °C, and 26 °C.

The denitrification process was tracked using Nitrate and Nitrite test strips (Quantofix, Macherey-Nagel GmbH & Co, Düren, Germany) and the presence of nitrogen gas formation by its entrapment in Durham chamber after 24 h, 48 h and 72 h of incubation time. Turbidity and total number of viable bacteria in the tests was determined by serial dilutions and colony forming units (UFU) $mL^{-1}$ counts [8,16,28]. As a negative control, tubes without bacteria were also analyzed. Tests were conducted in triplicate.

### 2.3. Biomass Production

*Pseudomonas stutzeri*, strain 930T (CECT) nitrate-reducing biomass production was carried out in 2 × 5 L of sterile Nitrate Broth (Scharlau SL) cultivation media. Five vol.% of seed culture was inoculated and incubated for 48 h at 28 °C under aerobic conditions, adapting the protocol of Vidakovik et al. (2019) [28]. The Nitrate Broth cultivation media consisted of: meat extract (3 g/L), peptone (5 g/L), and potassium nitrate (1 g/L). The 10 L bacteria suspension was centrifugated (6000 rpm for 10 min, Beckman Coulter Avanti™ J-25 high performance centrifuge, Beckman instruments, Palo Alto, CA, USA, EEUU). The pellet was washed with NaCl (0.8% pH 7.0) and re-suspended in 6 tubes with 50 mL mineral water. Tubes were sent by 24 h refrigerated courier at 4–8 °C to the church for their application. Once in the church, 3 × 50 mL tubes were diluted with 4850 mL of mineral water to obtain the 5000 mL of application suspension with a final bacterial concentration of approximately $10^{10}$ colony forming units (CFU) $mL^{-1}$. A final volume of application suspension of 10 L was obtained, which allowed for 30 $m^2$ bio-desalination. In order to be able to bio-desalinate the complete pavement of 233 $m^2$, eight cycles of bacterial preparation and shipping were needed.

### 2.4. High Scale Application Protocol Implementation

In order to correctly conduct the bio-desalination on high dimension pavement, the bio-desalination application protocol was implemented.

This was based on the assessed optimal denitrification conditions of *P. stutzeri* (see Sections 2.1 and 3.1) and on the previously conducted in vitro and in situ preliminary bio-desalination small tests conducted in the lab and in the church and published previously [8]. The protocol was defined as follows: 48 h application time, 2% agar as delivery system, and 26 ± 4 °C as the optimal pavement temperature for bacterial denitrification.

Three delivery systems were tested: warm 2% agar (AgarArt, C.T.S. España S.L, Madrid, Spain), ground 2% agar (AgarArt, C.T.S. España S.L., Madrid, Spain), and Nevek® (C.T.S. España S.L., Madrid, Spain). Agar is a complex polysaccharide extracted from red algae, soluble in water, and able to produce gels with no liquid water release. It is widely used in restoration, including cleaning and salt extract processes because it is able to limit the water release onto artistic surfaces [30]. In order to use it in the restoration field it can be acquired commercially in two main consistencies: as a powder (AgarArt) or as a pseudo-gel (namely Nevek®). The first type must be mixed with water and boiled in order to be used and the second can be applied direct from the storage package.

The tested delivery systems preparation protocols are as follows:

- Warm 2% agar is prepared by mixing 2% agar powder in deionized water and then boiling it. Then, it must be tempered (between 40–45 °C) to become a semisolid, which is a good consistency for the application.
- Ground 2% agar is previously prepared in the lab following the warm agar protocol but leaving to become completely cold and, therefore, to jellify on a crystal recipient. Then it is ground and used or stored in a fridge for a few hours.
- Nevek® does not need any preparation because it is directly applied from the storage package.

Three thermic systems were tested: infrared heat lamps (Massive Halo R7s, Zenith CT, Italy; with 300 W halogen bulb, Lexman R7s, Leroy Merlin España S.L.U., Valencia, Spain), air heater (Honeywell 2.000 W, Charlotte, NC, USA), and heating electric mat (Al 140 W/m$^2$, CEILHIT SLU, Barcelona, Spain). These thermic systems selected augment the temperature in different ways. Infrared heat lamps contain an incandescent bulb that produces infrared radiation causing heat and increasing the temperature of the object that is being heated. Air heater works by passing an air flow across a heated element (coil of wire that by passing electrical current through it, heat is produced) that augments the air flow temperature. Heating electric mat is composed of a mat with a thin electric heating wire stuck on it. The wire heats the material directly in contact with the mat rather than warming the air.

The tested thermic systems preparation protocols are as follows:

- Infrared heat lamps are installed onto a tripod and oriented to face the treatment surface. They must be switched on for 14 h before treatment in order to reach the desired treatment temperature. The surface temperature is manually measured with an infrared thermometer (Parkside PTIA1, OWIM GmbH & Co. KG, Neckarsulm, Germany).
- The air heater must be inserted into a previously installed hothouse on the treated area in order to reduce the volume of air to be warmed. The air heater must be activated 24 h before treatment in order to obtain the adequate treatment temperature. The surface temperature is manually measured with an infrared thermometer.
- A heating electric mat (10 m long and 50 cm wide, Warmup, UK) is applied directly to the delivery system and, on top of the mat, a thermally insulating material (such as wood panels) must be added in order to avoid temperature loss to the ambient air. The heating electric mat must be activated 24 h before treatment in order to obtain the desired temperature. It has a temperature control thermostat.

In order to identify the optimal protocol, in situ application trials were conducted (Figure 1) and were analyzed in terms of: performance difficulty including: preparation difficulties, supporting equipment, preparation and application time; efficiency application and removal complexity; temperature and humidity (moisture meter, Stanley® STHT 77030, Stanley Black & Decker, Victoria, Australia) efficacy, and homogeneity; economic evaluation (material costs, energy consumption). After their evaluation, the best performing delivery systems and the best performing thermic systems were applied in situ to the whole chapel pavement (Section 2.4). Bio-desalination was monitored over time (after 1 and 12 months) by efficiency evaluation (Section 2.5) and visual cleaning was measured by digital image analysis (Section 2.6).

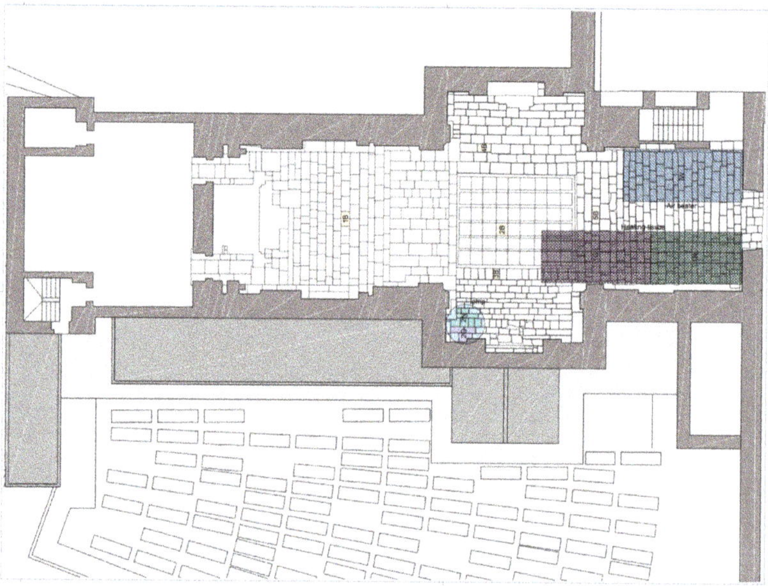

**Figure 1.** Cristo chapel plan (provided by Raquel Otero and Lourdes Pérez) showing the granite slabs. In situ preliminary test areas are marked. Tested thermic systems (lamp, air heater, and heating mat) and tested delivery systems; N: Nevek®; W: warm 2% agar; G: ground 2% agar. Selected slabs (1B, 2B, 3B, 4B, 5B) used for final bio-desalination monitoring tests are marked in yellow.

### 2.5. High Scale Optimized Protocol

Bio-desalination of the whole chapel pavement was conducted with *Pseudomonas stutzeri* CECT 930T (prepared as described in Section 2.2) and with the optimized application protocol selected according to the best performing and most efficient delivery system and thermic system (Section 2.3).

The application protocol consisted of application of the bacterial suspension directly to the pavement with a sterile brush. Then, Japanese paper was applied to facilitate the removal of the delivery system. After the delivery system was placed onto the surface, a plastic film (food plastic wrap) was later applied to cover the application treatment to avoid water evaporation and drying of the agar. Finally, the thermic system was applied, and the bio-desalination treatment was left for 48 h.

After treatment, the thermic system was switched off and removed. The plastic film, delivery system, and Japanese paper were removed, and the treated surface was cleaned using a sponge humidified with deionized water to remove bacterial cells and agar rests from the treated surface. Then, the treated surface was left to dry naturally [16]. Finally, the bio-desalination treatment was monitored over time (as described in Sections 2.5 and 2.6).

All the processed were carried out with the use of personal protection equipment such as working gloves and lab coat. All the residuals from the bio-desalination treatment (including bacterial suspensions, supported equipment in direct contact with bacteria, etc.) were sanitized by addition of bleach or cleaning with ethanol 70%.

*2.6. Bio-Desalination Monitoring*

The high dimension bio-desalination of the church was monitored on 5 slabs corresponding to north (3B), south (4B), east (1B), west (5B), and center (2B) of the chapel (Figure 1).

Bio-desalination efficiency monitoring consisted of measuring, on each selected slab, salt reduction, nitrate–nitrite reduction, and microbial presence. Monitoring was evaluated 1 and 12 months after the bio-desalination treatment in order to evaluate short-term and long-term effects of the treatment.

Salt reduction was measured by ionic conductivity (HANNA DiST3 H198303 EC Tester, Hanna Instruments S.L., Eibar, Spain). A circular (7 cm diameter) 50 g cellulose paper (Arbocel® BC 1000, Kremer Pigmente, Germany) was applied to each slab. Deionized water (4:1 wt/wt) was added and left for 10 min. Then, the cellulose paper was introduced into a recipient, and deionized water was added (until a final weight of 90 g), left for 1 h, and liquid conductivity was measured [34]. The conductivity of deionized water (5.8 µS/cm media values approximately) was subtracted from the test data. Each ion conductivity measurement was completed in triplicate.

Nitrate reduction was measured by semi-quantitative colorimetric nitrate–nitrite tests (Quantofix, Macherery-Nagel GmbH & Co, Düren, Germany). Nitrate–nitrite tests were introduced in the abovementioned recipient for 1 s. Then, tests were left at room temperature for 1 min, and the strip color was compared to the manufacturer's color scale, allowing for the nitrate and nitrite content (ppm) to be determined. The Quantofix tests sensibility values are between 10–500 mg/L for nitrates and between 1–80 mg/L for nitrites.

Microbial presence was analyzed by sterile swab sampling before and after the bio-desalination treatment. Each swab was then suspended in sterile 0.9% NaCl and 0.5 mL of the suspension was placed on a petri dish with nutrient agar (for bacteria) and sabouraud chloramphenicol agar (for fungi) and incubated at 28 °C for 48 h and 5 days, respectively. Bacteria and fungi colonies were counted and expressed in CFU per 25 $cm^2$.

*2.7. Digital Analysis Cleaning Study*

In order to confirm that the bio-desalination treatment corresponds with pavement cleaning, the relative percentage of cleanliness achieved was estimated by digital analysis of the images obtained before and after treatment.

The digital image analysis was based on the specific software scheme and method in MatLab (Matrix Laboratory) [8] with the necessary modifications to estimate the percentage of cleanliness from a cleaned slab selected as ideally clean.

For each five selected slabs (1B, 2B, 3B, 4B, 5B), three images were analyzed: before, after, and follow-up (one year after) treatment.

The images were captured as standardized as possible with a digital camera (NIKON D5100) under fixed conditions (ISO100, f/8.0, 1/250 s, focal length 19 mm) on a tripod and the same blue cardboard template, as a reference for the lighting correction. An automatic segmentation by thresholding of the Hue, Saturation, and Value (HSV) components, a morphological processing to obtain the masks of the area of interest and reference, as well as a correction of the luminosity of the images by homogenizing the images after the treatment with those before it, were carried out.

Finally, the results obtained were shown by means of different graphical representations, calculating the median, standard deviation, and outliers for each image. A cleaned slab, representing an ideally clean slab, and a corresponding slab before treatment were chosen as a common reference of 100% and 0% cleanliness, respectively.

## 3. Results

### 3.1. Selection of Best Bacterial Denitrification Conditions

As shown in Table 1, the process of denitrification was absent at 4 °C temperature at the three studied times, because bacterial growth was almost inhibited. At 20 °C, the denitrification process needed 48 h incubation time to occur, while at 26 °C the process started after a 24 h incubation period. This denitrification process was noticed by the reduction of nitrates and nitrite content and the formation of gas in Durham chambers.

**Table 1.** Denitrification of *P. stutzeri* 930T (CECT) at 4 °C, 20 °C, and 26 °C.

| Time (h) | Temperature (°C) | Turbidity (+/−) | CFU $mL^{-1}$ | Nitrates (mg $L^{-1}$) | Nitrites (mg $L^{-1}$) | Gas (+/−) |
|---|---|---|---|---|---|---|
| 24 | 4 | − | $2.0 \times 10^1$ | 500 | 23 | − |
| | 20 | − | $2.0 \times 10^2$ | 500 | 18 | − |
| | 26 | + | $8.0 \times 10^3$ | 0 | 77 | + |
| 48 | 4 | − | $1.0 \times 10^2$ | 500 | 24 | − |
| | 20 | + | $3.0 \times 10^6$ | 0 | 63 | + |
| | 26 | + | $1.0 \times 10^5$ | 0 | 64 | + |
| 72 | 4 | − | $1.9 \times 10^2$ | 500 | 26 | − |
| | 20 | + | $3.0 \times 10^6$ | 0 | 66 | + |
| | 26 | + | $3.0 \times 10^6$ | 0 | 67 | + |

+, presence of turbidity in the 10 mL nitrate broth medium tubes; or presence of nitrogen gas formation in the Durham chamber. −, absence of turbidity in the 10 mL nitrate broth medium tubes; or absence of nitrogen gas formation in the Durham chamber. CFU $mL^{-1}$, colony forming units of *P. stutzeri* in the nitrate broth medium tubes.

These results agree with other authors' results that demonstrate that *P. stutzeri* needs between an 8 to 24 h incubation period to start denitrification and then was able to complete nitrate elimination of 3 g/L (as $KNO_3$) in just 48 h [28].

Negative controls at the three studied temperatures and times showed constant nitrate and nitrite levels (500 ± 0 mg $L^{-1}$ and 20 ± 6 $mg^{-1}$, respectively) and absence of gas and microbial growth (negative turbidity and absence of CFU $mL^{-1}$).

### 3.2. Optimal High Scale Application Protocol Determination

The bio-desalination application protocol was implemented in order to be able to efficiently treat the 233 $m^2$ of the pavement. This was based on the bio-desalination application protocol described in the preliminary small tests conducted in the church [8]. That protocol defined the application time (48 h), the delivery system (2% agar), and the optimal pavement temperature for bacterial denitrification (26 ± 4 °C). While up-scaling the process, two critical points were identified and optimized: (i) the delivery system was applicable and easy to use on big, in situ surfaces; (ii) thermic systems were constant and adequate for maintaining homogeneous pavement temperature during treatment.

According to previous studies, agar was selected as a delivery system [8,30], but three different types of agar were tested: warm 2% agar, ground 2% agar, and Nevek® (Figure 2A, Figure 2B, and Figure 2C, respectively).

From the three delivery systems tested, warm 2% agar was the first to be tested, according to previous results conducted in the chapel on a smaller scale [8]. When up-scaling the in situ application dimensions, warm 2% agar showed the highest high dimension in situ performance difficulties (Table 2). This implies very high preparation difficulties that included the need for many supporting pieces of equipment. Once the agar is weighed and diluted in water, it must be boiled; therefore, a microwave or electric heater is needed. Then, before adding the warm agar to the pavement, the warm agar must be adjusted to about 45 °C, the application temperature. Adjusting the agar temperature is a very delicate task because there is a small temperature range where it can be correctly used in biocleaning; if it is too hot, bacteria will die, and if it gets too cold, it will jellify and will not be correctly adapted to the rough granite surface. Therefore, this process must be constantly monitored by a thermometer, and warm agar must be mixed constantly in order to allow for a correct homogeneous temperature. This is a very personnel-consuming process because two persons must work together in order to properly prepare and apply the warm agar. This high time-consuming process can be accelerated by cooling the agar with a cooled (with ice and salt) water bath. During this process (heating and cooling of the agar), a lot of water vapor is generated, which is detrimental to the environment of the church, which already has a high level of humidity. The application is also a delicate step because it must be quickly applied (in order to avoid the agar from jellifying) and, due to its liquid state, it can go over the treatment area. In order to limit the treatment area and avoid the warm agar from entering the gaps between the slabs, cellulose can be added to restrict the border (see Figure 2A), but it must be removed just after the agar application because if it is left on the pavement for a few days, fungi will grow on it due to the high air humidity. All these performance difficulties significantly reduced the quantity of square meters able to be treated in a day, up to 15 $m^2$, and augmented the economic costs. Personnel costs were the highest evaluated because there was a need for two persons working together for 2.4 h in order to bio-desalinate 1 $m^2$. Material costs included higher agar requirements (152 $g/m^2$) compared to 2% ground agar (52 $g/m^2$) because a significant quantity of agar jellified on the different recipients used in the preparation and application process. Due to all of the high in situ performance difficulties observed, warm 2% agar was discarded for high dimension bio-desalination, and ground 2% agar and Nevek® were tested next.

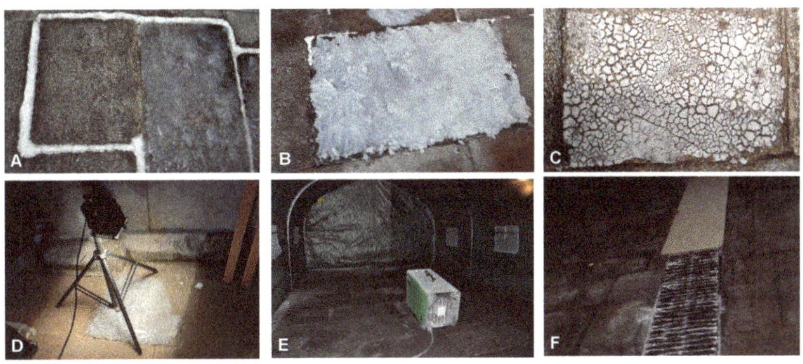


**Figure 2.** Image showing the delivery systems and thermic systems tested. (**A**): warm 2% agar; (**B**): ground 2% agar; (**C**): Nevek®; (**D**): infrared heat lamps; (**E**): air heater; (**F**): heating electric mat.

**Table 2.** High scale onsite bio-desalination application protocol evaluation.

| | | In Situ Performance Difficulties | | | | | | Economic Evaluation | |
|---|---|---|---|---|---|---|---|---|---|
| | | Preparation Difficulties | Supporting Equipment | Time-Consuming (1) | Efficiency (2) | Complexity (3) | Treatment Area/Day | Costs €/$m^2$ (4) Material (5) | Personnel (6) |
| Delivery system | Warm 2% agar | very high | very high | very high | very high | medium | 15 $m^2$ | EUR 11.91 | EUR 104.98 |
| | Ground 2% agar | low | low | medium | very high | low | 30 $m^2$ | EUR 7.42 | EUR 18.37 |
| | Nevek | low | low | low | medium | low | 30 $m^2$ | EUR 70.66 | EUR 14.00 |
| Thermic systems | Infrared heat lamps | high | high | medium | very low | high | 15 $m^2$ | EUR 3.54 | EUR 2.53 |
| | Air heater | high | high | high | low | high | 15 $m^2$ | EUR 7.40 | EUR 1.56 |
| | Heating electric mat | medium | medium | medium | very high | medium | 25 $m^2$ | EUR 13.49 | EUR 0.98 |

Evaluation criteria: absence; low; medium; high; very high; (1) Includes preparation, application, and removal time (2) Efficiency is related to the correct, homogeneous and constant water and temperature needed to be achieved by the thermic systems and the delivery system (3) Includes application and removal complexity (4) Costs do not include company overheads, industrial profit, or VAT. (5) Refers only to the materials needed for the application, including the delivery and thermic systems and the complementary materials (cellulose, deionized water, plastic film, Japanese paper, wood panels, hothouse, etc. and does not include bacterial cells) for the delivery system; and in the case of the thermal systems, it refers to the cost of renting the equipment and its power supply. (6) Personnel costs for the whole process including preparation, application, and removal for the delivery system and assembly and disassembly of the thermic systems.

Ground 2% agar significantly reduces the high dimension, in situ performance difficulties because it is directly applied to the pavement without any in situ preparations (agar preparation is completed previously in the laboratory). There is no need for in situ supporting equipment, and application and removal is easy and with medium time-consumption. Higher areas, up to 30 $m^2$, can be treated in a day, and economic costs are reduced both in terms of material and personnel costs (0.84 h for 1 $m^2$). Similar performance occurs with Nevek®, which also has low preparation difficulties (because it is applied directly from the original package), low supporting equipment, low time-consumption, and low application and removal complexity, also allowing for a 30 $m^2$ treatment area in a day. Economic costs are higher because, even though the personnel costs are low (0.64 h for 1 $m^2$), the material costs are very high as 3.3 kg of product is needed to bio-desalinate 1 $m^2$.

In regard to the delivery system efficiency, both warm and ground agar showed very high efficiency by supplying high (100% humidity) and homogeneous water to the surface (as needed for the microorganisms to clean). Instead, Nevek®, after the 48 h treatment time, showed excessive drying and the presence of significant cracks that increased when thermic systems were added (see Figure 2C). This could interfere with the correct and homogeneous biocleaning process and even complicate the removal. In order to avoid this, a thicker layer of Nevek® could be added, which implies an increase in economic costs.

These preliminary in situ delivery system trials allowed us to discard the use of warm 2% agar for high dimension surfaces bio-desalination application due to the high in situ performance difficulties observed, and to discard Nevek® due to the high economic costs and excessive drying. Ground 2% agar was selected as the best performing delivery system.

In addition, three thermic systems were tested: infrared heat lamps, an air heater, and a heating electric mat (Figure 2D, Figure 2E, Figure 2F, respectively). Infrared heat lamps were selected because they have previously been used in biocleaning studies [8,13,15–17]. The other thermic systems, tested here for the first time, were selected due to their potential ability to warm high dimension areas. Again, in situ performance difficulties, thermal efficiency, and economic costs were evaluated and compared (Table 2).

From the three thermic systems tested, the air heater had the highest in situ performance difficulty (see Table 2). Preparation and removal was tedious, complex, time consuming, and significant supporting materials were needed (a hothouse had to be installed on the treated area in order to reduce the volume of air to be warmed, which complicated the application on non-orthogonal areas and reduced the quantity of square meters able to be treated in a day, up to 15 $m^2$ in this case). Thermal efficiency was very low because the system needed about 24 h to reach the required average air temperature of 26 ± 6 °C. Additionally, the temperature was not homogeneous on the treated area, with high temperatures (about 30 °C) on the areas near the air heater and low temperatures (about 20 °C) on the areas far from the air heater. In addition, it was found that the heating of the hothouse air caused an increase in its capacity to contain water vapor and caused an increase in soil evaporation, dragging the existing salts inside the slabs towards its surface, and producing, again, darkening of the slabs. Economic evaluation showed a medium material cost/$m^2$ and a lower personnel cost/$m^2$.

Similar high dimension drawbacks were observed with the infrared heat lamp. The lamps had to be switched on 14 h before treatment in order to get the pavement temperature to 26 °C, but after 24 h treatment, the pavement temperature increased to 29 °C, and at the end of the treatment (48 h), the pavement temperature was 35 °C. In addition, the temperature was not homogeneous on the area illuminated with the lamps because it depended on the light beam, which was stronger in the central area and weaker at the edges. This inhomogeneous temperature can influence the bacterial desalination (even though the bacteria are viable within those temperatures, the enzymatic kinetics augment with temperature) and the water content of the delivery system by drying it completely in some areas and making its removal much more difficult [28]. The bio-desalination process could also completely stop. The in situ preparation performance on high dimensions was also difficult, with medium time-consumption, and needing supporting equipment.

The system needed a structure to be assembled (like a mobile scaffolding) to allow for the assembly of different lamps that covered a bigger area. Both systems, due to their limitations, did not allow for a homogenous, high treatment area to be completed in a day. Economic evaluation shows a lower costs of material but a higher personnel costs.

The heating electric mat showed medium onsite performance difficulties because the pavement had to be covered with the heating mat 24 h previous to treatment, and it had to be raised in order to add the treatment. On the top of the mat, a thermal insulating material had to be added in order to avoid temperature loss. Nevertheless, it showed a very high thermic efficiency (constant 26 °C temperature was measured on the granite pavement surface) and allowed for coverage over large areas. The mat was easily and perfectly adapted to the church space because the mat could be cut according to the dimensions of the space (as long as the electric radiant wire was not cut). The mat was the most versatile, easy, and comfortable to use in high dimension treatments. The system allowed one person to perform a 25 $m^2$ treatment area per day. The total costs per square meter were about double the costs of the lamps and the air heater, but the thermic benefits and the convenience and simplicity of its in situ performance were considered worthwhile.

According to these results, the high dimension bio-desalination protocol was optimized by selecting ground 2% agar in combination with a heating electric mat as the best performing delivery system and thermic system. Therefore, the optimized protocol (Figure 3) was applied to the 233 $m^2$ of the chapel pavement.

### 3.3. *Efficiency of High Dimension Bio-Desalination*

Before and after the bio-desalination treatment, ion conductivity, nitrate–nitrite reduction, and microbial presence was measured for the selected five slabs.

One month after the bio-desalination high dimension treatment, a significant salt reduction was observed in all of the slabs (Table 3), with mean values of 31.1%, which was maintained in most of the slabs during the following 12 months (follow-up).

**Figure 3.** High dimension bio-desalination optimized protocol applied to the 233 $m^2$ church granite pavement. In lab preparation: (**a**) bacterial biomass production; (**b**) 2% ground agar preparation. In situ application: (**c**) bacterial application; (**d**) 2% ground agar application over the Japanese paper; (**e**) thermic mat applied and insulating wood panels. After treatment: removal of (**f**) wood panels; (**g**) thermic mat; (**h**) plastic film, ground agar, and Japanese paper; (**i**) final cleaning with deionized water and sponge.

**Table 3.** High dimension bio-desalination treatment Ion conductivity monitoring.

| Treated Slab | Ion Conductivity (µS/cm) | | |
|---|---|---|---|
| | Before | After [1] | Follow-Up [2] |
| 33 | 659 | 387 | 350 |
| 40 | 413 | 208 | 303 |
| 41 | 423 | 300 | 390 |
| 45 | 287 | 252 | 200 |
| 46 | 282 | 216 | 350 |
| Mean values | 412.8 | 272.6 | 318.6 |
| Standard deviation | 153.0 | 73.6 | 73.1 |

[1] One month after bio-desalination treatment. [2] Twelve months after bio-desalination treatment.

In order to determine if the salt reduction observed was correlated to a nitrate and nitrite reduction, an in situ analysis was performed on the same slabs before and one month after the bio-desalination treatment (Table 4). Initial values for all of the slabs show the maximal value measured by the test strips (500 mg/L for nitrates and 80 mg/L for nitrites). After the bio-desalination, all slabs showed a reduction in nitrate and nitrite values, with 30% and 60% statistically significant reduction rates, respectively ($p$-values of 0.04 and 0.0001, respectively). Twelve months after the treatment, the slabs showed an increased reduction in nitrate and nitrite values of up to 58% and 100% statistically significant reduction rates, respectively ($p$-values of 0.000044).

**Table 4.** High dimension bio-desalination treatment nitrate–nitrite reduction monitoring.

| Treated Slab | Nitrate–Nitrite Reduction (mg/L) | | | | | |
|---|---|---|---|---|---|---|
| | Before | | After [1] | | Follow-Up [2] | |
| | Nitrates | Nitrites | Nitrates | Nitrites | Nitrates | Nitrites |
| 33 | 500 | 80 | 250 | 20 | 250 | 0 |
| 40 | 500 | 80 | 500 | 40 | 250 | 0 |
| 41 | 500 | 80 | 250 | 40 | 250 | 0 |
| 45 | 500 | 80 | 250 | 20 | 50 | 0 |
| 46 | 500 | 80 | 500 | 40 | 250 | 0 |
| Mean values | 500 | 80 | 350 | 32 | 210 | 0 |
| Standard deviation | 0.0 | 0.0 | 122.3 | 9.8 | 80.0 | 0 |

[1] One month after bio-desalination treatment. [2] Twelve months after bio-desalination treatment.

The microbial (bacteria and fungi) presence was also evaluated on the five slabs before and after the bio-desalination treatment. The results showed absence of fungi before and after the bio-desalination treatments and bacteria mean values of 4.7 ± 2.3 UFC/25 cm$^2$ before treatment and 2 ± 2 UFC/25 cm$^2$ after treatment. The difference observed for the bacteria was not significant ($p > 0.05$), indicating that this biocleaning treatment did not significantly alter the CFU content that was present on the chapel pavement. These results agree with the preliminary analysis carried out on the same pavement where contact plates showed very low microbial counts [8].

### *3.4. High Dimension Bio-Desalination Visual Cleaning and Digital Imaging Analysis*

After the bio-desalination treatments, a cleaner and clearer surface was visually observed (Figure 4), and no damage of the stone surface was observed. In order to transfer these visual observations into quantifiable data, digital image analysis was used. Digital analysis showed median cleaning percentages of 18.12% when comparing before and 1 month after the bio-desalination treatment. The cleaning effect was augmented after the 12 months follow-up, reaching median cleaning percentages of 37.78% (Table 5, Figure 5).

**Figure 4.** Detail of a church granite pavement 25 $m^2$ treated area following the high dimension bio-desalination optimized protocol. Area before treatment (**left image**), area during treatment (**middle image**), and area after treatment (**right image**).

**Table 5.** Cleaning percentage of the five slabs selected for high dimension bio-desalination monitoring compared with a 100% cleaned (ideally cleaned) slab.

| Median ± std (%) | 1B | 2B | 3B | 4B | 5B | Media |
|---|---|---|---|---|---|---|
| After | 17.80 ± 21 | 8.73 ± 28 | 24.08 ± 34 | 26.30 ± 28 | 13.72 ± 22 | 18.12 ± 27 |
| Follow-up | 28.70 ± 18 | 37.05 ± 25 | 48.92 ± 32 | 56.41 ± 33 | 17.85 ± 22 | 37.78 ± 26 |

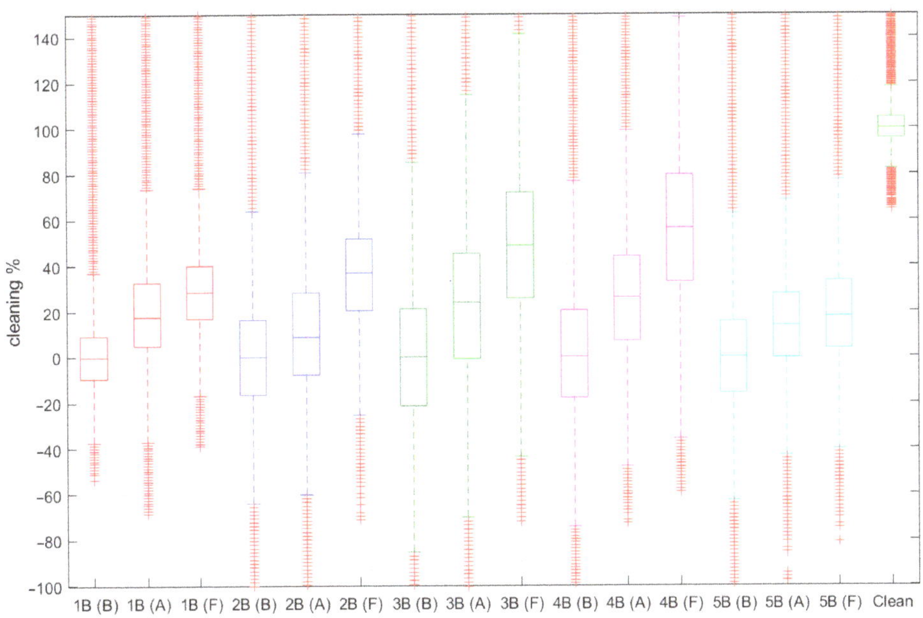


**Figure 5.** Boxplot showing median, standard deviation, and outliers of the cleaning percentage of the five slabs selected for high dimension bio-desalination of the church monitoring. All values have been adjusted with the control median before treatment (0%) and with the clean slab selected as ideally clean (100%). For each slab, three measurements were completed: before treatment (B), 1 month after treatment (A), and 12 months after treatment follow-up (F).

## 4. Discussion

Salt weathering is a global and important deterioration problem for high dimension CH porous building materials that causes both physical and aesthetic damage and leads to significant cultural and economic consequences. Therefore, in the field of conservation and restoration, many multidisciplinary research teams are focused on understanding this complex and extensive problem in order to find ways to reduce damage caused by salts [34,35]. The most logical and common solution to reduce salt weathering is to reduce the stone salt content [36]. In this sense, bio-desalination has emerged as one possible green conservation solution for this widespread problem by showing real CH applications with good results on small dimension samples.

Due to the fact that the salt problems usually affect high dimension CH surfaces, an in situ scale-up implementing study was conducted here.

When working in situ in a CH building, a church in this particular case study, many important limitations are encountered that do not typically appear when restorations are performed in a laboratory. For example, room temperature cannot be fixed, energy supply/power is frequently low, supporting equipment must be moved in situ, etc. These issues lead to an increase in preparation, application, and removal difficulties, an increase in process time, a reduction in efficiency, a reduction in the treatment area/day, and an increase in costs.

In this study, the scale-up problems that appear when bio-desalination is applied on- site to high dimension CH surfaces have been identified, evaluated, and overcome. The scale-up problems here identified and evaluated are related to: in situ performance difficulties, efficiency, and economic costs.

In order to reduce in situ high dimension bio-desalination application difficulties, the following suggestions can be followed:

- Reduce in situ material preparation. Any needed preparations must first be completed in the lab to reduce in situ preparation time and equipment.
- Use versatile materials, allowing them to be adapted to the building architectural needs.
- Select materials and protocols that allow a high treatment area per day and self-handling in order to reduce economic costs.
- Use methods that modify the interior environmental conditions of the monument as little as possible.
- Use electricity generation methods that are compatible with the low electrical power generally supported by the monuments and avoid impact on the surrounding environment (noise, gases, etc.).

In this particular case study, after the evaluation of the critical points identified on the in situ high dimension application protocol, the optimized protocol selected was the use of 2% ground agar and a heating electric mat (Figure 4).

This protocol has low–medium preparation difficulty, use of supporting materials, application complexity, time consumption, and economic costs, and high thermic and water system efficiency as well as good bio-desalination efficacy.

The optimized protocol was applied to the 233 m$^2$ granite pavement, allowing a general salt reduction of up to 31.1%. In particular, as denitrified bacteria were used, nitrates were evaluated, and a 30% reduction was observed. The visual cleaning observations and measurements by digital analysis when compared with an "ideally clean" slab, showed an 18.12% media cleaning percentage, which is slightly lower than the salt removal. This agrees with previous results that showed that there is no direct relationship between salt reduction and visual cleaning [8]. It is important to indicate that high variability (high standard deviation) was observed on the treated slabs for all of the conducted analyzes. This agrees with other authors who indicated that building material desalination typically results in non-proportional, non-linear, and non-homogeneous salt removal [37,38]. Other authors indicated that controlling desalination is a very difficult task that can achieve results that are "contradictory and sometimes even undesirable on large monuments" [39].

In order to evaluate the long-term effects of the bio-desalination treatment, the same slabs were analyzed one year after the treatment. Ion conductivity results after one year showed very similar media values; this agrees with other authors who indicated that constant salt content was a sign that the desalination process had terminated [40]. But it is interesting that the analyzed slabs showed, one year after the bio-desalination treatment, an increase in nitrate reduction (up to 58% median values) and an increase in visual cleaning percentage (up to 37.78% median values). This visual cleaning and nitrate reduction increase over a one year period could be related to environmental conditions, slab salt mixture variations after the bio-desalination process, or even by long-term dry bio-desalination.

Salt behaviour is a very complex system, and multiple variables, such as salt type and quantity (supersaturation), substrate properties (e.g., porous size), and environment (temperature and humidity), are involved [36]. It is remarkable that the one year period coincided with the COVID-19 pandemic that resulted in the closing of the church and chapel for a few months and then reopened with an important restriction on the number of visitors. This could have influenced the chapel environment and, therefore, the surface salt content. Some authors have demonstrated that visitors' exhalations and corporal temperatures are able to modify salt crystallization patterns on artistic materials [41,42]. In addition, different field observations have shown how environmental variations in non-heated rooms are associated with salt crystallization behaviours [43,44].

On the other hand, the bio-desalination process modifies the slab surfaces by removing an important percentage of nitrates and, therefore, modifies the previously existing salt mixtures. This slab superficial salt variation could have altered the salt behaviour in relation with the environment and produced the observed cleaning increase and nitrate reduction after one year. As described by other authors, in stone materials with salt mixtures, the particular salt combination composition is important to surface decay [45]. Some authors have demonstrated that changing stone surface parameters, such as surface tension using surfactants, directly affects salt crystallization [46]. Diverse species of the genus *Pseudomonas* have been reported to produce bio-surfactants, which might contribute to this long-term effect [47].

Another explanation of this phenomena could be related to remaining denitrifying bacteria that are active and produce a low-rate and long-term dry bio-desalination process. Recent publications have shown how other microorganisms (*Saccharomyces cerevisiae*) can be used to bioclean in situ stonework by using only environmental humidity and temperature [32].

Future research is needed to investigate the particular mechanisms that could be related to this observed increase in long-term nitrate–nitrite reduction, taking special consideration of the slab surface: microbiological analysis, thermohygrometric analysis, and salts analysis.

This work has contributed to the improvement and quantification of desalination evaluation by the use of non-destructive cleaning techniques based on digital imaging. Digital imaging is a non-destructive tool that is easy to use in situ and that allows for the process to be recorded and followed. Diverse efforts are shown in the field to use digital image analysis as a cheap non-destructive method for quantification of what experts can see by the naked eye [8,48–50]. In this work, a step forward has been achieved by standardizing a protocol that allows, by comparing with an "ideally cleaned" slab image, a relative percentage of cleaning to be obtained. Digital imaging is shown to be a useful method for verifying the effectiveness of biocleaning, which is complementary to the traditional salts and nitrates measurements. This digital image analysis could be used for other types of cleaning techniques and other types of cleaning areas in order to establish the degree of cleaning.

## 5. Conclusions

This application oriented research has taken into account the practical aspects conservators face in their in situ daily work, such as materials choice, the practical needs of the high dimension CH building, selecting the most efficient, ease of use application and removal, and good cost/benefit ratios. This practical analysis has enabled, for the first time, the development and application of a bio-desalination protocol that is able to reduce the salt content of the high dimension granite pavement of Santa Maria del conxo Chapel. This innovative application protocol could be transferable to other large scale in situ biocleaning strategies. Furthermore, digital image analysis tools have been improved in order to convert the visual observations of cleaning into a quantitative cleaning percentage.

New frontiers of intervention are opening, and the obtained results contribute by giving conservators a new choice of a green desalination treatment procedure suitable for high dimension restoration sites.

**Author Contributions:** Conceptualization, P.B.-R. and L.P.-C.; Experimental procedures and methodology on bio-desalination P.B.-R. and Á.F.-S.; bio-desalination investigation, formal analysis, and data curation P.B.-R.; Digital image software design, validation, and analysis I.B.; writing—original draft preparation and visualization P.B.-R., L.P.-C., Á.F.-S., and I.B.; writing—review and editing, P.B.-R.; supervision, project administration, P.B.-R. and L.P.-C.; funding acquisition, P.B.-R., L.P.-C., and I.B. All authors have read and agreed to the published version of the manuscript.

**Funding:** This research was funded by CONSORCIO DE LA CIUDAD DE SANTIAGO, which promotes intervention studies in this building.

**Acknowledgments:** The authors wish to thank the Instituto Universitario de Restauración del Patrimonio (IRP) and Instituto de tecnologías y Aplicaciones Multimedia for their support. Authors wish to give special thanks to Mari Cruz Rochina Peñalver from the Instituto de Biología Molecular y Celular de Plantas (IBMCP) from the Universitat Politècnica de Valencia for their technical support and for allowing us to use their high-performance centrifuge for this project.

**Conflicts of Interest:** The authors declare no conflict of interest.

## References

1. Bosch-Roig, P.; Ranalli, G. Biocleaning of Cultural Heritage treasures. In *Biodeterioration and Preservation of Cultural Heritage Treasures;* Mitchel, R., Ed.; Archetype Publications: London, UK, 2018; pp. 169–183.
2. Bosch-Roig, P.; Ranalli, G. The safety of biocleaning technologies for Cultural Heritage. *Front. Microbiol.* **2014**, *5*, 155. [CrossRef]
3. Ranalli, G.; Zanardini, E. Biocleaning on Cultural Heritage: New frontiers of microbial biotechnologies. *J. Appl. Microbiol.* **2021**, 1–21. [CrossRef]
4. Lombardi, E.; Balloi, A.; Troiano, F.; Gulotta, D.; Polo, A.; Gioventù, E.; Sorlini, C.; Cappitelli, F.; Daffonchio, D. Strategies for increasing the scale of biocleaning treatment through sulfate crust bioremoval. In Proceedings of the Built Heritage 2013 Monitoring Conservation Management, Milano, Italy, 18–20 November 2013; Politecnico di Milano: Milano, Italy, 2013; pp. 1424–1430.
5. Romano, I.; Abbate, M.; Poli, A.; D'Orazio, L. Bio-cleaning of nitrate salt efflorescence on stone samples using extremophilic bacteria. *Sci. Rep.* **2019**, *9*, 1–11. [CrossRef]
6. Soffritti, I.; D'Accolti, M.; Lanzoni, L.; Volta, A.; Bisi, M.; Mazzacane, S.; Caselli, E. The Potential Use of Microorganisms as Restorative Agents: An Update. *Sustainability* **2019**, *11*, 3853. [CrossRef]
7. Gioventù, E.; Lorenzi, P.F.; Villa, F.; Sorlini, C.; Rizzi, M.; Cagnini, A.; Griffo, A.; Cappitelli, F. Comparing the bioremoval of black crusts on colored artistic lithotypes of the Cathedral of Florence with chemical and laser treatment. *Int. Biodeterior. Biodegrad.* **2011**, *65*, 832–839. [CrossRef]
8. Bosch-Roig, P.; Allegue, H.; Bosch, I. Granite Pavement Nitrate Desalination: Traditional Methods vs. Biocleaning Methods. *Sustainability* **2019**, *11*, 4227. [CrossRef]
9. Cappitelli, F.; Zanardini, E.; Toniolo, L.; Abbruscato, P.; Ranalli, G.; Sorlini, C. Bioconservation of the marble base of the Pietà Rondanini by Michelangelo Buonarroti. *Geophys. Res. Abstr.* **2005**, *7*, 06675.
10. Polo, A.; Cappitelli, F.; Brusetti, L.; Principi, P.; Villa, F.; Giacomucci, L.; Ranalli, G.; Sorlini, C. Feasibility of Removing Surface Deposits on Stone Using Biological and Chemical Remediation Methods. *Microb. Ecol.* **2010**, *60*, 1–14. [CrossRef] [PubMed]
11. Troiano, F.; Gulotta, D.; Balloi, A.; Polo, A.; Toniolo, L.; Lombardi, E.; Daffonchio, D.; Sorlini, C.; Cappitelli, F. Successful combination of chemical and biological treatments for the cleaning of stone artworks. *Int. Biodeterior. Biodegrad.* **2013**, *85*, 294–304. [CrossRef]
12. Martino, M.; Schiavone, S.; Palla, F.; Pellegrino, L.; De Castro, E.; Balloi, A. Bioremoval of sulphate layer from a 15th century polycrome marble artifact. *Conserv. Sci. Cult. Herit.* **2015**, *15*, 235–243.

13. Bosch-Roig, P.; Regidor-Ros, J.L.; Soriano-Sancho, P.; Doménec-Carbó, M.T.; Montes-Estelle s, R.M. Ensayos de biolimpieza con bacterias en pinturas murales. *Arché* **2010**, *4–5*, 115–122.
14. Lustrato, G.; Alfano, G.; Andreotti, A.; Colombini, M.P.; Ranalli, G. Fast biocleaning of mediaeval frescoes using viable bacterial cells. *Int. Biodeterior. Biodegrad.* **2012**, *69*, 51–61. [CrossRef]
15. Bosch-Roig, P.; Regidor-Ros, J.L.; Soriano-Sancho, P.; Montes-Estellés, R.M. Biocleaning of animal glue on wall paintings by *Pseudomonas stutzeri*. *Chem. Today* **2013**, *31*, 50–53.
16. Roig, P.B.; Ros, J.L.R.; Montes-Estelles, R.M. Biocleaning of nitrate alterations on wall paintings by *Pseudomonas stutzeri*. *Int. Biodeterior. Biodegrad.* **2013**, *84*, 266–274. [CrossRef]
17. Metaldi, S. Applicazione sperimentale di batteri solfato riduttori sul pigmento azzurrite. In Proceedings of the I batteri nel restauro, Thiene, Italy, 28 September 2013.
18. Mazzoni, M.; Alisi, C.; Tasso, F.; Cecchini, A.; Marconi, P.; Sprocati, A.R. Laponite micro-packs for the selective cleaning of multiple coherent deposits on wall paintings: The case study of Casina Farnese on the Palatine Hill (Rome-Italy). *Int. Biodeterior. Biodegrad.* **2014**, *94*, 1–11. [CrossRef]
19. Balloi, A.; Lombardi, E.; Troiano, F.; Polo, A.; Cappitelli, F.; Gulotta, D.; Toniolo, L.; Lucchini, A. Sulfate reducing bacteria as bio-cleaning agents: Development of new methodologies and study cases. *Conserv. Sci. Cult. Herit.* **2015**, *15*, 109–119.
20. Panella, E.; Giovannone, C.; Bartolini, M.; Fondi, V. Biotechnology in the conservation field: Removal of sulphates using bacteria and bioconsolidation of paintings and stuccos. *Conserv. Sci. Cult. Herit.* **2019**, *19*, 157–175.
21. Ranalli, G.; Zanardini, E.; Rampazzi, L.; Corti, C.; Andreotti, A.; Colombini, M.P.; Bosch-Roig, P.; Lustrato, G.; Giantomassi, C.; Zari, D.; et al. Onsite advanced biocleaning system for historical wall paintings using new agar-gauze bacteria gel. *J. Appl. Microbiol.* **2019**, *126*, 1785–1796. [CrossRef]
22. Alfano, G.; Lustrato, G.; Belli, C.; Zanardini, E.; Cappitelli, F.; Mello, E.; Sorlini, C.; Ranalli, G. The bioremoval of nitrate and sulfate alterations on artistic stonework: The case-study of Matera Cathedral after six years from the treatment. *Int. Biodeterior. Biodegrad.* **2011**, *65*, 1004–1011. [CrossRef]
23. Rampazzi, L.; Andreotti, A.; Bressan, M.; Colombini, M.P.; Corti, C.; Cuzman, O.; D'Alessandro, N.; Liberatore, L.; Palombi, L.; Raimondi, V.; et al. An interdisciplinary approach to a knowledge-based restoration: The dark alteration on Matera Cathedral (Italy). *Appl. Surf. Sci.* **2018**, *458*, 529–539. [CrossRef]
24. Ranalli, G.; Alfano, G.; Belli, C.; Lustrato, G.; Colombini, M.P.; Bonaduce, I.; Zanardini, E.; Abbruscato, P.; Cappitelli, F.; Sorlini, C. Biotechnology applied to cultural heritage: Borestoration of frescoes using viable bacterial cells and enzymes. *J. Appl. Microbiol.* **2005**, *98*, 73–83. [CrossRef] [PubMed]
25. García Morales, S.; Otero Ortiz de Cosca, R.; Allegue Castelos, H. *Investigación Sobre El Oscurecimiento Húmedo Que Afecta Al Enlosado De La Capilla Del Cristo De Santa María De Conxo*; Cuadernos técnicos; Consorcio de Santiago Oficina Técnica: Santiago de Compostela, Spain, 2016. Available online: https://issuu.com/consorciodesantiago/docs/conxo_oscurecimiento_humedo (accessed on 16 April 2021).
26. Ranalli, G.; Chiavarini, M.; Guidetti, V.; Marsala, F.; Matteini, M.; Zanardini, E.; Sorlini, C. The use of microorganisms for the removal of nitrates and organic substances on artistic stoneworks. In Proceedings of the 8th International Congress on Deterioration and Conservation of Stone, Berlin, Germany, 30 September–4 October 1996.
27. May, E.; Webster, A.M.; Inkpen, R.; Zamarreno, D.; Kuever, J.; Rudolph, C.; Warcheid, T.; Sorlini, C.; Cappitelli, F.; Zanardini, E.; et al. The BIOBRUSH Project for bioremediation of heritage stone. In *Heritage Microbiology and Science. Microbes, Monuments and Maritime Materials*; May, E., Jones, M., Mitchell, J., Eds.; RCS Publishing: Cambridge, UK, 2008; pp. 76–93.
28. Vidakovic, A.; Sovljanski, O.; Vucurovic, D.; Racic, G.; Djilas, M.; Curcic, N.; Markov, S. Novel denitrifying bacteria *Pseudomonas stutzeri* strain D1—From isolation to the biomass production. *Chem. Ind. Chem. Eng. Q.* **2019**, *25*, 403–415. [CrossRef]
29. Noble, R.C.; Overman, S.B. *Pseudomonas stutzeri* infection a review of hospital isolates and a review of the literature. *Diagn. Microbiol. Infect. Dis.* **1994**, *19*, 51–56. [CrossRef]
30. Atlas, R.M.; Chowdhury, A.N.; Gauri, K.L. Microbial calcification of gypsum-rock and sulfated marble. *J. Stud. Conserv.* **1988**, *33*, 149–153.
31. Bosch-Roig, P.; Lustrato, G.; Zanardini, E.; Ranalli, G. Biocleaning of Cultural Heritage stone surfaces and frescoes: Which delivery system can be the most appropriate? *Ann. Microbiol.* **2014**, *65*, 1227–1241. [CrossRef]
32. Ranalli, G.; Bosch-Roig, P.; Crudele, S.; Rampazzi, L.; Corti, C.; Zanardini, E. Dry biocleaning of artwork: An innovative methodology for Cultural Heritage recovery? *Microb. Cell* **2021**, *8*, 91–105. [CrossRef]
33. Yang, J.; Feng, L.; Pi, S.; Cui, D.; Ma, F.; Zhao, H.-P.; Li, A. A critical review of aerobic denitrification: Insights into the intracellular electron transfer. *Sci. Total. Environ.* **2020**, *731*, 139080. [CrossRef] [PubMed]
34. Franco, B.; Gisbert, J.; Mateos, I.; Navarro, P. Deterioro de los materiales pétreos por sales: Cinética del proceso, cartografía y métodos de extracción. In Proceedings of the I Congreso GEIIC, Valencia, Spain, 25–27 November 2002; pp. 287–294. Available online: https://www.ge-iic.com/wp-content/uploads/2006/06/Franco_Belen.pdf (accessed on 20 May 2021).
35. Randazzo, L.; Montana, G.; Castiglia, A.; La Russa, M.F. Salt extraction from lime-based mortars: An experimental study using different poultice formulations. Constr. *Build. Mater.* **2020**, *255*, 119391. [CrossRef]
36. Doehne, E. Salt weathering: A selective review. *Geol. Soc. Lond. Spéc. Publ.* **2002**, *205*, 51–64. [CrossRef]

37. Bromblet, P.; Verges-Belmin, V. The removal of sulfates from calcareous stone outdoor statuary: A questionable practice. In Proceedings of the Le dessalement des matériaux Poreux. 7es journées d'études de la SFIIC, Poitiers, France, 9–10 May 1996; pp. 55–63.
38. Charola, A.E. Salts in the deterioration of porous materials: An overview. *J. Am. Inst. Conserv.* **2000**, *39*, 327–343. [CrossRef]
39. Siedel, H. Experiences from desalting of tuffstone and sandstone monuments by compresses. In Proceedings of the Le dessalement des matériaux poreux. 7es journées d'études de la SFIIC, Poitiers, France, 9–10 May 1996; pp. 191–198.
40. Zornoza-Indart, A. Técnicas de desalación. In *La Conservación de los Geomateriales utilizados en el Patrimonio*; Programa Geomateriales: Madrid, Spain, 2012; pp. 143–154.
41. Wust, R.; McLane, J. Rock deterioration in the Royal Tomb of Seti I, Valley of the Kings, Luxor, Egypt. *Eng. Geol.* **2000**, *58*, 163–190. [CrossRef]
42. Wüst, R.A.; Schlüchter, C. The Origin of Soluble Salts in Rocks of the Thebes Mountains, Egypt: The Damage Potential to Ancient Egyptian Wall Art. *J. Archaeolog. Sci.* **2000**, *27*, 1161–1172. [CrossRef]
43. Arnold, A.; Zehnder, K. Decay of stony materials by salts in humid atmosphere. In Proceedings of the 6th International Congress on Deterioration and Conservation of Stone, Torun, Poland, 12–14 September 1988; Ciabach, J., Ed.; Copernicus University Press: Torun, Poland, 1988; pp. 138–148.
44. Cantón, Y.; Solé-Benet, A.; Queralt, I.; Pini, R. Weathering of a gypsum-calcareous mudstone under semi-arid environment at Tabernas, SE Spain: Laboratory and field-based experimental approaches. *Catena* **2001**, *44*, 111–132. [CrossRef]
45. Williams, R.B.G.; Robinson, D.A. Experimental frost weathering of sandstone by various combinations of salts. *Earth Surf. Process. Landf.* **2001**, *26*, 811–818. [CrossRef]
46. Rodriguez-Navarro, C.; Doehne, E.; Sebastian, E. Influencing Crystallization Damage in Porous Materials through the Use of Surfactants: Experimental Results Using Sodium Dodecyl Sulfate and Cetyldimethylbenzylammonium Chloride. *Langmuir* **2000**, *16*, 947–954. [CrossRef]
47. Mukherjee, A.K.; Das, K. Microbial Surfactants and Their Potential Applications: An Overview. In *Advances in Experimental Medicine and Biology*; Springer: New York, NY, USA, 2010; Volume 672.
48. Gazzano, C.; Favero-longo, S.E.; Matteucci, E.; Piervittori, R. Image analysis for measuring lichen colonization and within stonework. *Lichenologist* **2009**, *41*, 299–313. [CrossRef]
49. Miller, A.Z.; Rogerio-Candelera, M.A.; Dionísio, A.; Macedo, M.F.; Saiz-Jimenez, C. Microalgae, as biodeteriogens of stone cultural heritage: Qualitative and quantitative research by non-contact techniques. In *Microalgae: Biotechnology, Microbiology and Energy*; Johanssen, M.N., Ed.; Nova Science Publishers: New York, NY, USA, 2011; pp. 345–358.
50. Coutinho, M.L.D.; Miller, A.Z.; Gutierrez-Patricio, S.; Hernandez-Marine, M.; Gómez-Bolea, A.; Rogerio-Candelera, M.A.; Philips, A.; Jurado, V.; Saiz-Jimenez, C.; Macedo, M.F. Microbial communities on deteriorated artistic tiles from Pena National Palace (Sintra, Portugal). *Int. Biodeterior. Biodegrad.* **2013**, *84*, 322–332. [CrossRef]

*applied sciences*

MDPI

*Article*

# Varnishes with Biocidal Activity: A New Approach to Protecting Artworks

**Julio Romero-Noguera [1,*], Rafael Bailón-Moreno [2] and Fernando Bolívar-Galiano [3,*]**

[1] Department Painting, Faculty of Fine Arts, University of Seville, Laraña 3, 41003 Seville, Spain
[2] Department Chemical Engineering, Faculty of Sciences, University of Granada, Avenida de Fuente Nueva, s/n, 18071 Granada, Spain; bailonm@ugr.es
[3] Deparment Painting, Faculty of Fine Arts, University of Granada, Avenida de Andalucía s/n, 18014 Granada, Spain
* Correspondence: juliorn@us.es (J.R.-N.); fbolivar@ugr.es (F.B.-G.)

Received: 29 July 2020; Accepted: 12 October 2020; Published: 19 October 2020

**Abstract:** Keeping agents that cause biodeterioration under control is a common task in restoring and conserving artworks. However, there are very few studies, specifically those concentrating on pictorial works, that provide solutions compatible with the complex mix of organic and inorganic materials to be found in such cultural assets. This study aims to use biocide products that are commonly used in this field (benzalkonium chloride, o-phenylphenol, and tributyltin naphthenate) and which are associated with natural varnishes with a long tradition in artistic practices, in order to protect paintings and polychrome sculptures from environmental fungi and bacteria without altering the original materials or the visual appearance of the artworks. The effectiveness or innocuousness of the treatments was tested chemically and visually via optical microscopy, gas chromatography–mass spectrometry (GC-MS), and a multidimensional scaling analysis (MDS); the treatments produced good results as regards the inhibition of microbial growth and scarce interaction with the artistic materials being studied.

**Keywords:** biocides; terpenoid varnishes; painting; fungi; bacteria; biodeterioration; GC-MS; MDS

## 1. Introduction

Biological agents are one of the main causes of deterioration in cultural assets, and biocide products are among the solutions used to control them [1–4]. The use of these substances has been studied in significant papers focusing principally on treating ornamental stone, and these have been the main source of information on the matter until now [5–9]. Different products' action, toxicity, chemical stability, and interactions with the original materials have been evaluated in order to apply the treatments more effectively, meaning an approach that is innocuous, has a low-cost, and has low human and environmental toxicity for the artistic materials [10–15].

Studies focusing on paintings show the problems in applying biocide treatments to different types of polychromies, as well as the difficulty in eliminating microorganisms without altering the pictorial layer [16,17]. There are few products available for use on these cultural assets due to a lack of data about their necessary harmlessness towards the pigments and film-forming materials (binders and varnishes) that make up pictorial artworks, as well as the long-term effects on them [18]. The most commonly used biocides were not created specifically for use on cultural heritage; rather, they are disinfectants and pesticides for broad industrial usage. They fall within groups of chemicals such as the organometallic and nitrogen compounds, phenols, and quaternary ammonium derivatives [14,19–21].

There is also a rather limited range of methods for applying paint treatments [14]. The most common ones usually involve applying biocides in liquid form and fumigation with gases. One possible alternative that has been explored very little is to use materials with biocide incorporated. Restorers often carry out these tasks empirically, mixing them with water repellents or consolidants, but only in the area of mortars for restoration have serious attempts been made to obtain products with biocidal activity [22–24].

This study raises the possibility of using varnishes with incorporated biocide to give artworks protection against environmental microorganisms. Terpenic varnishes have traditionally been used as the final layer of protection so that they can act as a barrier between the pictorial layer and the external environment. It should be noted that they are susceptible to microbial colonization, although they are a restrictive substrate only suitable for microorganisms with certain metabolic and physiological capabilities (production of spores, resistance to a shortage of nutrients, resistance to xeric environments, etc.). *Penicillium* and *Aspergillus* are the most commonly described fungi associated with biodeterioration in polychrome artworks, and the genus *Bacillus* and *Arthrobacter* among bacteria [25–29]. Biodeterioration of these materials has been studied in past publications, which have described different analytical markers for these processes [30–33].

Nevertheless, terpenic varnishes naturally suffer from ageing processes that lead to phenomena such as yellowing and a loss of transparency. A common restoration task is to completely or partially eliminate them and revarnish the artworks. This can be done with products with incorporated biocide, which can add extra protection to the pictorial layer against microbial biodeterioration of paintings and polychrome sculptures.

The use of varnishes with biocidal activity has been studied very little until now. Souza and Gaylarde [34] evaluated the colonization of pinewood varnished with an unspecified alkyl resin with an incorporated biocide based on 2,3,5,6-tetrachloro-4 (methyl sulphonyl)-pyridine at a concentration of 0.3%. They concluded that the effects on the varnish layer, measured using the changes on the surface observed microscopically, were significantly lower than in the same varnish not treated with biocide.

Our research group also contributed some results in this area [35], which we shall look at in greater detail in this work. We propose using biocides associated with natural diterpene resins such as colophony (rosin), sandarac, and Manila copal, which are widely used in fine arts as the final layer of protection against biodeterioration in artworks (mainly paintings and polychrome sculptures). Their effectiveness against selected bacteria and fungi is tested in this study using optical microscopy, gas chromatography–mass spectroscopy (GC-MS), and a multidimensional scaling analysis (MDS). On the whole, we found that the treatments gave good results in inhibiting microbes from developing and in chemical innocuousness for varnishes.

## 2. Materials and Methods

### 2.1. Varnishes and Biocides

The standard colophony obtained by distillation from pine resin was supplied by RCM (Madrid, Spain). The varnish was prepared in the proportion of 100 g of resin per 175 mL of purified turpentine essence, supplied by Titan (Barcelona, Spain). Sandarac of *Tetraclinis articulata* and Manila Copal of *Agathis dammara* were supplied by Caremi Pigmentos (Seville, Spain). These varnishes were prepared in the proportion of 100 g of resin per 200 mL of absolute ethanol (Merck, Barcelona, Spain).

The biocides selected for the study were benzalkonium chloride (alkyl dimethyl benzyl ammonium chloride; CTS, Madrid, Spain), sodium-2-phenylphenolate, a solid alkaline formulation of o-phenylphenol (Preventol® ON Extra, Kremer Pigmente, Aichstetten, Germany), and tributyltin naphthenate (tributyl-mono (naphthenoyloxy) stannane; Metatin N 5810/10, Acima Chemical Industries, Buchs, Switzerland). They all are widely used in restoration and are representative of significant chemical groups of substances used in this field [14,19,21].

### 2.2. Microorganisms

The microorganisms listed here were selected on the basis of their capacity for growing and causing chemical changes in resins, as reported in earlier works [30–33].

- Colophony: *Chrysonilia sitophila* and *Bacillus amyloliquefaciens*
- Sandarac: *Chrysonilia sitophila*, *Penicillium chrysogenum*, and *Streptomyces celluloflavus*
- Manila copal: *Chrysonilia sitophila*, *Phoma herbarum*, and *Arthrobacter oxydans*

The fungus *Penicillium chrysogenum* (Pc) (CECT-2306), and the bacteria *Streptomyces celluloflavus* (Sc) (CECT-3242), *Bacillus amyloliquefaciens* (Ba) (CECT-493), and *Arthrobacter oxydans* (Ao) (CECT-386) came from stock collections belonging to the Spanish Type Culture Collection (CECT, Colección Española de Cultivos Tipo, Universidad de Valencia, Spain). The fungi *Chrysonilia sitophila* (Cs) and *Phoma herbarum* (Ph) were isolated from the surface of the oil on the canvas paintings of the Fine Arts Museum of Granada (Spain) entitled *Allegory of Death* (by P. Toma) and *St. Francis of Assisi* (anonymous), both from the 17th century and are severely affected by biodeterioration processes (Figure 1). Samples were taken from the painting using a noninvasive sampling procedure [36] of rubbing a dry sterile cotton bud, suitable for isolations in culture media (Class IIa) (Eurotubo, Deltalab, Barcelona, Spain), on the surface of the painting over an area of 2 cm$^2$. Cultivation assays were performed immediately on these samples. The cotton swabs were inoculated directly onto Sabouraud-chloramphenicol Agar plates (Scharlab, Barcelona, Spain), which were incubated at 28 °C for two weeks. During this period, colonies exhibiting a different morphology and appearance were btransferred to PDA culture plates (Scharlab, Barcelona, Spain) to obtain pure strains. All purified strains were stored in 70% glycerol at −80 °C for conservation.

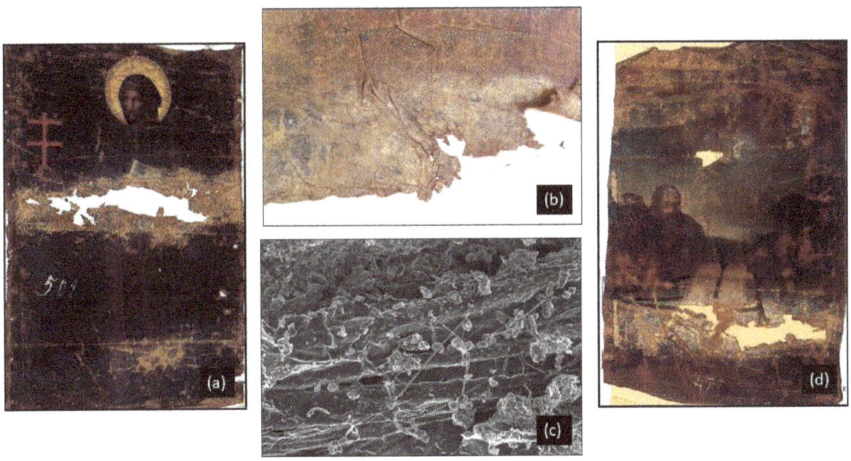


**Figure 1.** Overview of two paintings colonized by fungi, the *St. Francis of Assisi* (anonymous) (**a**) and the *Allegory of Death* (by P. Toma) (**b**), a detail of the biodeterioration pattern observed on the *St. Francis of Assisi* painting (**c**), and a SEM image of *C. sitophila* (**d**).

In order to obtain fungal spores, lyophilized collection stocks were hydrated in CM broth (Oxoid, Hampshire, UK) and incubated for one week (28 °C, 75% RH). Afterwards, cultures were spread on a solid malt agar medium (Oxoid, Hampshire, UK) and incubated for 15 days. Sporulated cultures were resuspended in 2 mL of Tween 80 at 0.1% (Aldrich, Madrid, Spain). After centrifugation, pellets were washed and resuspended in 2 mL of distilled water. The suspensions were then filtered through glass wool to eliminate any remains of mycelium. After a count in a Neubauer chamber, the concentration was adjusted to 10$^6$ spore mL$^{-1}$.

As for bacteria, the lyophilized strains collected were rehydrated in a TSB medium (Oxoid, Hampshire, UK), and seeded after 48 h of incubation in TSA medium (Oxoid, Hampshire, UK). Using these cultures, inocula were prepared in a TSB medium and kept incubated for 24 h. Afterwards, they were centrifuged and washed with distilled water to eliminate any remains of the culture medium, and resuspended in distilled water. Using the growth curve, the inocula were adjusted to $10^7$–$10^8$ cells $mL^{-1}$.

### 2.3. Sample Preparation

In order to prepare varnish samples with biocide incorporated, solutions in ethanol (Sigma-Aldrich, Spain) at 10% (*w*/*v*) of o-phenylphenol and at 10% (*v*/*v*) of tributyltin naphthenate and benzalkonium chloride were mixed with the varnishes at a concentration of 3% (*v*/*v*). To help prepare the mixtures, a small amount (0.1% *v*/*v*) of the nonionic surfactant Tween 80 (Sigma-Aldrich, Spain) was also added in the case of colophony. These varnishes with biocide were brush-applied to the surface of standard-sized glass slides (24 × 80 mm) and were allowed to dry in the dark for 90 days. It was not possible to obtain a homogeneous, perfectly transparent mixture of o-phenylphenol with the sandarac and Manila copal varnishes, so those samples were not used for this study.

Solutions of the biocides were also prepared at 3% *v*/*v* (tributyltin naphthenate and benzalkonium chloride) and at 3% *w*/*v* (o-phenylphenol) in distilled water, which were applied by brush onto the surface of samples, namely standard-sized glass slides (24 × 80 mm) varnished with the resins under study, which were allowed to dry in the dark for 90 days before applying the biocides on them.

In summary, two variants of samples were prepared: samples with biocide (3%) dried in darkness for 90 days, and samples of varnish dried in darkness for 90 days before having a biocide applied to them (3%). Similarly, control samples of the varnishes without biocide were also prepared to follow the progress of microorganisms and possible chemical changes with no biocidal treatment.

The Manila copal and sandarac samples treated with o-phenylphenol and the sandarac ones treated with tributyltin naphthenate (spread on) were discarded for this study because an increase in the opacity of the varnish was observed when they interacted with these products, which is incompatible with their use in artworks (Table 1).

**Table 1.** Samples included in the study.

| Varnish | Colophony | Sandarac | Manila Copal |
|---|---|---|---|
| **Microorganisms** | **Cs, Ba** | **Cs, Pc, Sc** | **Cs, Ph, Ao** |
| Biocides | Benzalkonium chloride (mixed, spread)<br>Metatin N (mixed, spread)<br>Preventol ON (mixed, spread) | Benzalkonium chloride (mixed, spread)<br>Metatin N (mixed) | Benzalkonium chloride (mixed, spread)<br>Metatin N (mixed, spread) |

Cs = C. sitophila; Pc = P. chrysogenum; Ph = Phoma herbarum; Ao = Arthrobacter oxydans; Ba = B. amyloliquefaciens.

In summary, five groups of test specimens were prepared so as to compare the possible chemical changes produced by the microorganisms in the varnishes studied in samples treated and not treated with biocides: type 1, varnish only (blank samples); type 2, varnish treated with biocides but not inoculated with microorganisms; type 3, varnish not treated with biocides but inoculated with microorganisms; type 4, varnish treated with biocides (mixed) and inoculated with microorganisms; type 5, varnish treated with biocides (spread) and inoculated with microorganisms.

### 2.4. Monitoring Growth

Each type of fungus and bacteria were inoculated separately in samples with several drops (75 µL) of spore or bacterial suspensions respectively ($10^7$–$10^8$ cells $mL^{-1}$), as described in Section 2.2. Previous experiments helped us to establish the optimal conditions for incubating the model varnish specimens. The specimens were incubated for 30 days in darkness at 28 °C and 90% RH, water activity (aw) = 0.85.

The fungal cultures were monitored by microscope to study the production of mycelium from the initial suspension of spores in distilled water both in the presence and absence of biocides. All of the samples were observed after a time of 0 h, 24 h, 48 h, and five, ten, fifteen, and thirty days using a Nikon Eclipse TS100 microscope equipped with a Nikon DS-5M digital camera.

### 2.5. Gas Chromatography–Mass Spectrometry (GC-MS)

The GC-MS technique was used to evaluate the chemical changes caused by the proposed treatments and the microorganisms under study. A sterile scalpel was used to scrape the samples from the glass slides, providing about 5 mg per sample. These were dissolved in 25 μL of benzene in test tubes, into which 500 μL of chloroform was added for 5 min to extract the organic components. Once filtered, the supernatants were subjected to a vacuum to evaporate the solvent in 2 mL vials, then the dry residue obtained was processed by adding 30 μL of benzene and 30 μL of (m-trifluoro-methylphenyl) trimethylammonium hydroxide (Meth-Prep II, Alltech, Stamford, UK) to each vial. The reaction mixture was shaken for 4 min at room temperature, and 2 μL of the sample solution was injected into an Agilent 7890A gas chromatograph (Agilent Technologies, Santa Clara, CA, USA) with a Combi-xt PAL sample injector, coupled to a Waters mass spectrometer, Quattro micro GC (Waters, Milford, MA, USA).

Chromatographic separations were achieved in a capillary column (WCOT; wall coating open tubular) ZB-5MS, 30 m × 0.25 mm I.D. with 0.25 μm film thickness. The chromatographic conditions for the GC-MS analysis were as follows: injector temperature = 250 °C, transfer line temperature = 250 °C, oven temperature = 140 °C (2 min.), 4 °C/min to 300 °C then isothermal for 5 min. The carrier gas was helium at a flow-rate of 1 mL/s. The samples were injected in splitless mode. Mass spectra were performed in total ion monitoring mode (mass range 50–550 *m/z*), and ions were generated by electron impact ionization (70 ev). The source temperature was 210 °C. A Masslynx v.4.0 data system was used for data acquisition and processing, and the peak area (TIC) data were used to obtain the peak area percentage value.

Calculations were based on values of the normalized peak area $Ni$, defined as the percentage of peak area of each individual compound relative to the set of peak areas of the $n$ components found in the model varnish:

$$Ni = \frac{Ai}{\sum_{i=1}^{n} Ai} 100$$

where $Ai$ is the peak area of each of the n compounds found in the chromatogram. Control samples inoculated with distilled water and incubated under the same conditions as those of the samples inoculated with microorganisms were also analyzed to evaluate the composition of the varnishes studied in the absence of microorganisms. The changes in composition were determined by establishing a $\Delta Ni$ parameter, defined as the increase or decrease in $Ni$ for each analyte in relation to the values corresponding to the control sample $Nio$.

$$\Delta Ni = Ni - Nio$$

Three replicates were used for each sample. The repeatability found under the experimental conditions and the relative standard deviations, calculated using the formula RSD = (standard deviation/mean of the normalized peak area) × 100%, fell within the range 1–5% for all the compounds analyzed. A Student's *t*-test was used for a comparison of the differences between mean values of inoculated and noninoculated blank samples. Only the values with a $p < 0.05$ were considered statistically significant.

### 2.6. MDS Analysis

In order to evaluate the analytical data obtained by GC-MS, an MDS (multidimensional scaling) statistical analysis was carried out. Each sample included in the study can be represented by a point in a multidimensional space in which the $n$ dimensions in the space are the $Ni$ values of each of the $n$ components of the varnish. The analogous chemical composition samples will be found near this multidimensional space, whereas those with a different chemical composition will be further away. Since it is not possible to visualize spaces of more than three dimensions, it is common to resort to techniques to decrease the scale and project the $n$ dimensions of a multidimensional space onto a bidimensional space by means of an MDS analysis, which enables the chemical changes produced by the microorganisms and treatments tested to be represented in a simplified diagram. In this study, the MDS analysis was carried out using the ASCAL command from the IBM SPSS Statistics® 26.0 software for 64 bits. For each varnish, a matrix was constructed of Euclidean distances between the different samples. For each pair of samples $i$ and $j$, the Euclidean distance $d_{ij}$ was calculated according to the equation:

$$d_{ij} = \sqrt{(x_i - x_j)^2 + (y_i - y_j)^2 + (z_i - z_j)^2 + \cdots}$$

where the values $x$, $y$, $z$, etc., are the $Ni$ values of each component of the varnish in samples $i$ and $j$. To improve the quality of the MDS graphs obtained with IBM SPSS Statistics, the resulting data from the analysis were processed with a Microsoft Excel 2013 spreadsheet.

## 3. Results and Discussion

### 3.1. Monitoring Growth

In the control samples not treated with biocides (type 3), the samples inoculated with *C. sitophila* in the three varnishes studied, the sandarac samples inoculated with *P. chrysogenum* and the Manila copal samples inoculated with *P. herbarum*, fungal growth was observed with mycelium developing from the spore suspensions after the incubation period. The growth observed appeared mainly in the first five days of incubation in all cases, which agrees with previous studies [30,33]. In the varnishes treated with surface biocides, only very little growth was observed in the colophony samples inoculated with *C. sitophila* (Figure 2), and in the samples prepared with biocide incorporated, there was no growth observed in any case.

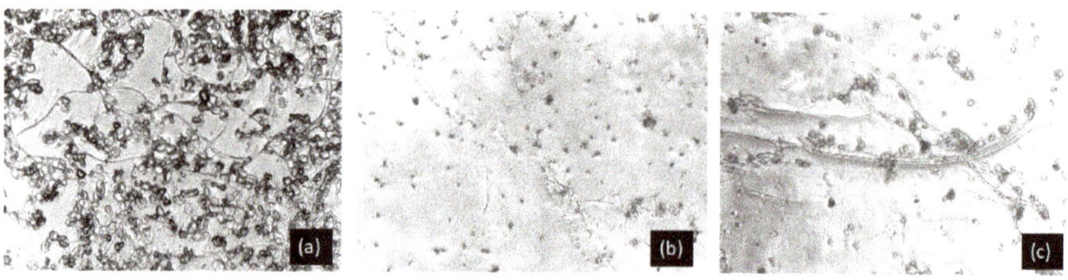


**Figure 2.** *C. sitophila* in colophony not treated with biocides (**a**) (400×) and treated with (**b**) Metatin N (spread, 100×) and (**c**) Preventol® ON (spread, 400×) after five days of incubation.

### 3.2. Colophony

The results from the GC-MS analysis are shown in Table 2. They include data from the blank samples (type 1); the samples not treated with biocides but inoculated with microorganisms (type 3) and the samples treated with biocides (types 4 and 5): fungal samples with growth and the samples inoculated with bacteria. The biocide treatments did not lead to any significant change in the original composition of the varnish. The GC-MS results for the type 2 samples were very similar to those obtained in the blank sample (type 1), so they have not been included in the table. The data from fungal samples in which no growth was recorded were also left out to simplify the study, since the data are very similar to the data from blank samples.

**Table 2.** GC-MS analysis ($p < 0.05$ in all cases) of the diterpenes from the colophony varnish and the samples inoculated with microorganisms (values expressed in *Ni*).

| Test | Sample Type | DHA | AA | 7-OH-DHA | 15-OH-DHA | 7-oxo-DHA | 15-OH-7-oxo-DHA |
|---|---|---|---|---|---|---|---|
| Colophony | 1 | **56.5** | **10.3** | **3.0** | **9.8** | **5.8** | **1.4** |
| Colophony-Cs | 3 | 51.3 | 0.0 | 2.8 | 8.9 | 17.2 | 3.3 |
| Colophony-Ba | 3 | 24.4 | 0.0 | 6.1 | 12.2 | 36.5 | 14.8 |
| Colophony Benzalkonium chloride (sp) Cs | 5 | 57.2 | 0.9 | 3.1 | 6.1 | 9.6 | 0.0 |
| Colophony Tributyltin naphthenate (sp) Cs | 5 | 58.9 | 0.8 | 3.3 | 5.6 | 8.9 | 0.0 |
| Colophony Orthophenyl phenol (sp) Cs | 5 | 56.4 | 11.4 | 1.8 | 5.3 | 3.0 | 0.0 |
| Colophony Benzalkonyum chloride (sp) Ba | 5 | 56.0 | 1.6 | 2.4 | 8.5 | 10.1 | 0.0 |
| Colophony Tributyltin naphthenate (sp) Ba | 5 | 55.7 | 1.0 | 3.1 | 7.3 | 11.7 | 0.0 |
| Colophony Orthophenyl phenol (sp) Ba | 5 | 54.5 | 2.0 | 2.0 | 7.8 | 7.8 | 0.0 |
| Colophony Benzalkonium chloride Ba | 4 | 57.1 | 2.4 | 1.6 | 6.8 | 6.1 | 0.0 |
| Colophony Tributyltin naphthenate Ba | 4 | 43.7 | 36.8 | 1.0 | 4.5 | 0.6 | 0.0 |
| Colophony Orthophenyl phenol Ba | 4 | 53.2 | 11.5 | 1.1 | 3.3 | 0.6 | 0.0 |

DHA = dehydroabietic acid; AA = abietic acid; 7-oxo-DHA = 7-oxo-dehydroabietic acid; 15-OH-7-oxo-DHA = 15 hydroxy-7-oxo-dehydroabietic acid. Sample type: 1: varnish only (blank samples); type 3: varnish inoculated with microorganisms; types 4 and 5: varnish treated with biocides, mixed (4) and spread (sp), (5) and inoculated with microorganisms. Cs = *C. sitophila*; Ba = *B. amyloliquefaciens*.

In the type 3 sample corresponding to *B. amyloliquefaciens*, a clear decrease was recorded in the case of the relative amount of abietic and dehydroabetic acids compared to the original composition of varnish, and a sharp rise in oxidized diterpenes with respect to the type 1 sample, especially 7-oxo-dehydroabietic and 7 oxo-15-hydroxy-dehydroabietic reported in *C. sitophila* and *B. amyloliquefaciens* (example: Colophony-Cs, $\Delta Ni$ 7-oxo-DHA = +11.4; Colophony-Ba, $\Delta Ni$ 7-oxo-DHA = +30.7; $\Delta Ni$ 15-OH-7-oxo-DHA = 13.4). These oxidative changes coincide with the chemical degradation patterns described for this varnish in previously published studies, and they may be related to the biological activity of the microorganisms tested [30,37–40].

These results differ significantly from those obtained with biocides (Figure 3). There was very little growth of the fungus *C. sitophila* observed in the type 5 samples (biocide spread), and none at all in those with the biocide incorporated (type 4), where growth was totally inhibited. In chemical terms, the oxidative processes described in the samples not treated with biocides did not occur, nor in the case of the fungus, nor with the bacteria, although with *B. amyloliquefaciens* there were changes in composition recorded in the samples treated with Metatin N (Colophony tributyltin naphthenate Ba, $\Delta Ni$ AA = −12.84; $\Delta Ni$ DHA = + 26.44), suggesting biological activity.

Figure 4 shows the MDS diagram for the colophony samples. It can be seen that the samples without biocide (type 3) inoculated with *C. sitophila* and *B. amyloliquefaciens* are far from the nucleus formed by the blank sample and the samples inoculated and also treated with the biocides. This nucleus contains chemical compositions very similar to the blank sample with no inoculation or treatment. In other words, the biocides not only totally or partially prevented microbial growth, but they also ensured that the chemical composition of the varnish did not change substantially.

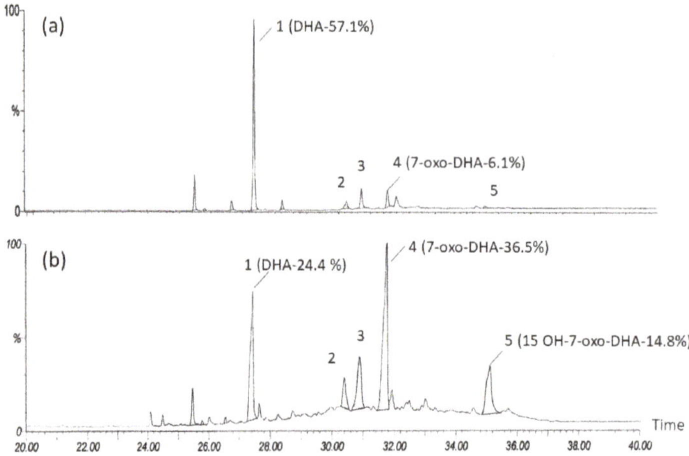


**Figure 3.** Chromatograms corresponding to (**a**) colophony varnish treated with Benzalkonium chloride and inoculated with *B. amyloliquefaciens* (type 4) and (**b**) colophony varnish not treated with biocides but inoculated with *B. amyloliquefaciens* (type 3). Values expressed in *Ni*. 1: DHA = dehydroabietic acid; 2: 7-OH-dehydroabietic acid; 3: 15-OH-dehydroabietic acid; 4: 7-oxo-dehydroabietic acid; 5: 15-OH-7-oxo-dehydroabietic acid.

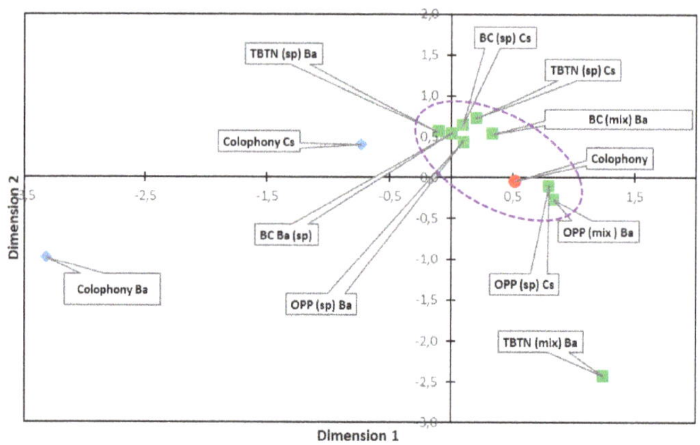


**Figure 4.** Multidimensional scaling (MDS) analysis of the colophony samples: blank sample (type 1, red circle); samples inoculated with microorganisms without biocide (type 3, blue diamonds); and samples inoculated with Cs and Ba, and treated with the studied biocides (type 4 (mix), and 5 (sp), green squares). BC = benzalkonium chloride, TBTN = tributyltin naphtenate, and OPP = ortophenyl phenol. Cs = *C. sitophila*; Ba = *B. amyloliquefaciens*.

The exception is the sample inoculated with *B. amyloliquefaciens*, the microorganism capable of producing the greatest chemical changes in the colophony, treated with the biocide tributyltin naphthenate mixed with the varnish. The biocide was not seen to be sufficiently effective, but nevertheless it managed to make the composition of the varnish somewhat closer to the blank sample than when it is inoculated with this bacteria type without the presence of the biocide. The results show the efficacy of the treatments tested, and they also resemble those obtained in a previous study carried out by our research group [35] that analyzed the changes produced by microorganisms inoculated in colophony resin samples treated on the surface with the biocides New Des (benzalkonium chloride), Biotin N, and Nipagine (methyl parahydroxybenzoate). The analyses by pyrolysis-GC-MS showed the efficacy of treatments with Biotin N on the growth of fungi and bacteria, without appreciable changes in the composition of the resins, and a lower effectiveness of the treatment with benzalkonium chloride and Nipagine.

### 3.3. Sandarac

The analytical results with the sandarac varnish are shown in Table 3. As in the case of colophony, the biocides studied did not lead to any significant change in the composition of the original varnish. The results from the GC-MS for the type 2 samples were very similar to those obtained in the blank sample, so they were not included in the table. However, the treatments tested inhibited the growth of the fungi *C. sitophila* and *P. chrysogenum* in all such samples studied. To simplify the study, the data for these latter samples are not included, since they are very similar to the data for the blank samples. The main chemical change observed in the sandarac inoculated with microorganisms but with no biocides (type 3) compared to the blank samples was a significant decrease in a hydroxylated derivative of sandaracopimaric acid to produce sandaracopimaric acid (example: Sandarac-Cs, $\Delta Ni$ OH-sandaracopimaric acid = +12.01; Sandarac-Pc/Sc, $\Delta Ni$ OH-sandaracopimaric acid= −14,54; Sandarac-Cs, $\Delta Ni$ OH-sandaracopimaric acid = −13,91). This has already been reported in previous studies [33,41], although this did not generally occur in samples treated with biocides, which may be related to a drop in the microorganisms' biological activity caused by the proposed methods.

**Table 3.** GC-MS analysis ($p < 0.05$ in all cases) of the diterpenes in the sandarac varnish and the samples inoculated with microorganisms (values expressed in *Ni*).

| Test | Sample Type | Manool | Sandaracopimaric Acid | Isopimaric Acid | OH-Sandaracopimaric |
|---|---|---|---|---|---|
| Sandarac | 1 | **25.79** | **47.34** | **8.02** | **14.54** |
| Sandarac-Pc | 3 | 28.33 | 61.72 | 9.95 | 0 |
| Sandarac-Cs | 3 | 27.47 | 64 | 7.9 | 0.63 |
| Sandarac-Sc | 3 | 26.29 | 65.27 | 7.74 | 0 |
| Sandarac-Benzalkonium chloride-Sc | 4 | 22.34 | 55.96 | 8.76 | 12.92 |
| Sandarac-Benzalkonium chloride (sp)-Sc | 5 | 24.54 | 49.94 | 10.41 | 15.1 |
| Sandarac-Tributyltin naphthenate-Sc | 4 | 18.43 | 54.33 | 13.58 | 13.65 |

Sample type: 1: varnish only (blank samples); type 3: varnish inoculated with microorganisms; types 4 and 5: varnish treated with biocides mixed (4) or spread (5) and inoculated with microorganisms. Cs = *C. sitophila*; Pc = *P. chrysogenum*; Sc = *S. celluloflavus*.

Figure 5 shows the MDS diagram for the samples studied. A nucleus can be seen very near to the blank sample, consisting of the samples inoculated with *C. sitophila* and treated with biocides. This indicates that there were no substantial chemical changes in the composition of the varnish. The type 3 samples (inoculated with a microorganism but with no biocide), however, appear very much grouped together and far from the control sample; this indicates the efficacy of the treatments tested in inhibiting microbial growth and also in preventing the appearance of chemical changes in the sandarac.

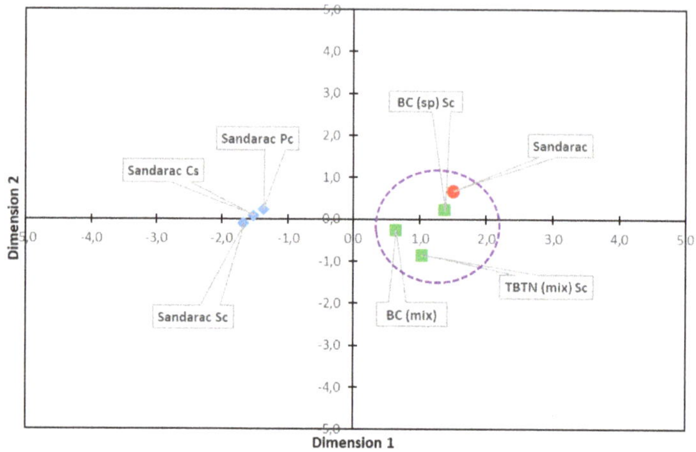


**Figure 5.** MDS analysis of the sandarac samples: blank sample (type 1, red circle); samples inoculated with microorganisms without biocide (type 3, blue diamonds); and samples inoculated with Pc, Cs and Sc, and treated with the studied biocides (type 4 (mix), and 5 (sp), green squares). BC = benzalkonium chloride, TBTN = tributyltin naphtenate, and OPP = ortophenyl phenol. Cs = *C. sitophila*; Pc = *P. chrysogenum*; Sc = *S. celluloflavus*.

### *3.4. Manila Copal*

Table 4 shows the analytical results for Manila copal. As with the other varnishes studied, the biocide treatments did not lead to any significant change in the composition of the original varnish, so the results for the type 2 samples are not included in the table. The study via microscope indicates that the treatments that were tested inhibited the growth of the fungi *C. sitophila* and *P. herbarum* in all such cases (Figure 6). As with the other two varnishes, the data corresponding to these latter samples are not included to simplify the study, since they are very similar to the data from blank samples.

As for the GC-MS analysis, in the samples inoculated with microorganisms but not treated with biocide, changes of little relevance occurred compared to the simple blank, most notably a decrease in the relative amount of agathalic acid, which oxidizes to agathic acid. Other trends seen were the moderate but generalized increase in the small peak for 19-norlabda-8 (20), 13-dien-15-oic acid and a decrease in acetoxy agatholic acid in *C. sitophila*, which agrees with the results obtained in previous studies [33].

The MDS statistical study (Figure 7) shows the results from all the samples somewhat grouped together to scale, indicating that in general the chemical changes occurring in the samples inoculated with microorganisms were not as clear as in the other varnishes studied. The results in the samples treated with biocides and inoculated with *A. oxydans* also appear in an area of the graph close to the uninoculated blank samples, indicating that the composition of the Manila copal was not substantially affected by the microorganisms tested and that the biocides therefore play a minor role in terms of the varnish's chemical composition. However, the treatments tested have shown to be efficient in controlling growth by inhibiting the development of mycelium in the samples inoculated with fungi.

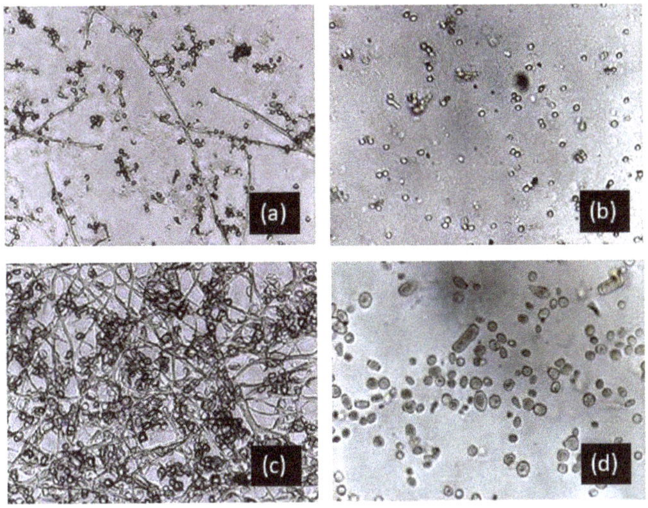


**Figure 6.** Microphotography (400×) showing fungal growth of (**a**) *P. chrysogenum* (sandarac) and (**c**) *C. sitophila* (Manila copal) after 5 days of incubation, images (**b**) and (**d**) show samples of sandarac treated with tributyltin naphthenate (mixed) and benzalkonium Chloride (mixed) in which no fungal growth was observed after the incubation period.

**Table 4.** GC-MS analysis ($p < 0.05$ in all cases) of the diterpenes free of Manila copal and the samples inoculated with microorganisms (values expressed in *Ni*).

| Test | Sample Type | 19-Norlabda-8 (20), 13-Dien-15-oic Acid | Agathalic Acid | Agathic Acid | Acetoxy Agatholic Acid |
|---|---|---|---|---|---|
| Manila copal | 1 | **3.24** | **7.35** | **41.66** | **23.28** |
| Copal-Cs | 3 | 3.92 | 3.19 | 46.07 | 17.03 |
| Copal-Ph | 3 | 3.92 | 1.93 | 39.07 | 19.81 |
| Copal-Ao | 3 | 2.88 | 3.58 | 48.14 | 22.27 |
| Copal Benzalkonium Chloride Ao | 4 | 3.31 | 1.05 | 46.18 | 28.06 |
| Copal Benzalkonium Chloride (sp) Ao | 5 | 2.16 | 0.33 | 45.82 | 34.13 |
| Copal Tributyltin naphthenate Ao | 4 | 1.30 | 0.49 | 40.59 | 33.85 |
| Copal Tributyltin naphthenate (sp) Ao | 5 | 2.40 | 0.00 | 50.28 | 26.00 |

Sample type: 1: varnish only (blank samples); type 3: varnish inoculated with microorganisms; types 4 and 5: varnish treated with benzalkonium chloride and tributyltin naphthenate mixed (4) or spread (5) and inoculated with microorganisms. Cs = *C. sitophila*; Ph = *P. herbarum*; Ao = *A. oxydans*.

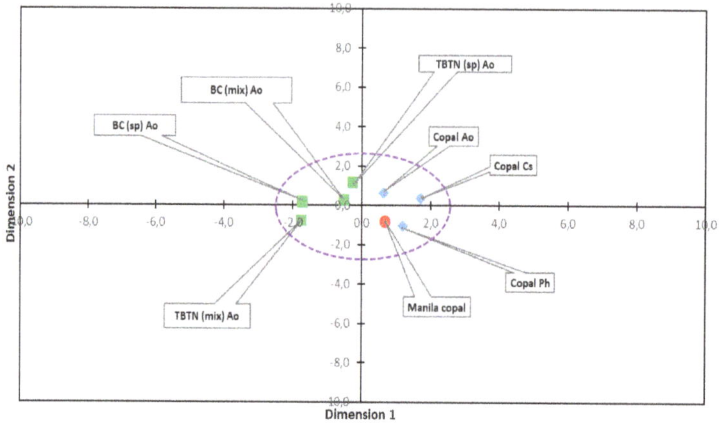


**Figure 7.** MDS analysis of Manila copal samples: blank sample (type 1, red circle); samples inoculated with microorganisms without biocide (type 3, blue diamonds); and samples inoculated with Ao and treated with the studied biocides (type 4 (mix), and 5 (sp), green squares). BC = benzalkonium chloride, TBTN = tributyltin naphtenate, and OPP = ortophenyl phenol. Cs = *C. sitophila*; Ph = *P. herbarum*; Ao = *A. oxydans*.

### 3.5. General Technical Considerations

As regards the biocides' physical and chemical properties and the possible practical applications of the treatments, the best properties were observed in the varnishes treated with benzalkonium chloride. This biocide enabled stable mixes formulas to be formed with the three varnishes without altering their transparency, which is in keeping with its properties as an emulsifier [21,42]. Another advantage of benzalkonium chloride's nature as a surfactant is that it tends to be heavily absorbed and also tends to spread easily over the surface where it is applied, forming an antiseptic layer that lends protection for a long time [19,42]. In the case of orthophenylphenol, transparent formulas dispersions were not achieved with the two varnishes of greatest polarity (sandarac and Manila copal), which also occurred with Metatin N spread on sandarac. This varnish, which is soluble in alcohol, gave the most problems in the mixes' formulas with biocides tested. As for the aesthetic appearance and transparency of the treated varnishes, in general good results were achieved visually, except in the aforementioned cases, although this factor has not been studied in the long term using quantitative colorimetric tests.

The innocuousness of treatments for painting materials (especially for binders and pigments) is an important matter that remains to be studied in future. Perhaps the most tried and tested product on polychrome artworks among those tested is benzalkonium chloride, which is widely used for fresco paintings, whether in an aqueous solution or in mixtures for cleaning poultices such as "AB 57", commonly used in restoration [21,42]. It can therefore be considered with some confidence for pigments. It is also compatible with metals, which is not the case with Preventol ON. The latter is corrosive to them, thereby restricting its use in artworks with gold or silver [43].

Finally, human and environmental toxicity is another significant issue to take into account. Metatin N is the most toxic biocide tested here, since it contains tin (acute oral LD50 of 1300 mg/kg) [4], which restricts its use in practical applications. Orthophenyl phenol is the least toxic of the three, since the formulations used do not irritate the skin or mucous membranes [44], whereas benzalkonium chloride is found to be somewhere in the middle, not showing special toxicity problems at the proposed concentrations [21].

## 4. Conclusions

New treatments were proposed to protect works of art against biodeterioration by using varnishes with biocidal activity, and three products were tested: benzalkonium chloride, orthophenyl phenol

(Preventol ON), and tributyltin naphthenate (Metatin N) incorporated into the colophony, sandarac, and Manila copal varnishes, and also applied by spreading it onto them. In general, growth of the microorganisms tested was inhibited without causing significant chemical changes to the varnishes.

In the case of colophony resin, the biocidal treatments that were incorporated into the varnish inhibited the growth of the fungus *C. sitophila*. In the samples with the biocides spread onto the resin, a little fungal growth was observed, indicating less efficiency for treatments applied in this way. The GC-MS analytical results did not show significant changes in resins that were inoculated with *C. sitophila* and with *B. amyloliquefaciens* when compared to the uninoculated blank samples, since the increase observed in oxidized forms of the abietane diterpenes without biocides was not seen. However, an exception was seen in the treatment with tributyltin naphthenate mixed with the varnish, which was seen to be less effective against the bacteria.

In the sandarac varnish, the biocides that were tested (benzalkonium chloride and Metatin N incorporated into the varnish, and benzalkonium chloride spread on the varnish) inhibited the growth of the fungi *C. sitophila* and *P. herbarum*. The chemical changes observed in the varnish due to the microorganisms studied, including the bacterium *S. celluloflavus*, did not occur in the presence of biocides.

Lastly, in the Manila copal varnish, the biocides that were tested (benzalkonium chloride and Metatin N incorporated into the varnish and spread on it) inhibited the growth of the fungi *C. sitophila* and *P. herbarum*, although the GC-MS analysis did not show conclusive data as regards the efficacy of the treatments chemically.

The results from the study suggest that the addition of biocides to natural varnishes has significant potential as a factor in controlling the biodeterioration of cultural assets, though questions remain such as the permanence of these products on the artwork, the durability of the treatments and their efficacy, and innocuousness for the original materials in the long-term, which can be addressed in future publications.

**Author Contributions:** Conceptualization and work design, J.R.-N. and F.B.-G.; methodology, analyses, and investigation, J.R.-N. and R.B.-M.; writing, review, and editing, J.R.-N.; supervision, project administration, funding acquisition, F.B.-G. All authors have read and agreed to the published version of the manuscript.

**Funding:** This research was funded by the following projects: Desarrollo de Nuevas Sinergias Arte-Ciencia aplicadas a la Conservación y Restauración de los Palacios y Jardines de la Alhambra y el Generalife (VIRARTE), MINECO, with reference HAR2016-79886-P; Métodos sinérgicos Arte-Ciencia-Tecnología para la Conservación-Restauración de la Alhambra y otros Bienes Culturales (VIRARTE II), MICINN, with reference PID2019-109713RB-I00; La aplicación de las algas procedentes de la Alhambra y el Generalife en técnicas artísticas y de conservación-restauración (FICOARTE), Universidad de Granada, with reference A-HUM-279-UGR18; Aplicación avanzada de las algas procedentes de la Alhambra y el Generalife en técnicas artísticas y de conservación-restauración. Modalidad Frontera consolidado en el ámbito del Plan Andaluz de Investigación, Desarrollo e Innovación (PAIDI 2020), with reference P18-FR-4477.

**Conflicts of Interest:** The authors declare no conflict of interest. The funders played no part in designing the study or in collecting, analyzing or interpreting the data, in writing the manuscript, or in the decision to publish the results.

## References

1. Ciferri, O.; Tiano, P.; Mastromei, G. *Of Microbes and Art: The Role of Microbial Communities in the Degradation and Protection of Cultural Heritage*; Springer Science & Business Media: New York, NY, USA, 2000.
2. Allsopp, D.; Seal, K.; Gaylarde, C. *Introduction to Biodeterioration*; Cambridge University Press: Cambridge, UK, 2004.
3. Koestler, R.J.; Koestler, V.H.; Charola, A.E.; Nieto-Fernandez, F.E. *Art, Biology, and Conservation: Biodeterioration of Works of Art*; Yale University Press: New Haven, CT, USA, 2004.
4. Caneva, G.; Nugari, M.P.; Salvadori, O. *Plant Biology for Cultural Heritage. Biodeterioration and Conservation*; Getty Conservation Institute: Los Ángeles, CA, USA, 2009.
5. Tiano, P. Biodeterioration of monumental rocks: Decay mechanisms and control methods. *Sci. Technol. Cult. Herit.* **1998**, *7*, 19–38.

6. Warscheid, T.; Braams, J. Biodeterioration of stone: A review. *Int. Biodeterior. Biodegrad.* **2000**, *46*, 343–368. [CrossRef]
7. Scheerer, S.; Ortega-Morales, O.; Gaylarde, C. Microbial deterioration of stone monuments—An updated overview. *Adv. Appl. Microbiol.* **2009**, *66*, 97–139. [PubMed]
8. Salvadori, O.; Municchia, A.C. The Role of Fungi and Lichens in the Biodeterioration of Stone Monuments. *Open Conf. Proc. J.* **2016**, *7* (Suppl. 1), 39–54. [CrossRef]
9. Tiano, P. Biodeterioration of Stone Monuments a Worldwide Issue. *Open Conf. Proc. J.* **2016**, *7* (Suppl. 1), 29–38. [CrossRef]
10. Nugari, M.P.; D'Urbano, M.S.; Salvadori, O. Test methods for comparative evaluation of biocide treatments. In Proceedings of the International UNESCO/RILEM Congress on Conservation of Stone and Others Materials, Paris, France, 29 June–1 July 1993; Thiel, M.J., Ed.; E&FN Spon: London, UK, 1993; pp. 565–572.
11. Koestler, R.J.; Salvadori, O. Methods of evaluating biocides for the conservation of porous building materials. *Sci. Technol. Cult. Herit.* **1998**, *5*, 1963–1968.
12. Malagodi, M.; Nugari, M.P.; Altieri, A.; Lonati, G. Effects of combined aplication of biocides and protectives on marble. In *Proceeedings of the 9th International Congress on Deterioration and Conservation of Stone, Venice, Italy, 19–24 June 2000*; Fassina, V., Ed.; Elsevier: Amsterdam, The Netherlands, 2000; Volume 2, pp. 225–233.
13. Nugari, M.P.; Salvadori, O. Biocides and treatment of stone: Limitation and future prospects. In *Art, Biology and Conservation: Biodeterioration of Works of Art*; Koestler, R.J., Koestler, V.H., Charola, E.A.E., Nieto-Fernandez, F.E., Eds.; The Metropolitan Museum of Art: New York, NY, USA, 2003; pp. 518–535.
14. Sterflinger, K.; Piñar, G. Microbial Deterioration of Cultural Heritage and Works of Art. Tilting at Windmills? *Appl. Microbiol. Biotechnol.* **2013**, *97*, 9637–9646. [CrossRef]
15. Mitova, M.; Iliev, M.; Novakova, A.; Gorbushina, A.; Groudeva, V.; Martín-Sánchez, P. Diversity and biocide susceptibility of fungal assemblages dwelling in the Art Gallery of Magura Cave, Bulgaria. *Int. J. Speleol.* **2017**, *46*, 67–80. [CrossRef]
16. Koestler, R.J.; Parreira, E.; Santoro, E.D.; Noble, P. Visual effects of selected biocides on easel painting materials. *Stud. Conserv.* **1993**, *38*, 265–273.
17. Poyatos, F.; Morales, F.; Nicholson, A.W.; Giordano, A. Physiology of biodeterioration on canvas paintings. *J. Cell. Physiol.* **2018**, *233*, 2741–2751. [CrossRef]
18. Bravery, A.F. Biodeterioration of paint—A state-of-the-art comment. In *Biodeterioration 7*; Houghton, D.R., Smith, R.N., Eggins, H.O.W., Eds.; Elsevier Applied Science: London, UK, 1988; pp. 466–485.
19. Caneva, G.; Nugari, M.P.; Pinna, D.; Salvadori, O. *Il Controllo del Degrado Biologico*; Nardini Editore: Florence, Italy, 1998; pp. 1–180.
20. Schieweck, A. Occurance of organic and inorganic biocides in the museum environment. *Atmos. Environ.* **2006**, *41*, 3266–3275. [CrossRef]
21. Nugari, M.P.; Salvadori, O. Biodeterioration control of cultural heritage: Methods and products. In *Molecular Biology and Cultural Heritage*; Sáiz-Jiménez, C., Ed.; Rouledge: London, UK, 2017; pp. 233–242. [CrossRef]
22. Ferone, C.; Pansini, M.; Mascolo, M.C.; Vitale, A. Preliminari study on the set up of mortars displaying biocidal activity. In *Proceeedings of the 9th International Congress on Deterioration and Conservation of Stone, Venice, Italy, 19–24 June 2000*; Fassina, V., Ed.; Elsevier: Amsterdam, The Netherlands, 2000; Volume 2, pp. 371–378.
23. Urzi, C.; de Leo, F. Evaluation of the efficiency of water-repellent and biocide compounds against microbial colonization of mortars. *Int. Biodeterior. Biodegrad.* **2007**, *60*, 25–34. [CrossRef]
24. Barrionuevo, M.R.E.; Gaylarde, C.C. Biocide-containing varnish for the protection of sandstone: Comparison of formulations and laboratory test methods. *Curr. Microbiol.* **2011**, *62*, 1671–1676. [CrossRef] [PubMed]
25. Capodicasa, S.; Fedia, S.; Porcellia, A.M.; Zannon, D. The microbial community dwelling on a biodeteriorated 16th century painting. *Int. Biodeterior. Biodegrad.* **2010**, *64*, 727–733. [CrossRef]
26. López-Miras, M.M.; Martín, I.; Yebra, A.; Romero-Noguera, J.; Bolívar, F.; Ettenauer, J.; Sterflinger, K.; Píñar, G. Contribution of the microbial communities detected on an oil painting on canvas to its biodeterioration. *PLoS ONE* **2013**, *8*, e80198. [CrossRef]
27. López-Miras, M.M.; Piñar, G.; Romero-Noguera, J.; Bolívar, F.C.; Ettenauer, J.; Sterflinger, K.; Martín, I. Microbial communities adhering to the obverse and reverse sides of an oil painting on canvas: Identification and evaluation of their biodegradative potential. *Aerobiología* **2013**, *29*, 301–314.

28. de Carvalho, H.P.; Sequeira, S.O.; Pinho, D.; Trovao, J.; da Costa, R.M.F.; Egas, C.; Macedo, M.F.; Portugal, A. Combining an innovative non-invasive sampling method and high-throughput sequencing to characterize fungal communities on a canvas painting. *Int. Biodeterior. Biodegrad.* **2019**, *145*, 104816. [CrossRef]
29. Piñar, G.; Poyntner, C.; Lopandic, K.; Tafer, H.; Sterflinger, K. Rapid diagnosis of biological colonization in cultural artefacts using the MinION nanopore sequencing technology. *Int. Biodeterior. Biodegrad.* **2019**, *148*. [CrossRef]
30. Romero-Noguera, J.; Bolívar-Galiano, F.C.; Ramos-López, J.M.; Fernández-Vivas, M.A.; Martín-Sánchez, I. Study of biodeterioration of diterpenic varnishes used in art painting: Colophony and Venetian turpentine. *Int. Biodeterior. Biodegrad.* **2019**, *62*, 427–433. [CrossRef]
31. Romero-Noguera, J.; Martín-Sánchez, I.; López-Miras, M.M.; Ramos-López, J.M.; Bolívar-Galiano, F.C. Biodeterioration patterns found in dammar resin used as art material. *Electron. J. Biotechnol.* **2009**, *13*. [CrossRef]
32. Romero-Noguera, J.; López-Miras, M.M.; Martín-Sánchez, I.; Ramos-López, J.M.; Bolívar-Galiano, F.C. An approach to the study of the fungal deterioration of a classical art material: Mastic varnish. *Electron. J. Biotechnol.* **2010**, *13*. [CrossRef]
33. Romero-Noguera, J.; Doménech-Carbó, M.T.; Osete-Cortina, L.; Martín-Sánchez, I.; Bolívar-Galiano, F.C. Analytical characterisation of the biodeterioration of diterpenoid labdanic varnishes used in pictorial techniques: Sandarac and Manila copal. *Int. Biodeterior. Biodegrad.* **2014**, *90*, 99–105. [CrossRef]
34. De Souza, A.; Gaylarde, C.C. Biodeterioration of varnished wood with and without biocide: Implications for standard test methods. *Int. Biodeterior. Biodegrad.* **2002**, *49*, 21–25. [CrossRef]
35. Doménech-Carbó, M.T.; Osete-Cortina, L.; de la Cruz Cañizares, J.; Bolívar-Galiano, F.; Romero-Noguera, J.; Fernández-Vivas, M.A.; Martín-Sánchez, I. Study of the microbiodegradation of terpenoid resin-based varnishes from easel painting using pyrolysis–gas chromatography–mass spectrometry and gas chromatography–Mass spectrometry. *Anal. Bioanal. Chem.* **2006**, *385*, 1265–1280. [CrossRef]
36. Michaelsen, A.; Piñar, G.; Montanari, M.; Pinzari, F. Biodeterioration and restoration of a 16th-century book using a combination of conventional and molecular techniques: A case study. *Int. Biodeterior. Biodegrad.* **2009**, *63*, 161–168. [CrossRef]
37. Pastorova, I.; Van den Berg, K.J.; Boon, J.J.; Verhoeven, J.W. Analysis of oxidised diterpenoid acids using thermally assisted methylation with TMHA. *J. Anal. Appl. Pyrolysis* **1997**, *43*, 41–57. [CrossRef]
38. Van den Berg, K.J.; Boon, J.J.; Pastorova, I.; Spetter, L.F.M. Mass spectrometric methodology for the analysis of highly oxidized diterpenoid acids in Old Master paintings. *J. Mass Spectrom.* **2000**, *35*, 512–533. [CrossRef]
39. Scalarone, D.; Lazzari, M.; Chiantore, O. Ageing behaviour and pyrolytic characterisation of diterpenic resins used as art materials: Colophony and Venice turpentine. *J. Anal. Appl. Pyrolysis* **2002**, *64*, 345–361. [CrossRef]
40. Azemard, C.; Menager, M.; Vieillescazes, C. Analysis of diterpenic compounds by GC-MS/MS: Contribution to the identification of main conifer resins. *Anal. Bioanal. Chem.* **2016**, *408*, 6599–6612. [CrossRef]
41. Kononenko, I.; Viguerie, L.; Rochut, S.; Walter, P. Qualitative and quantitative studies of chemical composition of sandarac resin by GC-MS. *Environ. Sci. Pollut. Res.* **2017**, *24*, 2160–2165. [CrossRef]
42. GE International Institute for Conservation of Historic and Artistics Works. Available online: https://www.geiic.com/files/fichas%20productos/cloruro_de_benzalconio.pdf (accessed on 16 September 2020).
43. GE International Institute for Conservation of Historic and Artistics Works. Available online: https://www.geiic.com/files/fichas%20productos/preventol_ON.pdf (accessed on 16 September 2020).
44. Food and Agriculture Organization of the United Nations (FAO). Available online: http://www.fao.org/fileadmin/templates/agphome/documents/Pests_Pesticides/JMPR/Evaluation99/23Phenylphenol.pdf (accessed on 22 September 2020).


**Publisher's Note:** MDPI stays neutral with regard to jurisdictional claims in published maps and institutional affiliations.

*applied sciences*

MDPI

*Review*

# Present and Future Perspectives for Biocides and Antifouling Products for Stone-Built Cultural Heritage: Ionic Liquids as a Challenging Alternative

**Sandra Lo Schiavo, Filomena De Leo and Clara Urzì ***

Department of Chemical, Biological, Pharmaceutical and Environmental Sciences, University of Messina, Viale F. Stagno d'Alcontres, 31, 98166 Messina, Italy; sloschiavo@unime.it (S.L.S.); fdeleo@unime.it (F.D.L.)
* Correspondence: urzicl@unime.it

Received: 21 July 2020; Accepted: 15 September 2020; Published: 20 September 2020

**Featured Application: This review underlines the pro and cons of recent research aiming to counteract the growth of biodeteriogen microorganisms on stone cultural heritage. Ionic liquids, a special class of salts, are proposed as potential biocides engineered by Safe by Design concepts.**

**Abstract:** This review offers an overview of the most recent research activities on counteracting the biodeterioration process of stone monuments, underlining all those aspects regarding eventual procedural drawbacks and compliance with sustainable criteria. For this purpose, the definition of "green conservation of cultural heritage" has been proposed. Its basics have been utilized in the text to highlight the issues arising from the most common conservative procedures as well as guidelines for the development of innovative technologies. The review then deals with the most innovative antimicrobial approaches, among which nano- and bio-technologies play a main role. Ionic liquids are a special class of salts, which can be prepared by applying *Safe by Design* concepts, to meet the Green Conservation criteria.

**Keywords:** cultural heritage; biofilm; biodeteriogens; biocides; antifouling; ionic liquids; green conservation; SMART conservation

## 1. Introduction

The major challenge for researchers involved in the conservation of Cultural Heritage (CH) regards the development of innovative specific protocols that match sustainability criteria. This involves concepts such as nondestructive and/or reversible procedures, and safety for the environment and for restorers [1]. For these reasons, we refer to these strategies as "green conservation".

In the recent fourth conference of YOCOCU (Youth in Conservation of Cultural Heritage) held in Portugal in 2019. (http://artes.porto.ucp.pt/en/greenconservation2019), the attendees tried to clarify the meaning of Green in the context of CH Conservation (Andrea Macchia personal communication).

The acronym SMART summarizes the clarification.

<u>S</u>ustainability: conservation as an engine of local development, with the cultural heritage linked to the aspects of the local context in which it is found.

<u>M</u>anagement: planning conservation from restoration into the present and future, and accounting for the health of the operators and environment.

Actors: people including tourists, students, businesses, and such heritage professionals as museum employees and archeologists.

Risks: resilience, loss of economic resources, loss of integrated and specific skills.

Tools: innovative materials and methods.

From this, a new definition of green conservation of cultural heritage is given as: all the eco-sustainable practices to be used in the conservation and restoration of cultural heritage assets, as alternatives to traditional products and methods which are often toxic and harmful for the users and the environment. This definition includes the term of sustainable conservation of cultural heritage linked to environmental and socio-economical (green economy) aspects of the subject, to tourism development, to sustainable chemistry, and to conservation planning.

## 2. Biodeterioration

Biodeterioration of CH assets is a well-known issue that needs to be faced with methodologies with two aims. First, to correctly comprehend the causes leading to the uncontrolled growth of macro and/or microorganisms (diagnosis). Second, to develop appropriate methodologies to slow down or eliminate the unwanted biological growth (therapy).

One factor that should be considered dealing with biodeterioration as a consequence of the biocolonization process (biofilm formation) is the so-called bioreceptivity of a material. This term was first used by Guillitte [2] to define the ability of a given material to be colonized by living organisms. This depends on several parameters such as material composition, the status of conservation, eventual surface treatments as well as the environmental conditions in which the artifact is placed. Determining the bioreceptivity is an important premise for the development of any sustainable conservation procedure as it highlights the relationships between the surface, the specific environment, and the organisms that thrive in it. A community of (micro)organisms irreversibly attached to a surface and encased in an extracellular polymeric matrix (EPS) (biofilm) possesses features such as high resistance to environmental stress and treatments. It is also referred to as biological patinas or lichenic crusts [3,4].

## 3. How to Treat and Protect Cultural Heritage Assets against Biodeteriogens?

According to Pinna [1] and Caneva [5], to control biological colonization, indirect and direct methods may be applied. The choice is dependent on various factors, which include the biodeteriogen types and their degree of extension, and the nature and status of conservation of the artefact, as first suggested by the Italian Committee [6].

Indirect methods are intended to modify or limit the effects of some harmful environmental factors (pH, RH, T, light, etc.), which directly affect the manufact, contributing to the microorganisms growth (Figure 1).

As these do not act directly on the object, they may be considered fully green in terms of conservation. They are suitable for indoor interventions, while they are difficult to adopt in outdoor (Figure 2). It is pleonastic to add that the control of environmental factors is almost unachievable for outdoor cultural heritage assets, since these may cover a large area and, hence, be subjected to climate changes and pollution.

**Figure 1.** Example of physicochemical parameters affecting an indoor environment. The presence of visitors requires lights and produces an increase of temperature, relative humidity, and release of $CO_2$, dramatically changing the delicate equilibrium of these environments. The image shows the Crypt of Popes in St. Callistus Catacombs, Rome, Italy, during the CATS project (photo, C. Urzì).

**Figure 2.** Example of outdoor archeological area. Remains of the antique Roman town of Ercolano (Naples, Italy). In general, outdoor environmental parameters such as sun irradiation, rain, raising humidity, wind, etc. can only be partially controlled due to the different exposition and the extension of the area. Indirect methods aim to keep environmental parameters in a range of acceptability. Image taken during the project POR Calabria FESR, project "NANOPROTECH" (NANO PROtection Technology for Cultural Heritage, J24E07000380005). (photo, C. Urzì).

Indirect methods of changing harmful environmental factors have similar limitations in submerged conditions (Figure 3), as artifacts are subjected to continuous contacts with fresh or seawater, wave currents, and seafloor nature (rock or sand) [7–9].

**Figure 3.** An example of submerged archeological items lying on the seafloor subjected to several underwater environmental factors such as sea current and the nature of the seabed. Biodeterioration is mainly due to the activity of micro- and macro-foulers depending on the position of the items (Photo, Alfina and Mario Tassone, Diving Center Punta Stilo, Calabria, Italy).

In order to control the growth of organisms in most conservation procedures, the use of direct methods (mechanical, physical, chemical, and biological) is the only solution. In this scenario, whatever method is chosen, any intervention has to be carefully planned in order to reduce the risks of unwanted recolonization. To prevent this drawback, a multistep approach, which takes into account the whole of the factors influencing medium and/or long-term conservation, has to be considered.

Direct methods, acting on the material's surface, are potentially harmful to the artifact. For this, it is important to be aware of the advantages/disadvantages that the different methods entail.

### 3.1. Physical Methods

Monochromatic visible light [10] is a very safe, useful, and simple method to reduce or prevent the growth of photosynthetic microflora. The most used wavelengths are blue (λ 470–490 nm), green (λ 500–530 nm), and red (λ 680–700 nm) (Figure 4). If the intent is to block the photosynthetic process, it has to be considered that most of these microorganisms also have accessory pigments. Removal of these may require different wavelengths. In addition, they are ineffective against other kinds of microorganisms, unless photosensitizers (e.g., erythrosine-ERY) are applied [1]. In some cases, those that survive may grow on organic debris from dead cells that are left behind [5,11].

UV irradiation, especially UV-C (λ 254 nm) will kill most indoor colonizing microorganisms and are especially used in natural caves and hypogea [12–14] and show caves [15]. They are easy to handle, but some preliminary tests need to be performed to establish the distance of lamps and duration of treatment. As their target is DNA, they need to be carefully used (for example, they cannot be left on during tourists' visits), and any damage to the organic component and pigments should be considered.

Thermal treatments (microwave, heat irradiation, heat shock treatment) are considered an ecologically safe alternative to traditional conservation methods [16–19]. Thanks to their low-interaction with the substrate and penetration depth, they are safe for the cultural asset, for operators, and the environment. Further, they are effective against lichens and bryophytes and partially on green algae, but they do not have any effect on most of the environmental bacteria and some black fungi [1]. Some side effects, such as the possible diffusion of water into the substrate and/or the production of the organic debris of dead cells, which are readily used as a carbon source by other microorganisms (see above), have also been reported.

**Figure 4.** Experimental chamber used to assess the effectiveness of monochromatic wavelength against photosynthetic biodeteriogens (photo, L.Bruno, University Tor Vergata, Rome, Italy).

### 3.2. Chemical Methods

Chemical methods bring up other issues. It has to be considered that the products utilized in CH restoration processes were usually created to be utilized in different fields and, in the past, indiscriminately applied on different substrates and not contextualized in a whole restorative project. Most of them were pesticides or herbicides normally used in agriculture [20]. The application has to take into account the possible effects on the substrate. The low cost of the products is also a consideration.

The European Community directive [21] has introduced significant restrictions on the use of chemical biocides based on their potential harm for users and other environmental issues.

## 4. Concept of Biocides and Antifouling Products

The following definitions are given to better clarify the procedural characteristics that have to be applied to counteract biodeteriogens.

Biocide: according to European legislation, a biocide is defined as "a chemical substance or microorganism intended to destroy, deter, render harmless, or exert a controlling effect on any harmful organism" [21]. As far as CH artifacts are concerned, the term biocide refers to chemicals/formulations or microorganisms that kill any biodeteriogen organisms, usually organized to form a biofilm, that colonize a monument's surface.

Antifouling defines the ability of specifically designed material coatings to remove or prevent biofouling by any number of organisms on wet surfaces [22].

## 5. Choice of Suitable Chemical Methods

As reported by Mascalchi et al. [19], although the treatment of biodeteriogens is a part of common restoration practice, the development of efficient protocols still represents a challenge for restorers.

There are two ways to counteract biodeteriogens. The first one is the removal of the already existing biomass. This is usually performed by applications of biocides, which have to fulfill a series of requirements, such as being effective against the specific microorganisms. In this case, conventional biocides (par. 7) are still the best products to be used because they are effective against a broad spectrum of organisms, are not harmful to the treated item, can be easily removed, and after cleaning, do not show any chemical or aesthetic interference with the stone surface.

The second is to prevent recolonization. According to the current guidelines for conservation, after cleaning, prevention procedures have to be planned in order to avoid a new recolonization and/or to slow down the growth rate of biofilms for a time longer than a bare surface [23–25]. This effect can be reached by the application of coating products that generally combine, among others, consolidants, hydro repellents, and surfactants [24,25].

Two major antibiofilm (or antifouling) coating approaches are used. The first strategy involves the use of chemically active antimicrobial coatings. The second consists in the inhibition of organism settlement on the surface or in the release of settled organisms without involving chemical reactions [26–31]. Fouling-release systems have been particularly used to counter marine biocolonization [32].

## 6. Biologically-Derived Biocides

There is an increasing interest in the use of natural or biologically derived products that can act as biocides for cultural heritage purposes [33,34].

They may have different origins, including:

- microbial by-products such as zoosteric acid, capsaicin, extracellular enzymes, hydrolases, usnic acid, parietin, or bacterial extract (e.g., *Bacillus*) [34];
- whole microorganisms against microorganisms [11,35];
- plant extracts such as essential oils [33,36,37], alone or embedded in sol/gel matrix to enhance their performance (Urzì, unpublished data).

The practical application of whole or extracts of microorganisms is quite difficult, as these are hard to control. Risk assessment protocols are still missing [38].

The natural origin of plant extracts as biocides limits their outdoor application. Indoors or in hypogean environments, they have been successfully used to some extent [33]. However, as reported by Fidanza and Caneva, there is a lack of adequate information as to specific efficacy of biodeteriogens at low doses, the durability of the treatment, or eventual interference with material substrates [39].

## 7. Conventional Biocides

Although many biocides and application techniques have recently been developed, formulations containing quaternary ammonium compounds (QACs) are still widely used in CH stone conservation. QACs are used alone or in formulations with organic compounds that increase efficiency and range of action, often with lengthened stability. In fact, Preventol R180 (CTS), Rocima™ 103 (CTS), and Biotin R (CTS) still remain the chemical compounds most used by restorers and conservators in Italy [40]. Although some ingredients of these formulates are currently considered toxic, they are applied at low concentrations, which do not create significant risks for humans or the environment (Table 1).

**Table 1.** List of most common biocides used in Italy.

| Commercial Product and Active Ingredient | % of Use and Solvent | Spectrum of Action | References |
|---|---|---|---|
| **Preventol RI80 (CTS)<br>alchyl-dimethyl-benzilammoniumchloride (78–82%) (benzalkonium chloride) (9–11%)** | 2–10% in water | fungi, bacteria and algae | [1,40] |
| **Rocima™ 103 (CTS)<br>di-n-decyl-dimethylammoniumchloride 40–60%, 2-N-octil-2H-isotiazol-3-one 7–10%, Isopropanol 15–20%, formic acid (1–2.5%)** | 2% water | lichens, fungi, bacteria and algae | [1,40] |
| **Biotin R (CTS)<br>3-iodo-2-propynylbutyl carbammate 10–25% 2% 2-N-octil-2H-isotiazol-3-one 2.5–10%, 2-(2-butossietossi)etanolo 50–100%** | 2–5% ethanol | fungi, bacteria and algae | [40] |
| **Biotin T (CTS)<br>di-n-decyl-dimethylammoniumchloride 40–60%, 2-N-octil-2H-isotiazol-3-one 7–10%, Isopropanol 15–20%, formic acid (1–2.5%)** | 2% water | fungi, bacteria and algae | [1] |

Field experiments have recently been reported by Mascalchi et al. [19]; the authors compared the efficacy and duration of three common conservation methods. Simple mechanical removal of biodeteriorants allowed the rapid recolonization of stone. The use of Rocima™ 103 and microwave treatment produced a short term comparable effect for 15 months; the biocide was active for 5 years.

In a recent paper [41], treatment of a wall in the archeological site of Ercolano with Biotin R, (CTS, Italy), (5% in ethyl alcohol), alone or combined with Titania ($TiO_2$ and Ag-$TiO_2$), produced a comparable efficacy after 8 months.

## 8. New Technologies

When synthesizing new materials with biocidal activity, a green approach, and ecosustainability considerations should be a part of the design, synthesis, and selection of renewable feedstocks or raw materials, and degradation planning, with the subjection of the standards of LCA (Life Cycle Assessment) [42].

Many scientists, aware of the importance of safeguarding cultural heritage, have been transferring their skills to CH purposes by providing innovative solutions and green alternatives to old procedures. Concepts such as Safe by Design structural/functional relationships are being used in the development of innovative products and formulations for specific cleaning, preservation, and repair tasks. Bio- and nanotechnologies are significant contributions in this context.

### *8.1. Nanoparticles*

Nanotechnologies contribute in several ways to the development of innovative methodologies for CH conservation [43]. In particular, new strategies based on nanoparticles have provided some promising alternatives, even as biocides, in the consolidation and protection of damaged CH materials.

Due to their large variety and multifunctional addressing (mechanical, hydrophobic, biocides), nanoparticles (NPs), such as $SiO_2$, ZnO, $TiO_2$, $Mg(OH)_2$, $Ca(OH)_2$, $ZrO_2$, $TiO_2$, Ag, etc., in suspension, combined with appropriate resins or as precursor formulations, have been widely applied in stone material restoration [44–46]. Featuring good penetration properties, NPs significantly contribute to the consolidative/protective processes of the substrate.

By using proper strategies, for example, the combination with low surface energy polymer matrices, nanoparticles have demonstrated high capabilities in producing biomimetic micro/nanostructured surfaces on a variety of stone substrates with controlled wettability and roughness [47–49]. For this, NPs are useful as alternatives to synthetic organic polymers for restorations.

Among NPs, semiconductor $TiO_2$ photocatalysts have played the main role. $TiO_2$-based materials are currently used as self-cleaning and antibacterial coatings. Their application is limited because near-ultraviolet irradiation is required for photocatalytic activation. $TiO_2$ doping or formation of heterostructures produce useful means to access to VIS-light driven reactions [45].

Although very promising, the use of nanomaterials and nanotechnologies in the field of CH conservation, over time, has raised serious concerns both in terms of human health and environmental risks, as well as in terms of efficiencies against microorganisms, and long-term effects on the material [50,51].

### *8.2. Ionic Liquids*

In spite of the research to find new and sustainable materials and solutions for the restoration of CH stones, very little attention has been paid to ionic liquids (ILs). In the last two decades, this class of functional and multifunctional materials has become a "major subject of study for modern chemistry due to their high potential of applications in many industrial fields" [52,53].

ILs, in brief, are low melting point organic salts which exhibit a variety of tunable properties such as low vapor pressure, high thermal and chemical stability, recyclability, good electrical conductivity, and wide electrochemical range. They are largely used as alternative "green" solvents for conventional organic media (VOCs) in chemical synthesis, catalysis, separation processes, biomass processing, electrochemistry, and nano- and biotechnologies [54].

The chemical versatility of ILs stems from their ionic nature, which allows a synthetic control on their physicochemical and biological properties and functions through appropriate choice of the anion/cation couple, creating theoretically infinite possibilities. By anchoring to chosen substrates such as polymer networks, silica substrates, and nanosystems, ILs based materials can be built with specific thermal properties, mechanical stability, catalytic functions, and even new functions [55–57].

The majority of ILs are based on nitrogen or phosphorus cations (imidazolium, ammonium, piperidinium, phosphonium, etc.) (Figure 5a). The anions, which usually are employed as modulating agents of overall ILs properties (hydrophobicity, viscosity, etc.), may consist of simple halides ($Cl^-$, $Br^-$, $I^-$), and organic and inorganic species (Figure 5b). Their widespread use has posed serious environmental concerns. Numerous studies have been carried out to establish ILs toxicity and biodegradability as functions of the cation and/or anion species [58]. The green credentials of ILs have been revised, giving rise to the so called "ILs of third generation" based on naturally occurring and/or biodegradable ions (for example, cholinium, morpholinium, aminoacids, drugs ions, etc.), which may have intrinsic biological and pharmaceutical activities [59–61].

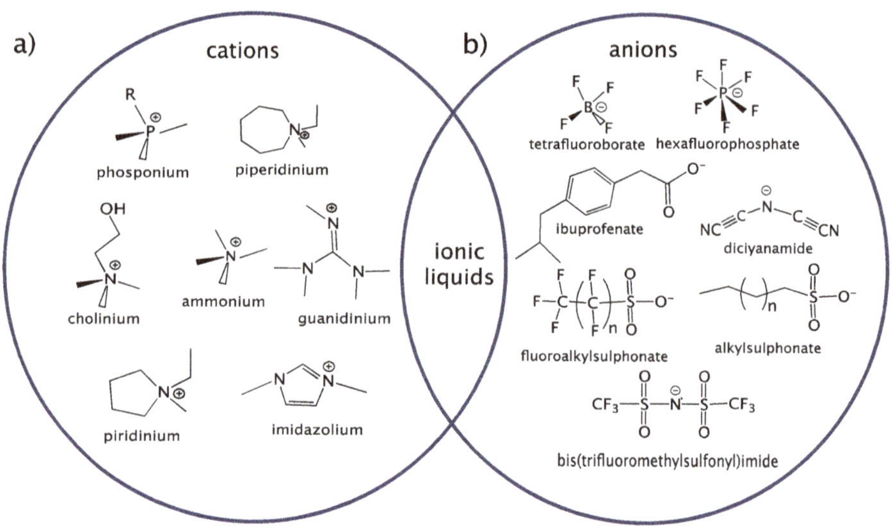


**Figure 5.** Representation of common classes of (**a**) cations and (**b**) anions, respectively, used as IL constituents.

IL surface activity has gained much attention. Surface active ionic liquids (SAILs) have been proposed, among others, as an intriguing and varied source of novel ionic surfactants, with a high impact in colloidal/surface/nano-science and biotechnology [62,63].

Due to their structural analogies with conventional amphiphilic quaternary ammonium (QACs) species, SAILs have an intense antimicrobial activity [64–66]. Their bioactivity is mainly correlated to their lipophilicity/hydrophobic behavior, with the cation playing the main role in that it is responsible for the primary electrostatic interactions with cell walls [67,68], while the role of the anion is not fully established [69]. Biodegradability of ILs is, among others, dependent on alkyl chain length. Too short or too long C atom chains make ILs less biodegradable [58–61].

By application of well-known techniques, such as polymerization, self-assembled monolayers, and so on, ILs grafted surfaces with controlled wettability, antifouling, and sensing functions have been obtained [70–72].

ILs were successfully applied in the CH field to clean calcium crust ($CaCO_3$, $CaSO_4$, $CaC_2O_4$) from stained glasses [73], and synthetic and natural varnishes [74]. Surfactant phosphonium ILs, based on the luminescent 8-anilino-1-naphthalenesulfonate and 1-pyrene carboxylate "metal-coordinating anions" (Figure 6), were tested for corrosion crust removal from medieval glasses. The results were more efficient than conventional EDTA [75].

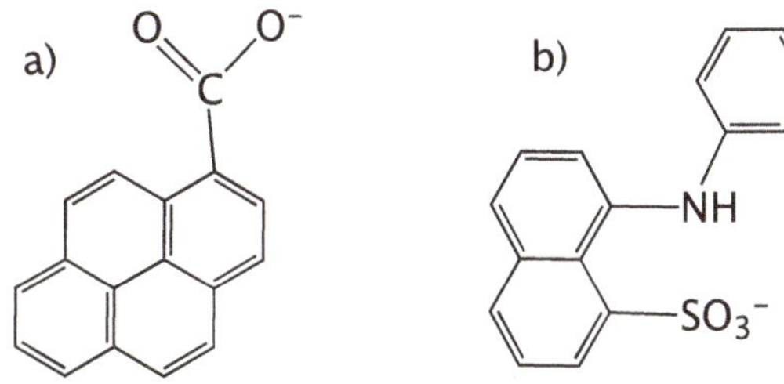


**Figure 6.** Luminescent anions: (**a**) 1-pyrenecarboxylate and (**b**) 8-Anilino-1-naphthalene sulfonate, respectively.

ILs have been widely exploited for the pretreatment, processing, and preservation of lignocellulose materials [76,77], displaying, among other characteristics, effective antibacterial and antifungal activities and wettability properties.

Protic ionic liquids, based on the 1-ammonium-2-propanol cation and a series of carboxylates anions, have been recently proposed as "green" cleaner agents, as an alternative to thymol [78]. Schmitz et al. [79] observed that imidazolium ILs used in "paper de-acidification and reinforcement processes" have high similarity with well-known antimicotic agents such as Clotrimazole and Miconazole. For this reason, the authors investigated the antifungal capabilities of chloride and acetate imidazolium ILs against common paper molds and the permanence of ILs after the treatment. The results of this study were positive and, as the authors suggest, may represent the basis of future research aimed at obtaining simultaneous repair and protection against microorganisms in treated documents.

Regarding the application of ILs in stone restoration, very little has been done. An interesting example is represented by ILs based on the multifunctional polyoxametalated $[\alpha\text{-}SiW_{11}O_{39}]^{8-}$ (POM) polyanion, a multifunctional class of metal cluster oxides never exploited in this frame. The two tetralkylammonium POM-ILs (Figure 7), namely $[(n\text{-}C_7H_{15})_4N]_8[\alpha\text{-}SiW_{11}O_{39}]$ (1) and $[(n\text{-}C_6H_{13})_3[(n\text{-}C_{14}H_{29})N]_8[\alpha\text{-}SiW_{11}O_{39}]$ (2), respectively, have been investigated both as antimicrobial and anticorrosive stone coatings [80].

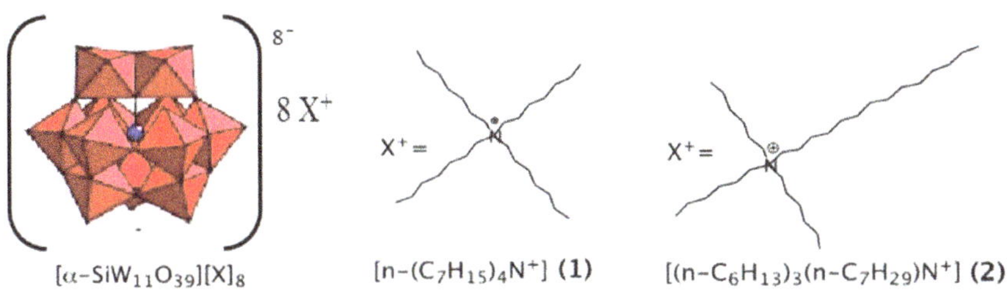


**Figure 7.** Schematic representation of the two POM-ILs, namely $[(n\text{-}C_7H_{15})_4N]_8$ $[\alpha\text{-}SiW_{11}O_{39}]$ (**1**) and $[(n\text{-}C_6H_{13})_3[(n\text{-}C_{14}H_{29})N]_8[\alpha\text{-}SiW_{11}O_{39}]$ (**2**), investigated both as antimicrobial and anticorrosive stone coatings.

Both 1 and 2 POM-ILs coatings proved very resistant to acid exposure. The degree of corrosion was dependent on stone features (> porosity = > damage), while the thickness of the coating was important to reduce the acidic corrosion. Their antimicrobial activity was tested by using Gram-positive strain (*B. subtilis*) and Gram-negative one (*E. coli*) as bacterial models [80]. The data obtained in vitro revealed a good activity for both POM-ILs against the two strains (particularly vs. *B. subtilis*) being better performing. They have also shown good antifouling activity on different stone substrates. However, as the same authors declared, the data needs further research, in that the tests were performed on strains not derived from stone, and over a short time (72 h).

Water wettability (absorbency/hydrophobicity) significantly contributes to the surface deterioration process of stone artifacts and buildings and also to biological colonization. ILs anion exchange constitutes a useful means to control the wettability of ILs based materials. Studies regarding a series of ammonium ILs species based on a variety of SA anions, both fluorinated and non-fluorinated, have shown that dodecylbenzenesulfonate (DBS) anion produced the most hydrophobic behavior [81,82]. These results prompted an investigation on the influence of surface-active anions on the antimicrobial activity of the IL series, named [QA][X], as reported by De Leo et al. [83] and in Figure 8.

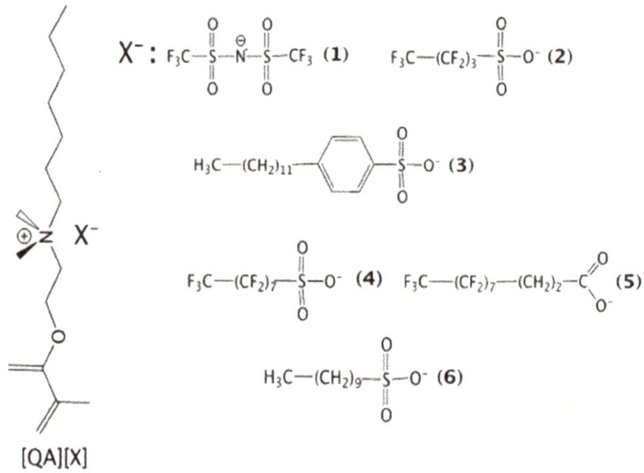


**Figure 8.** Schematic representation of the [QA][X], **1-6,** IL series. The IL numbering is the same used in the original paper [83].

The antimicrobial assessment of 1–6 was carried out on four microbial strains isolated from deteriorated monuments and compared with that of Preventol® 80 (CTS) (S1) and Rocima™ 103 (CTS) (S2) used as standards. The results obtained substantiate that lipophilicity of anions plays a pivotal role in determining the antimicrobial activities of these ILs, in that only the DBS derivative 3, and to some extent, the decanesulfonate one, 6, displayed significant bioactivity. In particular, the DBS based IL was remarkably active against the G (+) *Micrococcus* strain with an efficiency significantly higher than that of the two standards [83].

The interesting capabilities of DBS anion in inducing hydrophobic/lipophilic and antimicrobial properties as well, led to focused research on DBS based surface active anionic ionic liquids (SAAILs). Keeping in mind the mandatory principles of eco-sustainability, the selected cholinium cations (CH) were chosen as counterparts. The newly synthesized ILs include the heptyl and dodecyl [CH][DBS] derivatives, **2a** and **3a**, respectively, (Figure 9a) and the two germinal [CH]gem$[DBS]_2$, **4a**, and **5a**, respectively (Figure 9b). The latter two differ in the connecting methylene chain length (at 6 and 8 C atoms, respectively). For comparison, the bioactivity of the CH halide precursors (**2** and **3**, respectively) has also been investigated. [De Leo et al. in prep.] The precursors **4** and **5** have not been considered because, in preliminary tests performed at different concentrations, they did not show any antimicrobial activity.

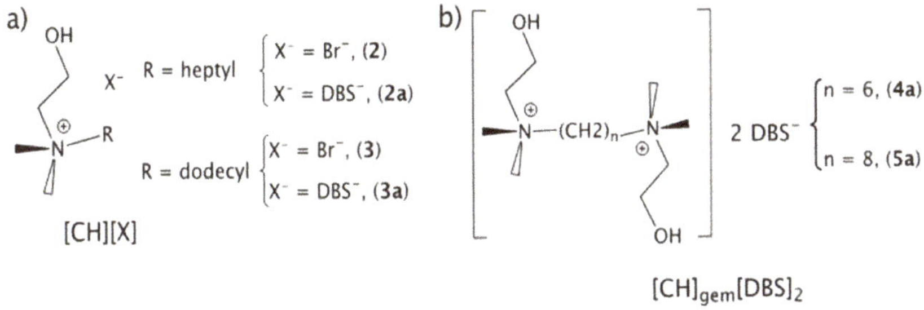


**Figure 9.** Novel (**a**) mono- and (**b**) germinal cholinium DBS based ILs. IL numbering is the same used for the antimicrobial test shown in Figure 8 and in the manuscript [De Leo et al., in prep.].

The tests have been carried out on a mixed stabilized microbial suspension (Gram+ and Gram-, yeasts, hyphomycetes, and unicellular algae) on untreated (CP) stone probes (marble and calcarenite), treated with consolidant only (NanoEstel, NP) or treated with the consolidant and ILs, namely **2**, **2a**, **3**, **3a**, **4a**, and **5a**, respectively. As shown in Figure 10, after 90 days of incubation in laboratory conditions (22 °C and at a constant illumination of ca. 1500 Lux), different behavior of the tested ILs was observed. In fact, **2** and **2a**, and, especially, **3** and **3a**, showed interesting antimicrobial colonization capabilities. On the contrary, **4a** and **5a**, promoted a remarkable microbial growth, even more pronounced than the untreated control sample. Such findings corroborate, once more, that is the IL lipophilicility that determines their bioactivity in that 3 and 3a, featuring longer alkyl chain lengths on the CH cation resulted in the most effective ones.

**Figure 10.** Laboratory tests carried out with ILs showed in Figure 9 on stone probes. On the left, soon after the inoculation of 30 μL of mixed suspension ($T_0$) and, on the right, after 90 days of incubation ($T_{90}$). Tests were carried on marble and tupha probes in double and incubating at 22 °C and a constant illumination of ca. 1500 Lux [De Leo et al., in prep.].

A brief summary of the direct methods currently used or proposed as innovative ones is listed in Table 2.

**Table 2.** Pros and cons of current direct methods and their functions.

| Method | When to Use | | Advantages | Disadvantages |
|---|---|---|---|---|
| | Removal | Preventing | | |
| **Monochromatic visible light** | X | X | Easy to use<br>Safe for the cultural asset and for humans | Effective only against target microorganisms<br>Possible production of organic debris to be used as a carbon source by other microorganisms |
| **Ultraviolet (UV-C)** | X | X | Bleaching effect on biofilms. Irreversible cellular and molecular damages | Cannot be used in the presence of visitors<br>Possible damages on painted surfaces |
| **Heat (microwave, heat treatment, Shock Heat treatment)** | X | - | Portability<br>Safe for the cultural asset and for the humans | Possible production of organic debris<br>Possible water diffusion into the substrate<br>Need to be used with other biocidal treatments |
| **Microorganisms against other microorganisms** | X | - | Effective against target microorganisms | Hard to control practical applications.<br>Risk assessment protocols are still missing |
| **Microbial by-products** | X | - | Effective against target microorganisms | Hard to control practical applications.<br>Risk assessment protocols are still missing |
| **Plant derived biocides (e.g., essential oils)** | X | X | Effective at low doses<br>Successfully employed in indoor environments | Not suitable for outdoor application<br>Missing adequate application information |
| **QAC*-based conventional biocides** | X | - | Effective against a broad spectrum of (micro)-organisms | Short or medium-term efficacy |
| **Nanoparticles** | - | X | Depend on the properties of the treated stone and of the inorganic or organic binder | Concerns of human health and environmental risks.<br>Some microorganisms are resistant (e.g., black fungi) |
| **Ionic Liquids** | X | X | Tunable solvent properties and antimicrobial and surface activity<br>Their biocidal/antifouling efficacy change depending on the length of IL chain, cation and anion used | Careful use of each molecule for dose and efficacy assessment. Some ILs may have an unwished effect incrementing microbial growth |

* = quaternary ammonium compound.

## 9. Final Remarks

Innovation, nowadays, has become synonymous with ecosustainability, and that, in the CH field, is well summarized by the term "green conservation". Starting with the acronym SMART (sustainability, management, risks, tools) from the "green conservation of cultural heritage" conference in 2019, comes the most all-inclusive term "green". Its basics were applied to a critical review of recent results about preventing biodeterioration of stone materials. The conclusion is that chemical treatment for biocidal and antifouling actions is currently the most commonly used and most versatile means toward developing innovative solutions. Despite the interesting results obtained by using biotechnology and nanotechnology, surface-active QAC–based formulations, adopting new sustainable protocols, are being used the most.

Attention should be paid to IL technologies, although currently, there are few data. Tunable solvent properties and antimicrobial and surface activity are just some of the most interesting features of ILs. As reported in the literature, IL is an enabling technology to access to advanced materials, which can contribute to the production of new formulations of antifouling and antimicrobial surface coatings, developed as gel materials and other forms.

Interest in ILs, which exploded in the early 2000s, was mainly based on their green characteristics such as low volatility and recyclability. Over time, studies of their bioactivity and biodegradation caused their greenness to be significantly reconsidered. Research on the design of biodegradable ILs has significantly progressed, leading to the implementation of Safe by Design strategies for ionic liquids developments [58,60,61].

**Author Contributions:** Conceptualization, S.L.S. and C.U.; methodology, S.L.S., C.U., and F.D.L.; resources, S.L.S., C.U., and F.D.L; data curation, S.L.S., C.U., and F.D.L.; writing—original draft preparation, S.L.S. and C.U.; writing—review and editing, S.L.S., C.U., and F.D.L. All authors have read and agreed to the published version of the manuscript.

**Funding:** This research was partially funded by the financial contribution of FFABR 2017 and FFABR 2018 (Italian fund of basic research activities).

**Acknowledgments:** Authors would like to thank Andrea Macchia (Yococu, Youth in Conservation of Cultural Heritage), for his input and definition of SMART conservation and Sherron Collins for her revision of the English text.

**Conflicts of Interest:** The authors declare no conflict of interest.

## References

1. Pinna, D. *Coping with Biological Growth on Stone Heritage Objects. Methods, Products, Applications, and Perspectives*; Apple Academic Press: Oakville, ON, Canada, 2017, ISBN 9781771885324.
2. Guillitte, O. Bioreceptivity: A new concept for building ecology studies. *Sci. Total Environ.* **1995**, *167*, 215–220. [CrossRef]
3. UNI. Beni Culturali—Materiali Lapidei Naturali Ed Artificiali—Descrizione della Forma di Alterazione—Termini e Definizioni. Available online: http://store.uni.com/catalogo/uni-11182-2006/ (accessed on 26 June 2020).
4. ICOMOS-ISCS. Illustrated Glossary on Stone Deterioration Patterns. Available online: http://iscs.icomos.org/glossary.html (accessed on 26 June 2020).
5. Caneva, G.; Nugari, M.P.; Salvadori, O. *Plant Biology for Cultural Heritage: Biodeterioration and Conservation*; Getty Publications: Los Angeles, CA, USA, 2008, ISBN 978-0-89236-939-3.
6. Commissione Normal. *Raccomandazioni Normal: 38/93. Valutazione Sperimentale Dell'efficacia Dei Biocidi*; C.N.R. I.C. R.: Rome, Italy, 1993.
7. García-Castrillo, G.; Lanuza, P.; López, G. El entorno Marino de los Restos Arqueológicos. In *La Conservación del Material Subacuático*; Ayuntamiento de Santoña Monte Buciero Spain: Santoña, Spain, 2003; pp. 95–109, ISSN 1138-9680 9.
8. Ruffolo, S.A.; Macchia, A.; La Russa, M.F.; Mazza, L.; Urzì, C.; De Leo, F.; Barberio, M.; Crisci, G.M. Marine antifouling for underwater archaeological sites: TiO2 and Ag-Doped TiO2. *Int. J. Photoenergy* **2013**, *2013*, 251647. [CrossRef] [PubMed]

9. Camara, B.; Alvarez de Buergo, M.; Bethencourt, M.; Fernandez-Montblanc, T.; La Russa, M.F.; Ricca, M.; Fort, R. Biodeterioration of marble in underwater environment. *Sci. Total Environ.* **2017**, *609*, 109–122. [CrossRef]
10. Bruno, L.; Valle, V. Effect of white and monochromatic lights on cyanobacteria and biofilms from Roman Catacombs. *Int. Biodeter. Biodegr.* **2017**, *123*, 286–295. [CrossRef]
11. Krakova, L.; De Leo, F.; Bruno, L.; Pangallo, D.; Urzì, C. Complex bacterial diversity in the white biofilms of the Catacombs of St. Callixtus in Rome evidenced by different investigation strategies. *Environ. Microbiol.* **2015**, *17*, 1738–1752. [CrossRef]
12. Van Der Molen, J.M.; Garty, J.; Aardema, B.W.; Krumbein, W.E. Growth control of algae and cyanobacteria on historical monuments by a mobile UV unit (MUVU). *Stud. Conserv.* **1980**, *25*, 71–77. [CrossRef]
13. Borderie, F.; Alaoui–Sossé, B.; Aleya, L. Heritage materials and biofouling mitigation through UV–C irradiation in show caves: State of the art practices and future challenges. *Environ. Sci. Pollut. Res.* **2014**, *6*, 4144–4172. [CrossRef]
14. Baquedano Estévez, C.; Moreno Merino, L.; de la Losa Román, A.; Durán Valsero, J.J. The lampenflora in show caves and its treatment: An emerging ecological problem. *Int. J. Speleol.* **2019**, *48*, 249–277. [CrossRef]
15. Pfendler, S.; Einhorn, O.; Karimi, B.; Bousta, F.; Cailhol, D.; Alaoui-Sosse, L.; Alaoui-Sosse, B.; Aleya, L. UV-C as an efficient means to combat biofilm formation in show caves: Evidence from the La Glacière Cave (France) and laboratory experiments. *Environ. Sci. Pollut. Res.* **2017**, *24*, 24611–24623. [CrossRef]
16. Tretiach, M.; Bertuzzi, S.; Candotto Carniel, F. Heat shock treatments: A new safe approach against lichen growth on outdoor stone surfaces. *Environ. Sci. Technol.* **2012**, *46*, 6851–6859. [CrossRef]
17. Riminesi, C.; Olmi, R. Localized microwave heating for controlling biodeteriogens on cultural heritage assets. *Int. J. Conserv. Sci.* **2016**, *7*, 281–294.
18. Bertuzzi, S.; Candotto Carniel, F.; Pipan, G.; Tretiach, M. Devitalization of poikilohydric lithobionts of open-air monuments by heat shock treatments: A new case study centred on bryophytes, Int. *Biodeter. Biodegr.* **2013**, *84*, 44–53. [CrossRef]
19. Mascalchi, M.; Orsini, C.; Pinna, D.; Salvadori, B.; Siano, S.; Riminesi, C. Assessment of different methods for the removal of biofilms and lichens on gravestones of the English cemetery in Florence. *Int. Biodeter. Biodegr.* **2020**, in press. [CrossRef]
20. Allsopp, C.; Allsopp, D. An updated survey of commercial products used to protect materials against biodeterioration. *Int. Biodeterior. Biodegrad.* **1983**, *19*, 99–146. [CrossRef]
21. AA.VV. Regulation (EU) No 528/2012 of the European Parliament and of the Council of 22 May 2012 Concerning the Making Available on the Market and Use of Biocidal Products. 2012. Available online: http://data.europa.eu/eli/reg/2012/528/2019-11-20 (accessed on 18 September 2020).
22. Yebra, D.M.; Kiil, S.; Johansen, K.D. Antifouling technology-past, present and future steps toward efficient and environmentally friendly antifouling coatings. *Progr. Org. Coat.* **2004**, *50*, 75–104. [CrossRef]
23. Delgado Rodrigues, J.; Vale Anjos, M.; Charola, A.E. Recolonization of Marble Sculptures in a Garden Environment. In *Biocolonization of Stone: Control and Preventive Methods*; Charola, A.E., McNamara, C., Koestler, R.J., Eds.; Smithsonian, Contributions to Museum Conservation; Smithsonian Institution Scholarly Press: Washington, DC, USA, 2011; pp. 71–85.
24. Urzì, C.; De Leo, F. Evaluation of the efficiency of water-repellent and biocide compounds against microbial colonization of mortars. *Int. Biodeter. Biodegr.* **2007**, *60*, 25–34. [CrossRef]
25. Pinna, D.; Salvadori, B.; Galeotti, M. Monitoring the performance of innovative and traditional biocides mixed with consolidants and water-repellents for the prevention of biological growth on stone. *Sci. Total Environ.* **2012**, *423*, 132–141. [CrossRef]
26. Liu, X.; Liang, Y.; Zhou, F.; Liu, W. Extreme wettability and tunable adhesion: Biomimicking beyond nature? *Soft Matter* **2012**, *8*, 2070–2086. [CrossRef]
27. Parkin, I.P.; Palgrave, R.G. Self-cleaning coatings. *J. Mater. Chem.* **2005**, *15*, 1689–1695. [CrossRef]
28. Zarzuela, R.; Moreno-Garrido, I.; Blasco, J.; Gil, M.L.A.; Mosquera, M.J. Evaluation of the effectiveness of CuONPs/SiO2-based treatments for building stones against the growth of phototrophic microorganisms. *Construct. Build. Mater.* **2018**, *187*, 501–509. [CrossRef]
29. La Russa, M.F.; Macchia, A.; Ruffolo, S.A.; De Leo, F.; Barberio, M.; Barone, P.; Crisci, G.M.; Urzì, C. Testing the antibacterial activity of doped TiO2 for preventing biodeterioration of Cultural Heritage building materials. *Int. Biodeter. Biodegr.* **2014**, *96*, 87–96. [CrossRef]

30. Artesani, A.; Di Turo, F.; Zucchelli, M.; Traviglia, A. Recent Advances in Protective Coatings for Cultural Heritage–An Overview. *Coatings* **2020**, *10*, 217. [CrossRef]
31. Zarzuela, R.; Luna, M.; Carrascosa, L.A.; Mosquera, M.J. Preserving Cultural Heritage Stone: Innovative Consolidant, Superhydrophobic, Self-Cleaning, and Biocidal Products. In *Advanced Materials for the Conservation of Stone*; Hosseini, M., Karapanagiotis, I., Eds.; Springer: Chan, Switzerland, 2018; pp. 259–275. [CrossRef]
32. Lejars, M.; Margaillan, A.M.; Bressy, C. Fouling release coatings: A nontoxic alternative to biocidal antifouling coatings. *Chem. Rev.* **2012**, *112*, 4347–4390. [CrossRef]
33. Palla, F.; Bruno, M.; Mercurio, F.; Tantillo, A.; Rotolo, V. Essential Oils as Natural Biocides in Conservation of Cultural Heritage. *Molecules* **2020**, *25*, 730. [CrossRef] [PubMed]
34. Bruno, L.; Rugnini, L.; Spizzichino, V.; Caneve, L.; Canini, A.; Ellwood, N.T.W. Biodeterioration of Roman hypogea: The case study of the Catacombs of SS. Marcellino and Pietro (Rome, Italy). *Ann. Microbiol.* **2019**, *69*, 1023–1032. [CrossRef]
35. Silva, M.; Rosado, T.; Teixeira, D.; Candeias, A.; Caldeira, A.T. Green mitigation strategy for Cultural Heritage: Bacterial potential for biocide production. *Environ. Sci. Pollut. Res.* **2017**, *24*, 4871–4881. [CrossRef]
36. Jurado, V.; del Rosal, Y.; Gonzalez-Pimentel, J.L.; Hermosin, B.; Saiz-Jimenez, C. Biological control of phototrophic biofilms in a show cave: The case of Nerja Cave. *Appl. Sci.* **2020**, *10*, 3448. [CrossRef]
37. Veneranda, M.; Blanco-Zubiaguirre, L.; Roselli, G.; Di Girolami, G.; Castro, K.A.; Madariaga, J.M. Evaluating the exploitability of several essential oils constituents as a novel biological treatment against cultural heritage biocolonization. *Microchem. J.* **2018**, *138*, 1–6. [CrossRef]
38. ECHA. *Guidance on the Biocidal Products Regulation: Volume V-Guidance on Active Microorganisms and Biocidal Products*; European Chemicals Agency: Helsinki, Finland, 2017. [CrossRef]
39. Fidanza, M.R.; Caneva, G. Natural biocides for the conservation of stone cultural heritage: A review. *J. Cult. Herit.* **2019**, *38*, 271–286. [CrossRef]
40. Bartolini, M.; Pietrini, A.M.; Ricci, S. Valutazione dell'efficacia di alcuni nuovi biocidi per il trattamento di microflora fotosintetica e di briofite su materiali lapidei. *Boll. ICR* **2007**, *14*, 101–111.
41. Ruffolo, S.A.; De Leo, F.; Ricca, M.; Arcudi, A.; Silvestri, C.; Bruno, L.; Urzì, C.; La Russa, M.F. Medium-term in situ experiment by using organic biocides and titanium dioxide for the mitigation of microbial colonization on stone surfaces. *Int. Biodeter. Biodegr.* **2017**, *123*, 17–26. [CrossRef]
42. Anastas, P.T.; Warner, J.C. *Green Chemistry: Theory and Practice*; Oxford University Press: New York, NY, USA, 1998.
43. Baglioni, P.; Carretti, E.; Chelazzi, D. Nanomaterials in art conservation. *Nat. Nanotechnol.* **2015**, *10*, 287–290. [CrossRef] [PubMed]
44. Sierra-Fernández, A.; Gómez-Villalba, L.S.; Rabanal, M.E.; Fort, R. New nanomaterials for applications in conservation and restoration of stony materials: A review. *Mater. Construct.* **2017**, *67*, e107. [CrossRef]
45. Toniolo, L.; Gherardi, F. The Protection of Marble Surfaces: The Challenge to Develop Suitable Nanostructured Treatments. In *Advanced Materials for the Conservation of Stone*; Hosseini, M., Karapanagiotis, I., Eds.; Springer: Cham, Switzerland, 2018; pp. 57–78. [CrossRef]
46. Ruffolo, S.A.; La Russa, M.F. Nanostructured Coatings for Stone Protection: An Overview. *Front. Mater.* **2019**, *6*, 147. [CrossRef]
47. Manoudis, P.N.; Karapanagiotis, I.; Tsakalof, A.; Zuburtikudis, I.; Kolinkeovà, B.; Panayiotou, C. Superhydrophobic films for the protection of outdoor cultural heritage assets. *Appl. Phys. A* **2009**, *97*, 351–360. [CrossRef]
48. Pinho, L.; Mosquera, M.J. Titania-silica nanocomposite photocatalysts with application in stone self-cleaning. *J. Phys. Chem. C* **2011**, *115*, 22851–22862. [CrossRef]
49. Martines, E.; Seunarine, K.; Morgan, H.; Gadegaard, N.; Wilkinson, C.D.W.; Riehle, M.O. Superhydrophobicity and superhydrophilicity of regular nanopatterns. *Nano Lett.* **2005**, *5*, 2097–2103. [CrossRef]

50. Reyes-Estebanez, M.; Ortega-Morales, B.O.; Chan-Bacab, M.; Granados-Echegoyen, C.; Camacho-Chab, J.C.; Pereañez-Sacarias, J.E.; Gaylarde, C. Antimicrobial engineered nanoparticles in the built cultural heritage context and their ecotoxicological impact on animals and plants: A brief review. *Herit. Sci.* **2018**, *6*, 52. [CrossRef]
51. Semenzin, E.; Giubilato, E.; Badetti, E.; Picone, M.; Volpi Ghirardini, A.; Hristozov, D.; Brunelli, A.; Marcomini, A. Guiding the development of sustainable nano-enabled products for the conservation of works of art: Proposal for a framework implementing the safe by design concept. *Environ. Sci. Pollut. Res.* **2019**, *26*, 26146–26158. [CrossRef]
52. Plechkova, N.V.; Seddon, K.R. Applications of ionic liquids in the chemical industry. *Chem. Soc. Rev.* **2008**, *37*, 123–150. [CrossRef]
53. Welton, T. Ionic liquids: A brief history. *Biophys. Rev.* **2008**, *10*, 691–706. [CrossRef] [PubMed]
54. Eftekhari, A. Lonic Liquid Devices. Royal Society of Chemistry. 2018. Available online: https://pubs.rsc.org/en/content/ebook/978-1-78801-183-9 (accessed on 18 September 2020).
55. Mecerreyes, D. Polymeric ionic liquids: Broadening the properties and applications of polyelectrolytes. *Prog. Polym. Sci.* **2011**, *36*, 1629–1648. [CrossRef]
56. Xin, B.; Hao, J. Imidazolium-based ionic liquids grafted on solid surfaces. *Chem. Soc. Rev.* **2014**, *43*, 7171. [CrossRef] [PubMed]
57. Cardiano, P.; Fazio, E.; Lazzara, G.; Manickam, S.; Milioto, S.; Neri, F.; Mineo, P.G.; Piperno, A.; Lo Schiavo, S. Highly untangled multiwalled carbon nanotube@polyhedral oligomeric silsesquioxane ionic hybrids: Synthesis, characterization and nonlinear optical properties. *Carbon* **2015**, *86*, 325–337. [CrossRef]
58. Jordan, A.; Gathergood, N. Biodegradation of ionic liquids—A critical review. *Chem. Soc. Rev.* **2015**, *44*, 8200–8237. [CrossRef]
59. Ferraz, R.; Branco, L.C.; Prudencio, C.; Noronha, J.P.; Petrovski, Z. Ionic Liquids as active pharmaceutical ingredients. *ChemMedChem* **2011**, *6*, 975–985; [CrossRef]
60. Petkovic, M.; Seddon, K.R.; Rebelo, L.P.N.; Pereira, C.S. Ionic liquids: A pathway to environmental acceptability. *Chem. Soc. Rev.* **2011**, *40*, 1383–1403. [CrossRef]
61. Santos, J.I.; Gonçalves, A.M.M.; Pereira, J.L.; Figueiredo, B.F.H.T.; Silva, F.A.; Coutinho, J.A.P.; Ventura, S.P.M.; Gonçalves, F. Environmental safety of cholinium-based ionic liquids: Assessing structure–ecotoxicity relationships. *Green Chem.* **2015**, *17*, 4657–4668. [CrossRef]
62. Marr, P.C.; Marr, A.C. Ionic liquid gel materials: Applications in green and sustainable chemistry. *Green Chem.* **2016**, *18*, 105. [CrossRef]
63. Brogan, A.P.S.; Clarke, C.J.; Charalambidou, A.; Loynachan, C.N.; Norman, S.E.; Doutche, J.; Hallet, J.P. Expanding the design space of gel materials through ionic liquid mediated mechanical and structural tuneability. *Mater. Horiz.* **2020**, *7*, 820–826. [CrossRef]
64. Pendleton, J.N.; Gilmore, B.F. The antimicrobial potential of ionic liquids: A source of chemical diversity for infection and biofilm control. *Int. J. Antimicrob. Agents* **2015**, *46*, 131–139. [CrossRef] [PubMed]
65. Egorova, K.S.; Gordeev, E.G.; Ananikov, V.P. Biological Activity of Ionic Liquids and Their Application in pharmaceutics and medicine. *Chem. Rev.* **2017**, *117*, 7132–7189. [CrossRef] [PubMed]
66. Sintra, T.E.; Vilas, M.; Martins, M.; Ventura, S.P.M.; Lobo Ferreira, A.I.M.C.; Santos, L.M.N.B.F.; Fernando, J.M.; Tojo, G.E.; Coutinho, J.A.P. Synthesis and characterization of surface-active Ionic Liquids used in the disruption of Escherichia coli cells. *ChemPhysChem* **2019**, *20*, 727–735. [CrossRef]
67. Benedetto, A.; Ballone, P. Room temperature Ionic Liquids Meet Biomolecules: A microscopic view of structure and dynamics. *ACS Sustain. Chem. Eng.* **2016**, *4*, 392–412. [CrossRef]
68. Benedetto, A.; Ballone, P. Room-Temperature Ionic Liquids and biomembranes: Setting the stage for applications in pharmacology, biomedicine, and bionanotechnology. *Langmuir* **2018**, *34*, 9579–9597. [CrossRef]
69. Guo, J.; Xu, Q.; Zheng, Z.; Zhou, S.; Hailei, M.H.; Wang, B.; Yan, F. Intrinsically antibacterial poly(ionic liquid) membranes: The synergistic effect of anions. *ACS Macro Lett.* **2015**, *4*, 1094–1098. [CrossRef]
70. Ye, Q.; Tingting Gao, T.; Wan, F.; Yu, B.; Pei, X.; Zhou, F.; Xue, Q. Grafting poly(ionic liquid) brushes for anti-bacterial and anti-biofouling applications. *J. Mater. Chem.* **2012**, *22*, 13123–13131. [CrossRef]
71. Jin, L.; Shi, Z.; Zhang, X.; Liu, X.; Li, H.; Wang, J.; Liang, F.; Zhao, W.; Zhao, C. Intelligent antibacterial surface based on ionic liquid molecular brushes for bacterial killing and release. *J. Mater. Chem.* **2019**, *7*, 5520–5527. [CrossRef]

72. Su-Yun, Z.; Zhuang, Q.; Zhang, M.; Wang, H.; Gao, Z.; Jian-Ke, S.J.; Jiayin, Y. Poly(ionic liquid) composites. *Chem. Soc. Rev.* **2020**, *49*, 1726–1755. [CrossRef]
73. Machado, A.S.; Redol, P.; Branco, L.; Vilarigues, M. Ionic Liquids for Medieval Stained-Glass Cleaning: A New Frontier. In Proceedings of the ICOM-CC Lisbon. Portugal: 16th Triennal Conference, Lisbon, Portugal, 19–23 September 2011.
74. Pacheco, M.F.; Pereira, A.I.; Branco, L.C.; Parola, A.J. Varnish removal from paintings using ionic liquids. *J. Mater. Chem. A* **2013**, *1*, 7016–7018. [CrossRef]
75. Delgado, J.M.; Nunes, D.; Fortunato, E.; Laia, C.A.T.; Branco, L.C.; Vilarigues, M. The effect of three luminescent ionic liquids on corroded glass surfaces—A first step into stained-glass cleaning. *Corros. Sci.* **2017**, *118*, 109–117. [CrossRef]
76. Przybysz, K.; Drzewinńska, E.; Stanisławska, A.; Wysocka-Robak, A.; Cieniecka-Rosłonkiewicz, A.; Foksowicz-Flaczyk, J.; Pernak, J. Ionic Liquids and paper. *Ind. Eng. Chem. Res.* **2005**, *44*, 4599–4604. [CrossRef]
77. Abushammala, H.; Mao, J. A review on the partial and complete dissolution and fractionation of wood and lignocelluloses using imidazolium Ionic Liquids. *Polymers* **2020**, *12*, 195. [CrossRef]
78. Dimitric, N.; Spremo, N.; Vraneš, M.; Belić, S.; Karaman, M.; Kovačević, S.; Karadžić, M.; Podunavac-Kuzmanović, S.; Crkvenjakovc, D.K.; Slobodan Gadžurić, S. New protic ionic liquids for fungi and bacteria removal from paper heritage. *RSC Adv.* **2019**, *9*, 17905–17912. [CrossRef]
79. Schmitz, K.; Wagner, S.; Reppke, M.; Maier, C.L.; Windeisen-Holzhauser, E.; Benz, J.P. Preserving cultural heritage: Analyzing the antifungal potential of ionic liquids tested in paper restoration. *PLoS ONE* **2019**, *14*, e0219650. [CrossRef]
80. Archismita, M.; Castillo, I.F.; Müller, D.P.; Gonzµlez, C.; Eyssautier-Chuine, S.; Ziegler, A.; de la Fuente, J.M.; Mitchell, S.G.; Streb, C. Polyoxometalate-Ionic Liquids (POM-ILs) as anticorrosion and antibacterial coatings for natural stones. *Angew. Chem. Int. Edit.* **2018**, *57*, 14926–14931. [CrossRef]
81. Cardiano, P.; Mineo, P.G.; Neri, F.; Lo Schiavo, S.; Piraino, P. A new application of ionic liquids: Hydrophobic properties of tetraalkylammonium based poly(ionic liquid)s. *J. Mater. Chem.* **2008**, *18*, 1253–1260. [CrossRef]
82. Mineo, P.G.; Livoti, L.; Giannetto, M.; Gulino, A.; Lo Schiavo, S.; Cardiano, P. Very fast $CO_2$ response and hydrophobic properties of novel poly(ionic-liquid)s. *J. Mater. Chem.* **2009**, *19*, 8861–8870. [CrossRef]
83. De Leo, F.; Cardiano, P.; De Carlo, G.; Lo Schiavo, S.; Urzì, C. Testing the antimicrobial properties of an upcoming "environmental-friendly" family of ionic liquids. *J. Mol. Liq.* **2017**, *248*, 81–85. [CrossRef]

*applied sciences*

MDPI

*Article*

# Biological Control of Phototrophic Biofilms in a Show Cave: The Case of Nerja Cave

**Valme Jurado [1], Yolanda del Rosal [2], Jose Luis Gonzalez-Pimentel [1,3], Bernardo Hermosin [1] and Cesareo Saiz-Jimenez [1,*]**

1 Instituto de Recursos Naturales y Agrobiologia, CSIC, 41012 Sevilla, Spain; vjurado@irnase.csic.es (V.J.); pimentel@irnas.csic.es (J.L.G.-P.); b.hermosin@csic.es (B.H.)
2 Instituto de Investigacion Cueva de Nerja, 29787 Nerja, Spain; yolanda@cuevadenerja.es
3 Current address: HERCULES Laboratory, Evora University, 7000-809 Evora, Portugal
* Correspondence: saiz@irnase.csic.es

Received: 20 March 2020; Accepted: 14 May 2020; Published: 16 May 2020

**Abstract:** Cyanobacteria and microalgae are usually found in speleothems, rocks and walls of show caves exposed to artificial lighting. These microorganisms develop as biofilms coating the mineral surfaces and producing aesthetic, physical and chemical deterioration. A wide number of physical, chemical and environmental-friendly methods have been used for controlling the biofilms with different results. Natural biological control has been suggested by some authors as a theoretical approach but without direct evidence or application. Here we report the finding of a natural biological control of phototrophic biofilms on the speleothems of Nerja Cave, Malaga, Spain. The formation of plaques or spots where the phototrophic microorganisms disappeared can be assumed on the basis of processes of predation of bacteria, amoebas and some other organisms on the phototrophic biofilms. This study aims at investigating the potentialities of the biological control of phototrophic biofilms in caves, but the originality of these data should be confirmed in future studies with a larger number of biofilm samples in different ecological scenarios.

**Keywords:** phototrophic biofilms; show caves; biological control; bacteria; cyanobacteria; amoeba

## 1. Introduction

Traditionally, caves have been considered as extreme environments characterized by low nutrient inputs. Although this may be possible in a few caves, most of the show caves cannot be considered extreme or oligotrophic environments because natural factors are often modified due to different anthropogenic impacts, being one of the most evident the installation of electrical lighting to facilitate visits. In fact, when caves are illuminated by natural or artificial lighting, phototrophic microorganisms become predominant. Under these conditions, complex communities develop on speleothems, walls and ground, mostly composed of cyanobacteria and algae, heterotrophic bacteria and fungi, and other eukaryotic organisms embedded in a matrix of extracellular polymeric substances. In further steps, mosses, ferns and other plants may also colonize caves. Examples of these communities have been studied in artificially illuminated show caves and catacombs, because their unconstrained growth can induce damage on valuable archeological or natural rock surfaces [1–3].

The biofilms cause aesthetic, physical and chemical damage [4,5], subject of great concern in general, and in particular in cases of caves with rock art paintings [6]. In addition, this invasion of foreign microorganisms induces great changes on the autochthonous microbial population, adapted to minimal inputs of C and N, and on the cave ecology.

A wide array of methods has been proposed to control and/or eliminate phototrophic biofilms in show caves, including physical, chemical and biological procedures. Physical methods comprise: (i) the mechanical removal of biofilms by brushing and subsequent washing with pressurized water;

(ii) the reduction of light intensity, and (iii) the use of ultraviolet radiation. Chemical methods include the use of sodium and calcium hypochlorite, hydrogen peroxide or commercial biocides. A detailed discussion of pros and cons were presented by Mulec and Kosi [7] and Baquedano et al. [8]. Some of these methods were tested in *Cueva del Tesoro*, a limestone cave in Rincon de la Victoria, Spain, depicting abundant biofilms coating the cave walls and speleothems [1]. The most effective treatments were sodium hypochlorite and hydrogen peroxide, but for the final cleaning hydrogen peroxide was adopted due to the absence of chemical residues on the rock [1].

Yet, there is no general consensus on the type of biological methods to be applied in cleaning processes. Albertano et al. [2] suggested the use of quorum-sensing molecules and siderophores, whereas Mulec and Kosi [7] proposed biological antagonists like genetically modified viruses. However, to our knowledge field-based experiments have not been conducted thus far.

Trophic relationships among components in phototrophic biofilms can be deduced by analyzing their composition; however, they have rarely been investigated in detail. Usually, most reports rely on a list of cyanobacteria and algae on speleothems but without establishing the trophic interactions among the different biofilm components [9–12]. Biofilms can suffer a loss of cells through different processes: grazing, erosion, abrasion, and sloughing [13]. The most relevant mechanisms to cave biofilms are likely grazing and sloughing, the first by predation of bacteria and protozoa, and the second by periodic loss of large patches of biofilms by drying processes.

By adopting advanced high throughput sequencing techniques, we aimed at investigating the composition of phototrophic biofilms within the Nerja Cave to highlight the potential role of natural biological control. To achieve this aim, we compared the composition of biofilms showing extensive detachments and/or rounded areas without photosynthetic microorganisms, with homogeneous biofilms on two speleothems, in order to understand the mechanisms involved in the process of detachment.

## 2. Materials and Methods

Nerja Cave was discovered in 1959 and open to public visits in 1960. The cave attracts about 450,000 visitors per year due to its location in Costa del Sol, a major tourist area, at 50 km from Malaga, Spain. The cave has an extension of almost 5 km, but the tourist trail is restricted to one-third of the cave. The cave is protected as a site of cultural interest due to its rock art paintings [3]. In addition to the cave entrance, this cave has two sinkholes that connect with the outside. The natural lighting that penetrates through the two sinkholes and the electric lighting installed in the tourist trail have allowed the development of photosynthetic-based biofilms on cave speleothems and walls. Currently, the electric lighting system consists of a trail, which runs along a circular route of about 600 m, a scenic lighting, and an emergency system. Some biofilms coating the speleothems close to light sources show extensive detachments and/or rounded areas without phototrophic microorganisms.

The biofilm samples were collected in November 2018 from two speleothems located in the so-called Tourist Galleries (Figure 1A). Unfortunately, sampling was restricted to a very limited number of samples due to the cataloging of the cave as a site of cultural interest. However, we believe that the data provided were representative of the cave microbiome and shed light on biofilms composition and trophic interactions.

Microclimatic data registered in the Tourist Galleries showed that cave air temperature ranges from 19.8 to 18.1 °C with a mean value of 19.0 °C, the relative humidity ranges from 98.6 to 67.5% with a mean value of 89.4% and the air $CO_2$ ranges from 1313 to 360 ppm with a mean value of 679 ppm [3]. Both biofilms studied in this work were exposed to white low pressure commercial fluorescent lamps (Roblan, Casarrubios del Monte, Spain, model ES27, 18W with a color temperature of 4845 K and 1030 lm), located 1 m away from speleothems. All the samples were collected at the same time by carefully scraping the biofilms with sterile scalpels; samples were kept at −80 °C until processing.

A total of six samples were collected from two speleothems displaying detachment patterns. In the sampling site 1 (Figure 1B), three areas were collected: with biofilms (1A), with and without biofilms

(1B), and a zone of rounded detachments (1C). Sampling site 2 (Figure 1C) comprised: a biofilm sample (2A), an extensive detachment (2B), and the interface between the biofilm and the detached zone (2C).

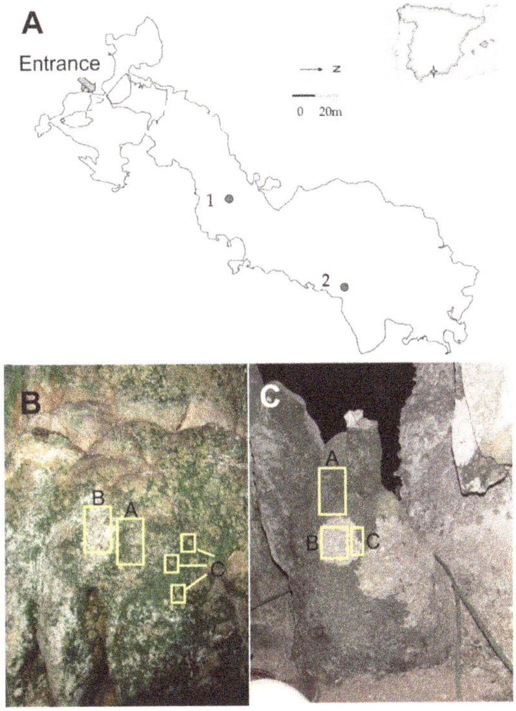


**Figure 1.** Sampling of biofilms in Nerja Cave speleothems. (**A**) Map of part of the cave and location of the sampling sites. (**B**) Sampling site 1. (**C**) Sampling site 2.

Genomic DNA was extracted from 250–500 mg of biofilms, using the FastDNA SPIN Kit for soil (MP Biomedicals, Illkirch, France) according to the manufacturer's instructions. The DNA concentrations, ranging from 6.0–104.0 ng/μL depending on each sample, were quantified by using a Qubit 2.0 fluorometer (Invitrogen, Carlsbad, CA, USA).

The extracted DNA of all samples was analyzed by Next Generation Sequencing (NGS). We investigated the bacterial V3 and V4 regions of the 16S rRNA gene and V4 region of the 18S rRNA gene using Illumina MiSeq and 2 × 250 paired-end sequencing, according to the Illumina metagenomic library preparation protocol used by Macrogen (Seoul, Korea). The primer sequences used in this study were 341F (CCTACGGGNGGCWGCAG) and 805R (GACTACHVGGGTATCTAATCC) for bacteria, Reuk454FWD1 (CCAGCASCYGCGGTAATTCC) and V4r (ACTTTCGTTCTTGAT) for eukaryotes [14].

Raw data were processed in QIIME 1.9.1. Quality control and trimming were performed using FASTQC (http://www.bioinformatics.babraham.ac.uk/projects/fastqc/) and Trimmomatic (0.36 version), respectively. Paired-end reads were assembled using PEAR. Operational Taxonomic Units (OTUs) were clustered at 97% cutoff using UCLUST. SILVA database for bacteria and eukaryotes was used to assign the taxonomic classification of each 16S rRNA and 18S rRNA gene sequences with a threshold of 80%.

Alpha_diversity.py command was employed for alpha-diversity. Taxonomic analyses of prokaryotic communities were represented through heat-maps built in R using gplots package [15]. Represented samples in heat-maps were reordered with dendrograms based on the row and column mean values as described by the authors. The raw reads were deposited into the NCBI Sequence Read Archive (SRA) database under accession number PRJNA611930.

## 3. Results

Table 1 indicates that the microbial communities of the samples taken from the speleothems coated by phototrophic biofilms collected in Nerja Cave were almost entirely composed of members of *Bacteria*, with percentages ranging between 100.0% and 99.7%. The *Archaea* were practically non-existent, reaching 0.2% in sample 1A and 0.3% in sample 1C. The percentage of unassigned community members was zero.

**Table 1.** Kingdoms distribution in phototrophic biofilms from Nerja Cave.

| Kingdom | 1A | 1B | 1C | 2A | 2B | 2C |
|---|---|---|---|---|---|---|
| *Bacteria* | 99.8 | 100.0 | 99.7 | 100.0 | 100.0 | 100.0 |
| *Archaea* | 0.2 | 0.00 | 0.3 | 0.0 | 0.0 | 0.0 |

The phototrophic communities of Nerja Cave were composed of 12 phyla with percentages greater than 1% (Figure 2). The most abundant phyla were *Firmicutes, Proteobacteria* and *Cyanobacteria*. Members of the phylum *Cyanobacteria* contribute to the green color of biofilms, due to the chlorophyll and its photosynthetic function, while other phyla are mainly composed of heterotrophic bacteria. *Firmicutes* was widely represented in sample 1 and *Proteobacteria* in sample 2. *Cyanobacteria* were abundant in samples 1A, 2A and 2C, which corresponded to the areas of epilithic phototrophic biofilms.

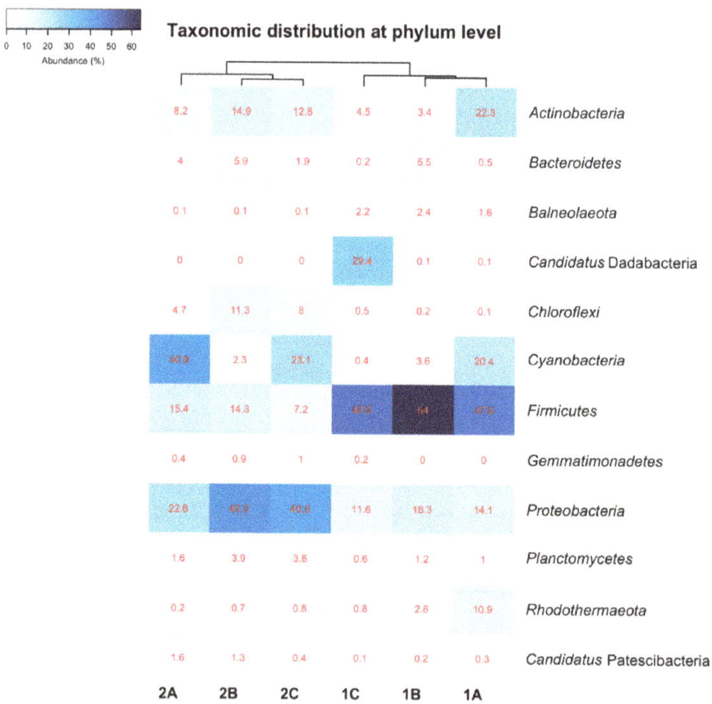


**Figure 2.** Heat-map analysis of Nerja Cave biofilms. Taxonomic identifications of *Bacteria* at phylum level. The phyla are described in the right column and their respective abundances included in the boxes.

In a range of lower abundance were the phyla '*Candidatus* Dadabacteria', which only reached significant representation in sample 1C, and *Actinobacteria*, in samples 1A, 2B and 2C. They were followed in abundance by the phylum *Chloroflexi* in sample 2B and *Rhodothermaeota* in sample 1A. The phylum *Bacteroidetes* was relatively well represented in samples 1B and 2B. *Planctomycetes* reached significant percentages in samples 2B and 2C, followed by *Balneolaeota* in samples 1B and 1C. Finally, the superphylum *Patescibacteria* reached percentages higher than 1% in samples 2A and 2B. The *Gemmatimonadetes* phylum was represented in sample 2 with percentages ≤1%.

Figures 3 and 4 illustrate the bacterial diversity at the class and order levels. The communities of the two speleothems showed a high heterogeneity in abundance, depending on the location area (biofilm, without biofilm, and interface area), with high percentages in the classes *Bacilli* and *Cyanophyceae* (Figure 3).

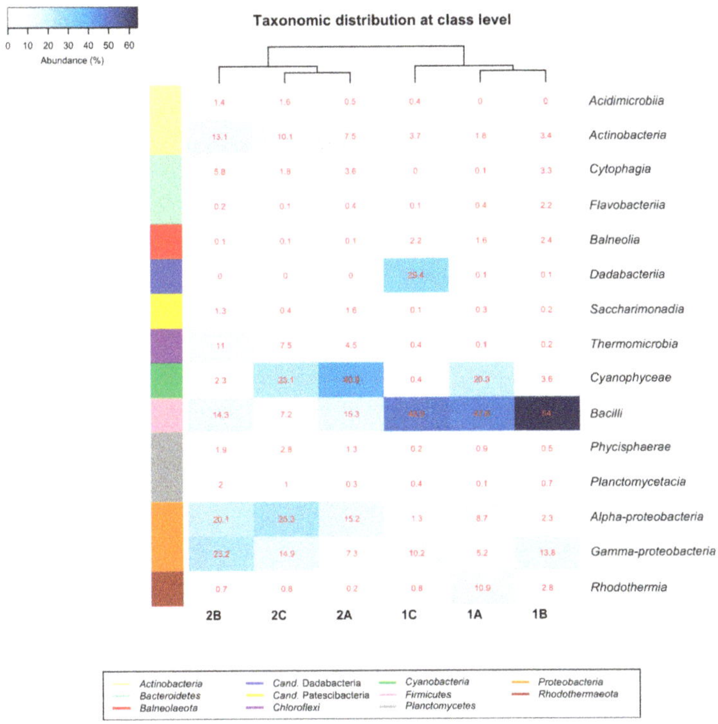


**Figure 3.** Heat-map analysis of Nerja Cave biofilms. Taxonomic identifications of *Bacteria* at class level. The classes are described in the right column and their respective abundances included in the boxes.

Considering all the six samples, the most abundant classes (those that at least reached a value greater than 5% in a sample) were *Bacilli, Cyanophyceae, Alphaproteobacteria, Gammaproteobacteria, Dadabacteriia, Actinobacteria, Thermomicrobia, Rhodothermia* and *Cytophagia*.

The phylum *Firmicutes* and the class *Bacilli* was one of the most abundant in samples 1 and 2. The order *Bacillales* comprised two genera: *Bacillus* and *Psychrobacillus*. The genus *Bacillus* was the most abundant in sample 1, while in sample 2 the abundance of *Bacillus* fell between six and three times, with respect to sample 1.

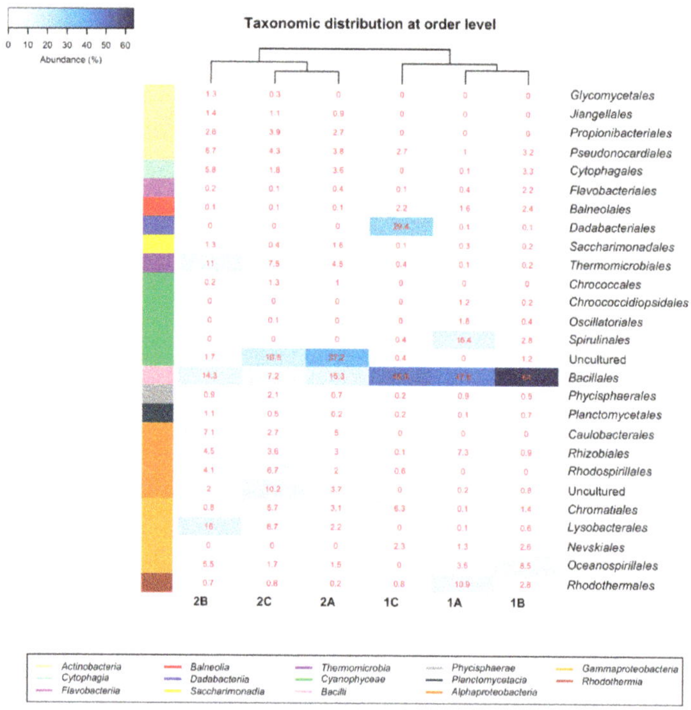


**Figure 4.** Heat-map analysis of Nerja Cave biofilms. Taxonomic identifications of *Bacteria* at order level. The orders are described in the right column and their respective abundances included in the boxes.

*Cyanobacteria*, the phylum that contained the phototrophic microorganisms, only included two identified genera: *Desertifilum* and *Spirulina* (Figure 5). In sample 1A, *Spirulinales* showed a remarkable abundance, and decreased in samples 1B and 1C (Figure 4). High abundances of uncultured members of the class *Cyanophyceae* were detected in samples 2A and 2C (Figure 5). It was interesting to find an abundance of *Spirulina* in sample 1 and a total absence in sample 2. However, a high abundance of uncultured members of the class *Cyanophyceae* was detected in sample 2, but lacking in sample 1. This pattern also occurred for the family *Chroococcidiopsaceae*, which was absent in sample 2, as well as for *Aphanothecaceae*, which was lacking in sample 1 (Figure 5).

The class *Alphaproteobacteria* was mainly composed of the orders: *Rhizobiales*, *Caulobacterales*, *Rhodospirillales*, and members of an uncultured order (Figure 4). In this class, only two genera were identified: *Brevundimonas* and *Mesorhizobium* (Figure 5).

The class *Gammaproteobacteria* was represented by the orders *Lysobacterales*, *Oceanospirillales* and *Chromatiales*, with high abundance in all samples, and by *Nevskiales*, solely present in sample 1 (Figure 4). The most representative genera within the *Gammaproteobacteria* were: *Lysobacter*, *Halomonas*, *Woeseia* and *Luteimonas*. *Thioalbus*, *Salinisphaera* and *Halopeptonella* were identified in low abundance (Figure 5).

The phylum *Actinobacteria* was composed of three classes (*Acidimicrobiia*, *Actinobacteria* and *Rubrobacteria*). However, only the class *Actinobacteria* attained a remarkable abundance in sample 2, with a noticeable richness of the order *Pseudonocardiales*. Regarding the genus level, *Amycolatopsis*, present in the samples 2B and 1B, and an uncultured genus of the family *Nocardioidaceae*, in samples 2C and 2A, showed relative abundance. The rest of genera (*Crosiella*, *Yuhushiella*, *Pseudonocardia*, *Actinophytocola*, *Glycomyces* and *Halactinopolyspora*) reached low abundances.

Within the *Chloroflexi* phylum, the class *Thermomicrobia* showed relatively high abundance in sample 2B. The orders *Phycisphaerales* and *Planctomycetales*, within the phylum *Planctomycetes*, reached scarce representation in the biofilm samples (Figure 4).

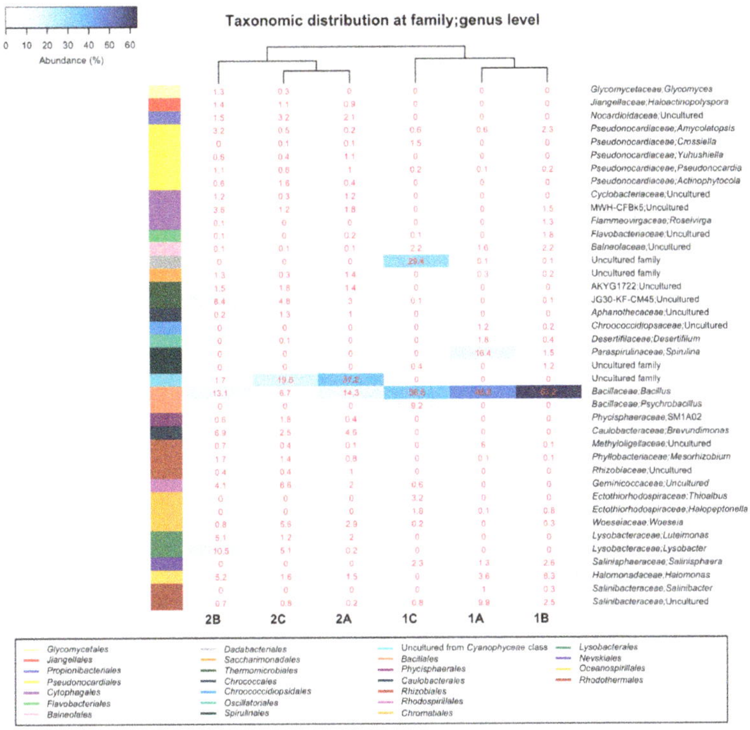


**Figure 5.** Heat-map analysis of Nerja Cave biofilms. Taxonomic identifications of *Bacteria* at families and genera level. The groups are described in the right column and their respective abundances included in the boxes.

A new phylum separated from *Bacteroidetes* is *Rhodothermaeota* [16], represented by the class *Rhodothermia* in sample 1A. This class comprises the genus *Salinibacter*, only found in sample 1A, together with other uncultured genera.

The phylum *Bacteroidetes* includes the class *Cythophagia* and the order *Flavobacteriales*, with uncultured genera, except for *Roseivirga*, with poor representation in all samples. The new phylum *Balneolaeota* [17], recently separated from *Bacteroidetes*, which includes the class *Balneolia*, only reached a representative abundance in sample 1.

The phyla '*Candidatus* Dadabacteria', *Gemmatimonadetes* and the superphylum *Patescibacteria* were commonly represented by uncultured bacteria, in low percentages of abundance which can only be identified using NGS. The abundance of the order *Dadabacteriales* was remarkable in sample 1C (Figure 4).

Most of the eukaryotic sequences retrieved in sample 1 were unknown, while in sample 2 the percentage of unassigned sequences was notably lower (Table 2).

**Table 2.** Relative abundance of eukaryotes, with higher taxonomic level in bold and lower taxonomic levels in italics.

| | | | Abundance of Sequences (%) * | | | |
|---|---|---|---|---|---|---|
| | | | 1A | 1B | 2A | 2B |
| | *Discosea* | *Dactylopodida* | 17.7 | 2.6 | 0.0 | 0.0 |
| | | *Acanthamoeba* | 0.0 | 0.0 | 1.2 | 6.0 |
| | | *Balamuthia* | 0.0 | 0.0 | 0.4 | 1.1 |
| ***Amoebozoa*** | *Schizoplasmodiida* | uncultured | 0.0 | 4.7 | 46.2 | 40.8 |
| | *Tubulinea* | *Echinamoebida* | 10.5 | 0.4 | 0.0 | 0.0 |
| | uncultured | uncultured | 6.8 | 8.3 | 0.0 | 0.0 |
| | Others | uncultured | 0.0 | 0.0 | 1.2 | 0.0 |
| ***Opisthokonta*** | *Holozoa* | *Codonosigidae* | 0.0 | 0.0 | 0.5 | 8.5 |
| | | *Rhabditida* | 0.0 | 0.0 | 12.4 | 19.1 |
| | *Fungi* | *Stachybotryaceae* | 0.0 | 0.0 | 0.0 | 3.2 |
| | | uncultured | 0.2 | 0.2 | 0.2 | 1.4 |
| **SAR** | *Alveolata* | *Gregarina* | 6.4 | 3.8 | 0.5 | 1.3 |
| | | uncultured | 0.0 | 0.0 | 1.7 | 0.0 |
| | | *Colpodea* | 0.0 | 0.0 | 4.2 | 3.3 |
| | *Rhizaria* | *Heteromita* | 0.0 | 0.0 | 12.2 | 8.2 |
| | | P34,6 | 0.0 | 0.0 | 0.0 | 2.9 |
| | *Stramenopiles* | *Sorodiplophrys* | 0.4 | 0.7 | 5.8 | 1.6 |
| | | *Chlamydomyxa* | 0.0 | 0.0 | 4.6 | 0.0 |
| **Unknown** | Others | Others | 58.0 | 79.3 | 8.9 | 2.6 |

* Relative abundance >1% in at least one of the samples.

The eukaryotic organisms identified in the four samples were affiliated to *Amoebozoa*, *Opisthokonta* and SAR. Table 2 shows the taxonomic composition of the samples where only the taxa represented with values greater than 1% were included.

Sample 1 was characterized by the abundance of *Amoebozoa* belonging to the order *Dactylopodida*, with uncultured members in sample 1A, an uncultured order in 1A and 1B, and members of the family *Echinamoebida* in 1A. The genus *Gregarina* (SAR) was relatively abundant in samples 1A and 1B.

The sampling site 2 showed very high abundances in samples 2A and 2B for uncultured *Schizoplasmodiida* of the *Amoebozoa* supergroup, followed by uncultured members of the order *Rhabditida*. The genus *Heteromita* (*Rhizaria*) also appeared with significant abundance in samples 2A and 2B. The genus *Acanthamoeba*, uncultured members of the *Codonosigidae* family and the genus *Sorodiplophrys* (*Stramenopiles*) in sample 2A were also noticeable. Cultivable genera such as *Balamuthia* and *Chlamydomyxa* were also identified.

To summarize, within *Bacteria*, significant differences were found between the two biofilm samples collected from speleothems displaying abundant growth of phototrophic microorganisms. Considering taxonomic groups in proportions greater than 1%, the analysis reflected a great diversity in the taxonomic composition of each sample, with a total of 27 orders (Figure 4). Comparatively, the samples presented similar orders, and shared common groups. Thus, samples 1 and 2 shared 22 orders. The greatest differences between each sample were in the percentage of abundance of certain groups, among which an uncultured order of *Alphaproteobacteria*, *Rhizobiales*, *Oceanospirillales*, *Chromatiales*, *Rhodothermales*, *Dadabacteriales*, *Bacillales* and the phototrophic *Spirulinales* were remarkable in sample 1, while the abundances of *Rhodospirillales*, *Caulobacterales*, *Lysobacterales*, *Thermomicrobiales*, *Pseudonocardiales*, *Cytophagales*, *Bacillales* and uncultured members of the phototrophic *Cyanophyceae* were noticeable in sample 2.

The *Eukaryota* showed a total of 19 groups, of which 7 were found in sample 1, and 16 in sample 2. Both sampling sites solely shared 5 groups of eukaryotes: *Schizoplasmodiida*, uncultured *Amoebozoa* uncultured *Fungi*, and the genera *Gregarina* and *Sorodiplophrys*. In addition, the high abundance of unassigned sequences in sample 1 indicated the existence of large unknown taxa in this community.

## 4. Discussion

Phototrophic microorganisms, such as cyanobacteria and algae, are usually found on speleothems, rocks and walls washed by percolation water or in the vicinity of light sources. The development of cyanobacteria and algae, which can be considered as primary colonizers, is usually influenced by the availability of light, water and dissolved mineral elements. The development of a photosynthetic-based biofilm may lead to the formation of a more complex ecosystem, because phototrophic microorganisms release extracellular compounds, which favor the establishment of heterotrophic bacteria. In later phases of the community, fungi, protozoa and metazoans will appear, giving rise to a complex community with important trophic interactions.

The formation of plaques or spots where the detachment of phototrophic microorganisms was evident, as found in Nerja Cave, can be assumed on the basis of processes of predation of some microorganisms on others. The samples labeled B and C in the two biofilms analyzed presented detachments indicative of some type of action (lysis, predation, etc.) on the cyanobacteria with respect to the intact biofilm (labeled A).

It is interesting to note that while the cyanobacterium *Spirulina* reached a remarkable abundance in sample 1A, it disappeared in sample 1C (complete detachment) and its concentration decreased six times in sample 1B (intermediary phase) with respect to 1A. This would indicate that the population of *Spirulina* has been predated by other microorganisms. Other groups and genera of cyanobacteria are minority in sample 1 (*Desertifilum*, *Chroococcidiopsaceae* and uncultured members of *Cyanophyceae*) and disappeared in the detached area (sample 1C).

The cyanobacterial composition of sample 2 is completely different. In this case, the most abundant cyanobacterium was not *Spirulina*, absent in these biofilms, but unidentified members of the class *Cyanophyceae*. Comparing the retrieved sequences with the NCBI database, we found that these unidentified members were related, although with a low percentage of similarity (87%) with *Loriellopsis cavernicola*. In sampling site 2, the detachment or loss of the biofilm occupies large areas, suggesting that the process of predation was much more pronounced. This is compatible with the high abundance of eukaryotes.

The richness of *Bacillus* in sample 1B was remarkable, although it was also present in considerable abundance in all other samples. Members of the phylum *Firmicutes*, and in particular the genus *Bacillus*, are well represented in caves [18,19].

The presence of *Bacillus* in Nerja Cave biofilms can be due to: (i) species of the genus *Bacillus* (*B. cereus*, *B. mycoides*) can lyse different species of cyanobacteria [20–23], and contributed to their disappearance (detached areas); or (ii) the periodic processes of humidification and desiccation suffered by the phototrophic biofilms. A similar process has been described by Karaoz et al. [23] for biological soil crusts dominated by cyanobacteria. In this case, the communities are well adapted to survive desiccation and quickly take advantage of rainfall periods, producing a bloom of *Firmicutes*, mainly of the order *Bacillales* (*Bacillus* spp.). This does not preclude a lytic activity.

The existence of a high number of *Dadabacteria* sequences in sample 1C is an unprecedented fact and reflects the poor existing knowledge about cave microbial ecology, and the microbial strategies necessary to growth and proliferate in cave conditions. *Dadabacteria* is a recently proposed candidate phylum, previously known as CSP1-2, and rarely present in environmental samples. These aerobic bacteria have been identified through metagenomic analysis [24]. The genome encodes glycolysis, a complete Krebs cycle and an electron transport chain for oxidative phosphorylation that includes a cytochrome c oxidase and possibly intervention in the nitrogen and sulfur cycles [24]. There is no information about a possible lytic activity on cyanobacteria or an explanation for their abundance.

Regarding eukaryotes, sample 1 showed abundance of uncultured *Amoebozoa* groups, such as the lobose amoebas of the order *Dactylopodida* that ingest cyanobacteria, algae, bacteria, yeasts and particulate matter, as well as *Echinamoebida*, and other group without assignment.

Protozoa are traditionally considered the main predators of bacteria, as this is a lifestyle common to this group. Predatory protozoa exist in both aquatic and terrestrial environments and represent a major component in nutrient remineralization or 'microbial loop.' Dryden and Wright [25] reported a list of protozoa (including *Amoebozoa*) that predated on a wide variety of cyanobacterial species (*Anabaena, Nostoc, Microcystis, Oscillatoria, Gloeocapsa, Phormidium, Lyngbya, Synechococcus,* etc.).

The genus *Gregarina* of the SAR supergroup was abundant in sampling site 1. According to Clopton et al. [26] there are more than 300 species of *Gregarina* that mainly infect coleoptera and orthoptera, but also colembola, diptera, lepidoptera, hymenoptera, etc. The presence of parasitic gregarins, in sample 1, must be related to the population of insects in the cave, since they have a great abundance of parasitic species.

Jurado et al. [27] identified in various samples of Nerja Cave sequences of the species *Gregarina blattarum*. *Gregarina blattarum* has been found in different species of cockroaches: *Blatta orientalis, Periplaneta americana, Periplaneta discoidales, Blattella germanica, Parcoblatta pennsylvanica* and *Blaberus craniifer* and is the most widely studied species of *Gregarina* [28].

A review of the literature reveals a series of predatory bacteria on cyanobacteria and algae, mainly the genera *Lysobacter, Bacillus, Cytophaga, Myxobacteria, Herpetosiphon, Ensifer, Vampirococcus, Vampirovibrio, Cupriavidus, Pseudomonas, Stenotrophomonas, Acinetobacter and Delftia* [22,29–31]. From all these genera, in sample 2 were present *Lysobacter, Bacillus* and uncultured members of the order *Cytophagales*, not being able to rule out the possible existence of other bacteria with similar capacities from the extensive list of uncultured members.

From the NGS analysis, the possible predatory activity and implication of *Lysobacter* in sample 2 was evident. Reichenbach [32] found that the densities of cyanobacterial and lytic bacteria populations correlated well and that bacteria attacked cyanobacteria in two ways, by excreting exoenzymes or by fixing them to the cell. Puopolo et al. [33] studied the *Lysobacter* genome confirming the ability to produce proteolytic enzymes. In addition, the genome revealed the presence of a high number of genes encoded for proteases, endopeptidases and metalloendopeptidases. The genus *Bacillus* was also abundant in sample 2 and could reinforce the predatory activity on this sample.

Considering the analysis of eukaryotes, the uncultured members of *Schizoplasmodiida* (*Amoebozoa*) stand out in sample 2. This group of protosteloid amoebas is predator of bacteria and fungi in terrestrial ecosystems, where they have a global distribution [34]. The genus *Acanthamoeba* (*Amoebozoa*) was present in sample 2B. Cyanobacteria are an adequate food source for amoebas, ciliates and flagellates, among the first is notable the genus *Acanthamoeba* and the species *A. castellanii* [35,36]. *Acanthamoeba castellanii* were found in some other caves [37,38].

Ma et al. [35] evaluated the capacity of five amoebas, among them *A. castellanii*, on the cyanobacteria *Synechococcus elongatus, Leptolyngbya* sp. and *Anabaena* sp., demonstrating a predatory activity. Wright et al. [36] showed that *A. castellanii* consumed unicellular and filamentous cyanobacteria. Predation was initiated by complete engulfment of the unicellular cyanobacterium and, in the case of the filamentous, by rupture of the filaments, followed by digestion in the vacuoles.

Important abundances reached the nematodes of the order *Rhabditida* in samples 2A and 2B, the uncultured members of the family *Codonosigidae* (*Choanoflagellida, Opisthokonta)*, flagellates of the genus *Heteromita* (*Rhizaria*), one of the most important predators of bacteria in aquatic ecosystems, and uncultured ciliates of the class *Colpodea*.

Finally, in sample 2A the amoeboid protist of the genus *Sorodiplophrys* (*Labyrinthulomycetes, Stramenopiles*) and *Chlamydomyxa* (*Ochrophyta, Stramenopiles*), a photophagotrophic ameboid alga, were noticeable.

Abolafia and Peña-Santiago [39] carried out a review of the order *Rhabditida* of nematodes in Spain, finding 80 species belonging to 40 genera and 12 families, of which six species were found in the province of Malaga. These nematodes are frequent in soils and waters, however, the sequences found in sample 2 belong to unidentified genera.

It is noteworthy the presence of ciliates, flagellates, amoebas and nematodes in sampling site 2, which implies that this biofilm conforms a complex ecosystem where a high dependence between prey and predators has been established. Apart from the microorganisms that act as predators, there are others that can act as saprophytes, feeding on decomposing organic matter.

## 5. Conclusions

The study of two phototrophic biofilms developed on the speleothems of Nerja Cave showed different structures and hence diverse predators. This indicated complex trophic interactions between the components of the biofilms. The question is how many different biofilms and predators are in this cave? This should be responded in future studies with a larger number of biofilm samples in different ecological scenarios, if further sampling is possible.

This study evidences that a natural biological control exists in caves and it could be a new method of control to be tested and used if the growth of predatory populations is enhanced. This research represents a first attempt to investigate the potentialities of the biological control of photosynthetic-based biofilms in caves.

**Author Contributions:** Conceptualization and work design, Y.d.R.; methodology, analyses and investigation, V.J., B.H. and J.L.G.-P.; writing, review and editing, C.S.-J. All authors have read and agreed to the published version of the manuscript.

**Funding:** This research was funded by Nerja Cave Foundation.

**Acknowledgments:** The authors acknowledge Nerja Cave Foundation for the support and facilities. The authors also acknowledge A.Z. Miller for the English language corrections.

**Conflicts of Interest:** The authors declare no conflict of interest.

## References

1. Jurado, V.; Hernández-Mariné, M.; Rogerio-Candelera, M.A.; Domínguez-Moñino, I.; Gutiérrez-Patricio, S.; Sáiz-Jiménez, C. La Cueva del Tesoro: Un paraíso microbiano. In *Cuevatur 2014, Iberoamerica Subterránea*; Calaforra, J.M., Durán, J.J., Eds.; Asociación de Cuevas Turísticas Españolas: Madrid, Spain, 2014; pp. 431–441.
2. Albertano, P.; Moscone, D.; Palleschi, G.; Hermosín Campos, B.; Saiz-Jimenez, C.; Sánchez-Moral, S.; Hernández-Mariné, M.; Urzì, C.; Groth, I.; Shroeckh, V.; et al. Cyanobacteria attack rocks (CATS): Control and preventive strategies to avoid damage caused by cyanobacteria and associated microorganisms in Roman hypogean monuments. In *Molecular Biology and Cultural Heritage*; Saiz-Jimenez, C., Ed.; Balkema Pub.: Lisse, The Netherlands, 2003; pp. 151–162.
3. Liñán, C.; del Rosal, Y.; Carrasco, F.; Vadillo, I.; Benavente, J.; Ojeda, L. Highlighting the importance of transitional ventilation regimes in the management of Mediterranean show caves (Nerja-Pintada system, southern Spain). *Sci. Total Environ.* **2018**, *631–632*, 1268–1278. [CrossRef]
4. Ariño, X.; Hernandez-Marine, M.; Saiz-Jimenez, C. Colonization of Roman tombs by calcifying cyanobacteria. *Phycologia* **1997**, *36*, 366–373. [CrossRef]
5. Saiz-Jimenez, C. Biogeochemistry of weathering processes in monuments. *Geomicrobiol. J.* **1999**, *16*, 27–37. [CrossRef]
6. Saiz-Jimenez, C.; Cuezva, S.; Jurado, V.; Fernandez-Cortes, A.; Porca, E.; Benavente, D.; Cañaveras, J.C.; Sanchez-Moral, S. Paleolithic art in peril: Policy and science collide at Altamira Cave. *Science* **2011**, *334*, 42–43. [CrossRef]
7. Mulec, J.; Kosi, G. Lampenflora algae and methods of growth control. *J. Cave Karst Stud.* **2009**, *71*, 109–115.
8. Baquedano Estévez, C.; Moreno Merino, L.; de la Losa Román, A.; Durán Valsero, J.J. The lampenflora in show caves and its treatment: An emerging ecological problem. *Int. J. Speleol.* **2019**, *48*, 249–277. [CrossRef]
9. Smith, T.; Olson, R. A taxonomic survey of Lamp Flora (Algae and Cyanobacteria) in electrically lit passages within Mammoth Cave National Park, Kentucky. *Int. J. Speleol.* **2007**, *36*, 105–114. [CrossRef]

10. Del Rosal Padial, Y.; Jurado Lobo, V.; Hernández Mariné, M.; Roldán Molina, M.; Sáiz Jiménez, C. Biofilms en cuevas turísticas: La Cueva de Nerja y la Cueva del Tesoro. In *El Karst y el Hombre: Las Cuevas como Patrimonio Mundial*; Andreo, B., Durán, J.J., Eds.; Asociación de Cuevas Turísticas Españolas: Madrid, Spain, 2016; pp. 103–114.
11. Cennamo, P.; Marzano, C.; Ciniglia, C.; Pinto, G.; Cappelletti, P.; Caputo, P.; Pilio, A. A survey of the algal flora of anthropogenic caves of Campi Flegrei (Naples, Italy) archeological district. *J. Cave Karst Stud.* **2012**, *74*, 243–250. [CrossRef]
12. Lamprinou, V.; Danielidis, D.B.; Economou-Amilli, A.; Pantazidou, A. Distribution survey of Cyanobacteria in three Greek caves of Peloponnese. *Int. J. Speleol.* **2012**, *41*, 267–272. [CrossRef]
13. Rittmann, B.E. Detachment from biofilms. In *Structure and Function of Biofilms*; Characklis, W.G., Wildered, P.A., Eds.; John Wiley & Sons: Chichester, UK, 1989; pp. 49–58.
14. Bradley, I.M.; Pinto, A.J.; Guest, J.S. Design and evaluation of Illumina MiSeq-compatible, 18S rRNA gene-specific primers for improved characterization of mixed phototrophic communities. *Appl. Environ. Microbiol.* **2016**, *82*, 5878–5891. [CrossRef]
15. Warnes, G.R.; Bolker, B.; Bonebakker, L.; Gentleman, R.; Liaw, W.H.A.; Lumley, T.; Maechler, M.; Magnusson, A.; Moeller, S.; Schwartz, M.; et al. *GPLOTS: Various R Programming Tools for Plotting Data. R Package*, version 2.17.0; GitHub, Inc.: San Francisco, CA, USA, 2015; Available online: http://CRAN.R-project.org/package=gplots (accessed on 10 February 2020).
16. Muñoz, R.; Rosselló-Móra, R.; Amann, R. Revised phylogeny of *Bacteroidetes* and proposal of sixteen new taxa and two new combinations including *Rhodothermaeota* phyl. nov. *Syst. Appl. Microbiol.* **2016**, *39*, 281–296. [CrossRef] [PubMed]
17. Hahnke, R.L.; Meier-Kolthoff, J.P.; García-López, M.; Mukherjee, S.; Huntemann, M.; Ivanova, N.N.; Woyke, T.; Kyrpides, N.C.; Klenk, H.-P.; Göker, M. Genome-based taxonomic classification of *Bacteroidetes*. *Front. Microbiol.* **2016**, *7*, 2003. [CrossRef] [PubMed]
18. Fernandez-Cortes, A.; Cuezva, S.; Sanchez-Moral, S.; Porca, E.; Jurado, V.; Saiz-Jimenez, C. Detection of human-induced environmental disturbances in a show cave. *Environ. Sci. Pollut. Res.* **2011**, *18*, 1037–1045. [CrossRef] [PubMed]
19. Saiz-Jimenez, C. The microbiology of show caves, mines tunnels and tombs: Implications for management and conservation. In *Microbial Life of Cave Systems*; Engel, A.S., Ed.; DeGruiter: Berlin, Germany, 2015; pp. 231–261.
20. Nakamura, N.; Nakano, K.; Sugiura, N.; Matsumura, M. A novel cyanobacteriolytic bacterium, *Bacillus cereus*, isolated from a eutrophic lake. *J. Biosci. Bioeng.* **2003**, *95*, 179–184. [CrossRef]
21. Gumbo, R.J.; Ross, G.; Cloete, E.T. Biological control of *Microcystis* dominated harmful algal blooms. *Afr. J. Biotechnol.* **2008**, *7*, 4765–4773.
22. Gumbo, R.J.; Cloete, E.T. Light and electron microscope assessment of the lytic activity of *Bacillus* on *Microcystis aeruginosa*. *Afr. J. Biotechnol.* **2011**, *10*, 8054–8063. [CrossRef]
23. Reim, R.L.; Shane, M.S.; Cannon, R.E. The characterization of *Bacillus* capable of blue green bactericidal activity. *Can. J. Microbiol.* **1974**, *20*, 981–986. [CrossRef]
24. Hug, L.A.; Thomas, B.C.; Sharon, I.; Brown, C.T.; Sharma, R.; Hettich, R.L.; Wilkins, M.J.; Williams, K.H.; Singh, A.; Banfield, J.F. Critical biogeochemical functions in the subsurface are associated with bacteria from new phyla and little studied lineages. *Environ. Microbiol.* **2016**, *18*, 159–173. [CrossRef]
25. Dryden, R.C.; Wright, S.J.L. Predation of cyanobacteria by protozoa. *Can. J. Microbiol.* **1987**, *33*, 471–482. [CrossRef]
26. Clopton, R.E.; Cook, T.J.; Cook, J.L. *Gregarina tropica* n. sp. (Apicomplexa: Eugregarinorida: Gregarinicae: Gregarinidae) parasitizing the brown-winged earwig, *Vostox brunneipennis* (Dermaptera: Labiidae), in the Texas Big Thicket. *Comp. Parasitol.* **2008**, *75*, 215–227. [CrossRef]
27. Jurado, V.; Del Rosal, Y.; Hernández-Mariné, M.; Galocha Zapata, I.; Domínguez-Moñino, I.; Rogerio Candelera, M.A.; Sáiz-Jiménez, C. Biología molecular de los biofilms fotosintéticos de la Cueva de Nerja. In *Análisis, Impacto y Evolución de los Biofilms Fotosintéticos en Espeleotemas. El caso de la Cueva de Nerja (Málaga, España)*; Del Rosal Padial, Y., Ed.; Fundación Pública de Servicios Cueva de Nerja: Nerja, Spain, 2017; pp. 229–270.

28. Clopton, R.E.; Gold, R.E. Host specificity of *Gregarina blattarum* von Siebold, 1839 (Apicomplexa: Eugregarinida) among five species of domiciliary cockroaches. *J. Invert. Pathol.* **1996**, *67*, 219–223. [CrossRef] [PubMed]
29. Christensen, P.; Cook, F.D. *Lysobacter*, a new genus of nonfruiting, gliding bacteria with a high base ratio. *Int. J. Syst. Bacteriol.* **1978**, *28*, 367–393. [CrossRef]
30. Rashidan, K.K.; Bird, D.F. Role of predatory bacteria in the termination of a cyanobacterial bloom. *Microb. Ecol.* **2001**, *41*, 97–105. [CrossRef] [PubMed]
31. Osman, O.A.; Beier, S.; Grabherr, M.; Bertilsson, S. Interactions of freshwater cyanobacteria with bacterial antagonists. *Appl. Environ. Microbiol.* **2017**, *83*, e02634-16. [CrossRef] [PubMed]
32. Reichenbach, H. The genus *Lysobacter*. In *The Prokaryotes*; Dworking, M., Falkow, S., Rosenberg, E., Schleifer, K.-H., Stackebrandt, E., Eds.; Springer: New York, NY, USA, 2006; Volume 6, pp. 939–957.
33. Puopolo, G.; Tomada, S.; Pertot, I. The impact of the omics era on the knowledge and use of *Lysobacter* species to control phytopathogenic microorganisms. *J. Appl. Microbiol.* **2017**, *124*, 15–27. [CrossRef] [PubMed]
34. Aguilar, M.; Lado, C. Check-list of microscopic protosteloid amoebae from the Southwest of Europe. *Anales Jard. Bot. Madrid* **2012**, *69*, 217–236. [CrossRef]
35. Ma, A.T.; Daniels, E.F.; Gulizia, N.; Brahamsha, B. Isolation of diverse amoebal grazers of freshwater cyanobacteria for the development of model systems to study predator prey interaction. *Algal Res.* **2016**, *13*, 85–93. [CrossRef]
36. Wright, S.J.L.; Redhead, K.; Maudsleys, H. *Acanthamoeba castellanii*, a predator of Cyanobacteria. *J. Gen. Microbiol.* **1981**, *125*, 293–300. [CrossRef]
37. Walochnik, J.; Mulec, J. Free-living amoebae in carbonate precipitating microhabitats of karst caves and a new vahlkampfiid amoeba, *Allovahlkampfia spelaea* gen. nov., sp. nov. *Acta Protozool.* **2009**, *48*, 25–33.
38. Garcia-Sanchez, A.M.; Ariza, C.; Ubeda, J.M.; Martin-Sanchez, P.M.; Jurado, V.; Bastian, F.; Alabouvette, C.; Saiz-Jimenez, C. Free-living amoebae in sediments from the Lascaux Cave in France. *Int. J. Speleol.* **2013**, *42*, 9–13. [CrossRef]
39. Abolafia, J.; Peña-Santiago, R. Rhabditid species (Nematoda, Rhabditida) recorded in peninsular Spain and Balearic islands. *Graellsia* **2001**, *57*, 113–131.

applied sciences

MDPI

Article

# Assessment of Bacterial Contamination of Air at the Museum of King John III's Palace at Wilanow (Warsaw, Poland): Selection of an Optimal Growth Medium for Analyzing Airborne Bacteria Diversity

**Mikolaj Dziurzynski [1], Karol Ciuchcinski [1], Magdalena Dyda [2,3], Anna Szych [1], Paulina Drabik [2], Agnieszka Laudy [4] and Lukasz Dziewit [1,*]**

1 Department of Environmental Microbiology and Biotechnology, Institute of Microbiology, Faculty of Biology, University of Warsaw, Miecznikowa 1, 02-096 Warsaw, Poland; mikolaj.dziurzynski@biol.uw.edu.pl (M.D.); k.ciuchcinski@student.uw.edu.pl (K.C.); aszych@biol.uw.edu.pl (A.S.)
2 Research and Development for Life Sciences Ltd., Miecznikowa 1/5a, 02-096 Warsaw, Poland; magdalena.dyda@rdls.pl (M.D.); paulinawdrabik@gmail.com (P.D.)
3 Department of Geomicrobiology, Institute of Microbiology, Faculty of Biology, University of Warsaw, Miecznikowa 1, 02-096 Warsaw, Poland
4 Museum of King John III's Palace at Wilanow, Laboratory of Environmental Analysis, Stanislawa Kostki Potockiego 10/16, 02-958 Warsaw, Poland; alaudy@muzeum-wilanow.pl
* Correspondence: ldziewit@biol.uw.edu.pl; Tel.: +48-22-554-14-06

Received: 3 September 2020; Accepted: 7 October 2020; Published: 13 October 2020

**Abstract:** There is no standardized protocol for the assessment of microbial air contamination in museums and other cultural heritage sites. Therefore, most museums conduct such assessments based on their own guidelines or good practices. Usually, microbial air contamination is assessed using only classical microbiology methods with the application of a single growth medium. Therefore, this medium should be carefully selected to limit any selective cultivation bias. Metabarcoding, i.e., a next-generation sequencing (NGS)-based method, combined with classical microbiological culturing was used to assess the effectiveness of various media applications in microbiological screening at the Museum of King John III's Palace at Wilanow (Warsaw, Poland). The obtained results indicated that when using a classical microbiology approach to assess the microbial air contamination at the museum, the selection of a proper growth medium was critical. It was shown that the use of rich media (commonly applied by museum conservators) introduced significant bias by severely underreporting putative human pathogens and the bacterial species involved in biodeterioration. Therefore, we recommend the use of other media, such as Frazier or Reasoner's 2A (R2A) medium, as they could yield more diverse communities and recovered the highest number of genera containing human pathogens, which may be suitable for public health assessments.

**Keywords:** air quality; cultural heritage; museum; next-generation sequencing; PCR amplicon

## 1. Introduction

Microbial indoor air contamination is attracting increasing attention from researchers. In 1955, in one of the first articles on this topic, Wells proposed that clean air should contain no more than $1.5 \times 10^3$ CFU/$m^3$ in summer and $4.5 \times 10^3$ CFU/$m^3$ in winter of microorganisms (bacteria and fungi) [1]. Other researchers proposed similar thresholds for clean, non-contaminated indoor air [2]. The common usage of such standards clearly indicates the importance of classical microbiology methods for indoor air quality assessments. Unfortunately, this approach is highly variable and strongly dependent on the institution performing the analysis and the applied methodology. The very first step of such an analysis may involve various approaches, i.e., active (oftentimes involving the usage of various air samplers and impactors) [3] and passive sampling [4,5]. Furthermore, various media and incubation conditions may be applied. Significantly different results may, therefore, be produced, even for the same indoor area.

Novel metagenomic approaches of microbial biodiversity assessment, based on next-generation sequencing (NGS), constitute an alternative to the classical microbiology analyses. NGS-based methods have many advantages over classical ones. First and foremost, they allow for the detection of uncultivable microorganisms. Furthermore, these high-throughput screening methods allow for the detection of minorities among microbial species. Despite their strengths, NGS-based methods bear some flaws, such as their relatively high cost, the need for expertise in bioinformatic data analysis, and a lack of standard operating procedures (SOPs) for air quality assessments [6,7]. Currently, the most widely adopted NGS-based technique is metabarcoding, which uses the sequencing of PCR-amplified marker genes (e.g., 16S rRNA for bacteria and ITS1 or ITS2 for fungi) for the assessment of microbial biodiversity [8].

Museums are specific open public spaces where cultural heritage monuments are presented to a broad audience. At the same time, exhibitions are strictly protected to prevent their destruction. Biodeterioration and degradation are critical concerns in the conservation and restoration of historical art pieces at museums all over the world [9–12]. To prevent and counteract the effects of these processes, it is of utmost importance to identify the microbial species inhabiting historical art pieces and discover their biological activities that are influencing abiotic surfaces. Classical microbiology methods have been used for many years to achieve this goal [13]. While beneficial for many reasons, including their low cost and the possibility for the further biochemical analysis of isolated microorganisms, these methods also bear some significant limitations, as it is estimated that less than 1% of all microorganisms are cultivable [14–16].

It is also worth mentioning that microbial air contamination in museums may be hazardous to visitors. Museums usually constitute enclosed, air-conditioned spaces that are visited by a multitude of people each day. This may lead to the accumulation of potentially pathogenic microorganisms, which could be easily transmitted between visitors via the circulating air [17–19]. Therefore, reliable monitoring systems for assessing the microbiological contamination of air at museums are needed.

The Wilanow Palace in Warsaw (Poland), which was analyzed in this study, was initially bought in 1677 for King Jan III Sobieski and was transformed into an art gallery in 1805. Nowadays, this unique property is the oldest art gallery in Poland that is open to the public and is visited by hundreds of thousands of visitors each year. The Palace itself hosts more than 9100 different artworks, including paintings, sculptures, decorative arts, prints, and textile arts from the XVII–XIX centuries, with most of them being of high importance for European and world heritage. Since the early XXI century, the facades and interiors of the palace, as well as the park and forefield, have undergone extensive revitalization works. Additionally, the Wilanow Palace is a member of the Association of the Royal Residences of Europe (ARRE), which enables the cooperation and exchange of knowledge and experience about the preservation and development of cultural heritage.

In this study, we used metabarcoding to select the best culture media combinations for the detection of the highest number of bacterial genera. Therefore, the goal of this study was to show how metagenomic methods can improve classic microbiology approaches that are commonly employed for microbial screening in cultural heritage institutions.

## 2. Materials and Methods

### 2.1. Sample Collection

Sampling was conducted over the autumn and winter seasons of 2017–2018. In total, six sampling campaigns were conducted, with 25 exhibition rooms sampled each time. Sampling took place in rooms housing various types of cultural heritage objects. This approach was applied to minimize the material-dependent bias and to maximize the observed biodiversity. The exact locations of the sampling rooms are shown in Figure 1.

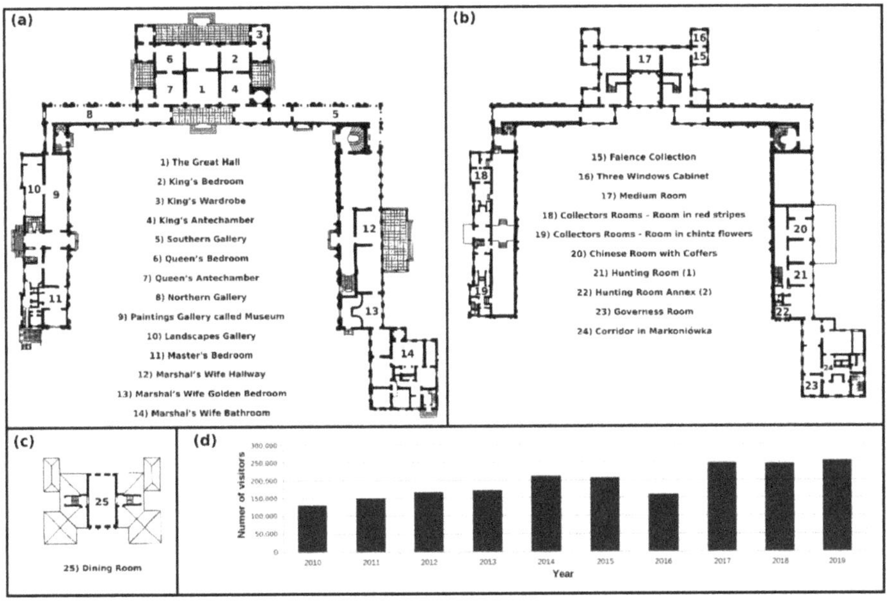


Figure 1. The plan of the Wilanow Palace rooms and the number of visitors in the last 10 years. Sections (**a–c**) show the locations of the sampling sites, i.e., rooms located on all three floors of the Wilanow Palace; (**d**) graph showing the number of visitors per year, starting from the year 2010.

Air samples were collected using the MAS-100 Eco device (Merck, Darmstadt, Germany) on agar plates with seven different bacterial growth media: Blickfeldt agar (BA), Brain Heart Infusion agar (BHI), Frazier agar (FA), lysogeny broth agar (LB), nutrient agar (NA), Reasoner's 2A agar (R2A), and yeast extract agar (YEA). Agar media were manufactured by BTL company (Lodz, Poland). At each sampling site, the device was positioned at 1.5 m above the ground using a tripod, and the airflow was set to 100 L per minute (±2.5%). Each plate was inoculated with 50 L of air. In total, 1050 agar plates (150 plates per medium) were inoculated and were later subjected to further analyses.

### 2.2. Bacterial Cultivation Experiments

The agar plates were incubated at 37 °C for 48 h, following the Museum's regular practice for the detection of bacterial pathogens. The bacterial colonies on each plate were counted and the number of colony-forming units (CFU) per cubic meter of air was calculated with the Feller's correction taken into account.

### 2.3. Metagenomic DNA Isolation and Sequencing

After incubation, each plate was washed with 2 mL of saline (0.85% NaCl solution in water). The obtained suspensions were merged by growth media, which gave seven mixed suspensions in total, one for each growth medium. The suspensions were centrifuged (12,000× *g*, 5 min), the supernatant was discarded, and bacterial pellets were subjected to further analyses. DNA was isolated from 100 mg of bacterial pellets for each of the seven pools separately using the FastDNA® SPIN Kit for Feces and the FastPrep® Instrument (MP Biomedicals, Santa Ana, CA, USA). Region V3/V4 of the 16S rDNA was PCR-amplified with primers: L16SV3V4 (5′-TCGTCGGCAGCGTCAGATGTGTATAAGAGACAGCCTACGGGNGGCWGCAG-3′), and R16SV3V4 (5′-GTCTCGTGGGCTCGGAGATGTGTATAAGAGACAGGACTACHVGGGTATCTAATCC-3′). The amplification was carried out in a Mastercycler Nexus GX2 thermocycler (Eppendorf, Hamburg, Germany) using KAPA HiFi DNA polymerase (KAPA Biosystems, Wilmington, MA, USA) with 5 ng of extracted DNA. For the DNA extracted from the bacteria grown on each type of medium, PCRs were conducted in triplicate and then mixed. The PCRs were prepared and performed as described previously [20]. The PCR products were quality-checked using electrophoresis in a 1.5% agarose gel. Seven PCR amplicons were then sequenced in a paired-end mode using a v3 chemistry kit with the Illumina MiSeq instrument (Illumina, San Diego, CA, USA) by the DNA Sequencing and Oligonucleotide Synthesis Laboratory—oligo.pl (Institute of Biochemistry and Biophysics, Polish Academy of Sciences).

### 2.4. Bioinformatics

The reads obtained from the MiSeq sequencer were first processed using fastp software in order to remove adaptor sequences [21]. Next, the reads were processed using the Qiime2 software package v. 2019.7 (http://qiime2.org) to perform quality-based trimming (–p-trunc-len-f 285 –p-trunc-len-r 240) and remove chimeras (–p-min-fold-parent-over-abundance 3) using the DADA2 library [22,23]. The remaining trimmed reads were treated as amplicon sequence variants (ASVs) [24]. Basic biodiversity indexes, such as Shannon's, Pielou's evenness, and Faith's indexes, were calculated for each of the samples using Qiime2 diversity (https://github.com/qiime2/q2-diversity), feature table (https://github.com/qiime2/q2-feature-table), and phylogeny (https://github.com/qiime2/q2-phylogeny) libraries. Taxonomies were assigned using a naive Bayes classifier from the feature-classifier (https://github.com/qiime2/q2-feature-classifier) library with the Silva database [25].

Other bioinformatic analyses were performed using custom Python scripts with matplotlib and seaborn libraries [26,27]. Statistical analysis was performed using R in the RStudio environment [28,29]. In order to determine significant differences between the colony-forming units (CFU) measured using methods of classical microbiology, the Kruskal–Wallis rank-sum test and the Mann–Whitney *U* test were used.

### 2.5. Systematic Literature Review for the Detection of Genera Hosting Human Pathogens

A systematic review of the available scientific literature was conducted in order to identify the bacterial genera hosting at least one species recognized as a human pathogen. The systematic review was conducted using the PubMed search engine according to the search scheme "(X) AND (pathogen) AND (human)," where X was a genus name. Additionally, each genus was checked for past name changes and reclassification. If such cases were found, previous genus names were also screened using the methodology described above. A genus was classified as "a genus containing pathogenic species" when at least one case study reported such a phenomenon.

### 2.6. Data Availability

The next-generation sequencing data was deposited at the European Nucleotide Archive under study number PRJEB38331 and link www.ebi.ac.uk/ena/data/view/PRJEB38331.

## 3. Results and Discussion

### *3.1. Assessment of Bacterial Contamination of Air-Quantitative Analysis Using the Classical Microbiology Approach*

There are no special protocols for assessing bacterial air contamination in places of cultural heritage significance. Moreover, the European standard for microbiological air contamination in cleanrooms, which are facilities that are required to maintain extremely low levels of particulates (ISO 14698-1:2003), does not recommend any specific culture media for microbiological assessments. The only requirement established in these norms is that a non-selective medium is used. In this study, seven different growth media (LB, NA, BHI, YEA, BA, FA, and R2A) were tested for their suitability for use in the screening of microbiological air contamination. We focused on airborne mesophilic bacteria, as this group provides the best estimate for contamination with human pathogens [30].

Figure 2 presents the overall air contamination that was measured using classical microbiology methods. Pairwise statistical analysis between yields from different media highlighted the YEA medium as having the lowest global recovery rate. The pairwise Wilcox rank-sum tests with YEA were always statistically significant, with $p$-values < $10^{-6}$ (Table S1). This highlighted the fact that the YEA medium enabled the formation of significantly fewer colonies (mean CFU/m$^3$ equaled $3.0 \times 10^2$) compared with the other investigated growth media (mean CFU/m$^3$ ranged between $5.2 \times 10^2$ and $9.3 \times 10^2$).

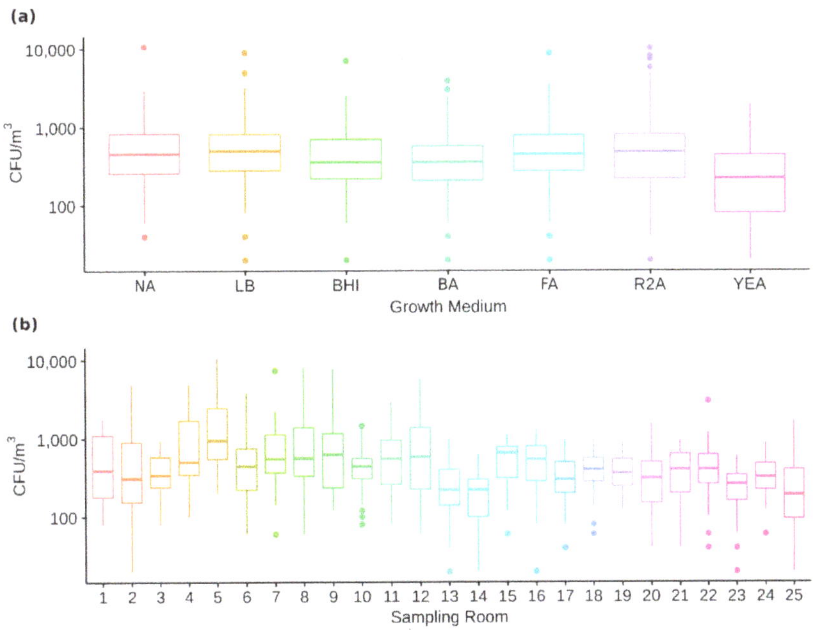


**Figure 2.** Box plots showing the measured CFU/m$^3$ distributions with a logarithmic scale. (**a**) Graph presenting the measured CFU/m$^3$ counts for each investigated medium. (**b**) Graph showing the CFU/m$^3$ counts by sampled rooms. The bold, central horizontal bars are the medians. The lower and upper limits of the box are the first and third quartiles, respectively. Points above and below the whiskers' upper and lower bounds are outliers. NA—nutrient agar, LB—lysogeny broth, BHI—brain heart infusion agar, BA—Blickfeldt agar, FA—Frazier agar, R2A—Reasoner's 2A agar, YEA—yeast extract agar.

The R2A, FA, and LB media showed the highest global recovery rates, with average CFU/m$^3$ values equal to $9.3 \times 10^2$, $7.1 \times 10^2$, and $7.0 \times 10^2$, respectively. It is worth mentioning that the highest average CFU/m$^3$ was obtained for the R2A medium, which was designed to promote the growth of slow-growing heterotrophic bacteria. This observation is in line with the findings of other studies, highlighting the potential for the application of R2A medium in air contamination assessment protocols. In this case, the R2A medium could replace commonly used rich media, such as NA or tryptone soya agar (TSA) [31–33]. The benefit of using the R2A medium relies on its specific features, as a wide range of cultivable bacteria can be obtained due to the promoted growth of slow-growing species. In contrast, the diversity of cultivable isolates observed on rich media was lower, as the promoted growth of fast-growing bacteria suppressed the appearance of slow-growing ones.

The results obtained for R2A, FA, and LB media were in line with other works describing indoor environments. Brągoszewska et al. investigated indoor air pollution in human dwellings in Upper Silesia, Poland. According to their case study, the average microbiological air pollution in human dwellings was $9.6 \times 10^2$ CFU/m$^3$ [34]. A more comprehensive analysis of microbiological air pollution in office buildings conducted in another study led by Brągoszewska reported recovery rates of $1.4 \times 10^3$ CFU/m$^3$ in naturally ventilated offices and $9.7 \times 10^2$ CFU/m$^3$ in offices with air conditioning [35]. A study by Kalwasińska et al. investigated microbiological air contamination in different rooms of a university library. By sampling different rooms with varying visitor accessibilities, the authors concluded that bacterial contamination is highly influenced by human presence. Rooms with high human activity showed a mean CFU/m$^3$ of $1.1 \times 10^3$ (library cafeteria), while the Old Prints Storeroom, which is not publicly accessible, showed a mean CFU/m$^3$ of $1.3 \times 10^2$ [36]. These results are in agreement with our observations since the Museum of King Jan III's Palace in Wilanow is a naturally ventilated indoor space with a high human presence.

### 3.2. Assessment of Bacterial Contamination of Air-Qualitative Analysis Using a Metagenomic Approach

PCR amplicons were obtained using bacterial material collected from 1050 Petri dishes. Metagenomic analysis of seven samples (with each sample being the sum of bacteria cultured on a specific medium) revealed 130 bacterial genera. The number of bacterial genera obtained from each medium is presented in Table 1. Sequences that could not be classified at the genus level constituted between 1.3% (YEA medium) and 4.4% (BHI medium) of sequences obtained from each medium type and were assigned to the "others" group (Figure 3).

**Table 1.** Number of bacterial taxa recovered on each growth medium.

| Taxonomic Rank | NA | LB | BHI | BA | FA | R2A | YEA |
|---|---|---|---|---|---|---|---|
| Genera | 54 | 55 | 45 | 49 | 71 | 63 | 27 |
| Families | 30 | 30 | 29 | 30 | 38 | 38 | 21 |
| Orders | 17 | 16 | 15 | 16 | 20 | 21 | 12 |
| Classes | 6 | 5 | 5 | 6 | 5 | 7 | 5 |

NA—nutrient agar, LB—lysogeny broth, BHI—brain heart infusion agar, BA—Blickfeldt agar, FA—Frazier agar, R2A—Reasoner's 2A agar, YEA—yeast extract agar.

The relative abundance analysis revealed that all used media reported similar dominant genera, i.e., *Acinetobacter*, *Bacillus*, *Enhydrobacter*, *Micrococcus*, and *Staphylococcus* (Figure 3). Amongst them, *Acinetobacter*, *Enhydrobacter*, and *Staphylococcus* contain mainly species associated with common human skin microbiota, including several pathogens [37], whereas *Bacillus* and *Micrococcus* can be included in the large group of so-called environmental bacteria.

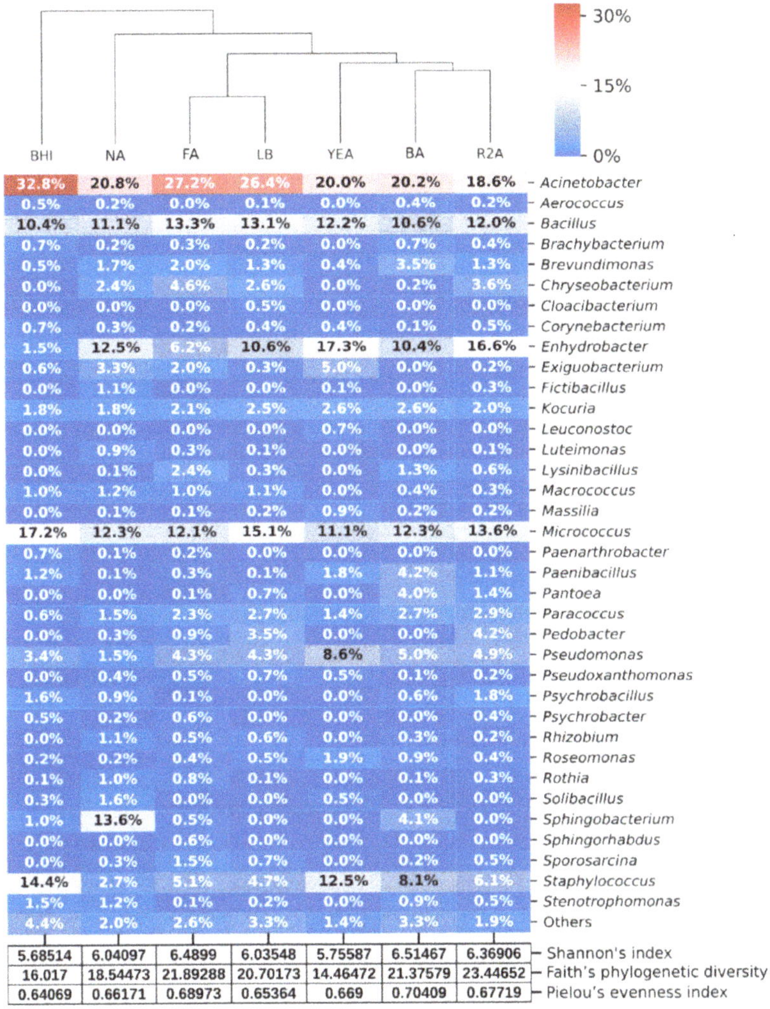


**Figure 3.** Cultivable bacteria taxonomy and their relative abundances. Heatmap with a dendrogram, depicting genera detected on the investigated media, and the following alpha-diversity indexes for each medium: Shannon's index, Faith's phylogenetic diversity, and Pielou's evenness index. Only the genera that constituted at least 0.5%, abundance-wise, of all genera on at least one medium are shown on the heatmap. Genera accounting for less than 0.5% across all samples were assigned to the "others" group. NA—nutrient agar, LB—lysogeny broth, BHI—brain heart infusion agar, BA—Blickfeldt agar, FA—Frazier agar, R2A—Reasoner's 2A agar, YEA—yeast extract agar.

While the dominant genera were similar on all media, the metabarcoding analysis showed that the total recovered bacterial taxa were different, depending on the type of medium. Shannon's biodiversity and Pielou's evenness indexes were similar for all media (Figure 3). On the other hand, Faith's phylogenetic diversity index indicated significant differences between the tested samples. The lowest Faith's phylogenetic diversity index was obtained for the YEA medium (14.46), while the R2A medium showed the highest value (23.45) (Figure 3). This result further supported the finding that rich media should not be used when trying to recover a high diversity of environmental microbiomes.

Based on the 16S rDNA amplicon analysis, we computed media combinations that would allow for the recovery of the highest bacterial diversity when sampling cultural heritage objects (Table 2). It was revealed that the FA medium enabled the growth of the most diverse community. It recovered 64.62% of cultivable bacterial genera detected on all tested media (Table 2). The addition of the R2A medium raised this value to 80.77% (by adding 21 new genera) (Table 2). Therefore, the combination of these two media could be used as a basis for microbiological assessments. The addition of other media subsequently increased the recovery rate, as shown in Table 2.

**Table 2.** Summary of the bacterial genera recovery using various media combinations.

| Row No. | Media Combination * | Percent of All Genera Detected ** | Genera Detected |
|---|---|---|---|
| 1 | FA | 64.62% | *Acinetobacter, Aerococcus, Agrococcus, Algoriphagus, Amaricoccus, Arthrobacter, Aureimonas, Bacillus, Bosea, Brachybacterium, Brevibacillus, Brevibacterium, Brevundimonas, Candidatus Paracaedibacter, Carnobacterium, Caulobacter, Chryseobacterium, Corynebacterium, Cupriavidus, Devosia, Dyadobacter, Enhydrobacter, Enterococcus, Exiguobacterium, Fictibacillus, Flavihumibacter, Flavobacterium, Glutamicibacter, Gordonia, Hydrogenophaga, Hymenobacter, Jeotgalicoccus, Kocuria, Kytococcus, Lactococcus, Luteimonas, Lysinibacillus, Lysobacter, Macrococcus, Marmoricola, Massilia, Methylobacterium, Microbacterium, Micrococcus, Mycobacterium, Nakamurella, Nocardioides, Novosphingobium, Oceanobacillus, Paenarthrobacter, Paenibacillus, Paeniglutamicibacter, Paenisporosarcina, Pantoea, Paracoccus, Pedobacter, Pseudomonas, Pseudorhodobacter, Pseudoxanthomonas, Psychrobacillus, Psychrobacter, Ralstonia, Rhizobacter, Rhizobium, Rhodobacter, Rhodococcus, Rhodoferax, Roseomonas, Rothia, Rummeliibacillus, Skermanella, Sphingobacterium, Sphingobium, Sphingomonas, Sphingopyxis, Sphingorhabdus, Sporosarcina, Staphylococcus, Stenotrophomonas, Streptococcus, Trichococcus, Truepera, Variovorax, Williamsia* |
| 2 | FA, R2A | 80.77% | Row no. 1 + *Aeromicrobium, Agromyces, Cellulomonas, Cloacibacterium, Cohnella, Curtobacterium, Deinococcus, Dermacoccus, Dietzia, Escherichia-Shigella, Microvirga, Ornithinibacillus, Parasegetibacter, Planomicrobium, Porphyrobacter, Pseudarthrobacter, Pseudoclavibacter, Sanguibacter, Shinella, Solibacillus, Timonella* |
| 3 | FA, R2A, LB | 89.23% | Row no. 2 + *Achromobacter, Clostridioides, Desemzia, Domibacillus, Enterobacter, Gracilibacillus, Kaistia, Leucobacter, Terribacillus, Vagococcus, Verticia* |
| 4 | FA, R2A, 0LB, NA | 93.85% | Row no. 3 + *Aneurinibacillus, Bergeyella, Cellulosimicrobium, Nosocomiicoccus, Phenylobacterium, Streptomyces* |
| 5 | FA, R2A, LB, NA, BA | 96.92% | Row no. 4 + *Acidovorax, Lechevalieria, Ochrobactrum, Promicromonospora* |
| 6 | FA, R2A, LB, NA, BA, BHI | 99.23% | Row no. 5 + *Dermabacter, Marinilactibacillus, Virgibacillus* |
| 7 | FA, R2A, LB, NA, BA, BHI, YEA | 100% | Row no. 6 + *Leuconostoc* |

* NA—nutrient agar, LB—lysogeny broth, BHI—brain heart infusion agar, BA—Blickfeldt agar, FA—Frazier agar, R2A—Reasoner's 2A agar, YEA—yeast extract agar. ** The percentage of all genera detected is equal to the number of genera detected on a given medium or media combination, divided by the summarized number of unique genera grown on at least one tested medium.

This result can be used as a guideline for selecting optimal growth media for microbiological air pollution assessments. Usually, due to budget and human resources limitations, it is impossible to use more than three different media during regular air contamination assessments in museums. Therefore, it is critical to establish reliable and comparable protocols (including medium selection) for classical analyses when employing microbial cultivation. In order to obtain the most comprehensive results, based on our findings, we suggest starting with the FA medium and then, depending on the available resources, adding other media in the following order: R2A, LB, NA, BA, BHI, and YEA.

### *3.3. Microbial Risk Assessment at the Museum Palace in Wilanow*

#### 3.3.1. Bacteria Potentially Involved in the Biodeterioration and Biodegradation of Historical Art Pieces

Microbial activity plays an important role in the biodeterioration of objects of cultural heritage. This is primarily due to the extraordinary diversity of bacterial metabolism and the involvement of bacteria in biogeochemical processes that affect various surfaces. Lignocellulolytic microorganisms, found in most bacterial phyla, are capable of degrading lignocellulose, which is the main component of paper, plant-based textiles, and wood [38]. Animal-derived materials, such as wool, silk, or hide, are more resistant to degradation, as they are mainly made of keratin, which is a complex of densely packed amino acid chains. However, members of *Alcaligenes*, *Bacillus*, *Proteus*, *Pseudomonas*, and *Streptomyces* exhibiting keratolytic activity were found to be able to degrade those as well [39]. Successful colonization of historical objects by autotrophic bacteria results in the formation of syntrophic chains, in which one microorganism thrives on the metabolites produced by another species. This phenomenon further accelerates the disruption of a given material by indirect actions, such as acidification or increased water retention [40,41].

Previously performed studies highlighted the fact that the biodeterioration mechanisms occurring on historical art pieces, as well as their pace, are determined by the material the object is made of. The Museum Palace in Wilanow hosts a wide range of historical objects made of various materials, including textiles, stone, and wood. Therefore, each object may be subjected to different kinds of deteriorating microbial activity. In this study, we performed a general assessment of the microbial contamination of air at the Museum Palace in Wilanow to recognize the total microbial diversity and how these bacteria may potentially affect historical objects.

As mentioned above, amongst the identified bacteria, the three dominant taxa were *Bacillus*, *Micrococcus*, and *Pseudomonas*. These are heterotrophic bacteria that are often detected as secondary colonizers in biofilms, which are primarily produced by filamentous *Actinobacteria*. Interestingly, such biofilms were found to be responsible for damaging plaster, marble, and textiles [42]. While *Actinobacteria* build the foundation of these biofilms, by providing simpler organic carbon sources and retaining moisture, secondary colonizers introduce cellulolytic, hemicellulolytic, and keratolytic activity that further accelerates the biodegradation [42].

Although the dominant taxa potentially involved in biodegradation were discovered on all tested media, some other bacteria that may also be involved in the destruction of cultural heritage objects were detected exclusively on some media. Examples of such taxa were: (i) *Flavobacterium*, which was previously recognized to be marble degrader [43], was detected only on FA; (ii) members of the *Cytophagales* order, damaging wooden objects [44], were recovered on FA, R2A, and LB; (iii) *Sphingomonas*, recognized as an important biodeterioration agent, previously detected on old manuscripts, wooden sculptures, and mural paintings [45–47], was found on all media except BHI and YEA. Generally, it was observed that the majority of genera (including *Rhodococcus*, *Achromobacter*, *Oceanobacillus*, *Cellulomonas*, *Arthrobacter*, and *Streptomyces*) previously recognized as being involved in the biodeterioration of historical objects was detected only on selected non-rich media. This proves that the rich media commonly used for microbiological contamination assessments in museums generate strongly biased results.

### 3.3.2. Putative Bacterial Pathogens

While human-to-human transmission of pathogenic bacteria is strongly dependent on a pathogen's biology, many studies emphasize that the environment and its characteristics, such as indoor/outdoor location, active/passive ventilation, heating, humidity, and temperature, play an important role in pathogen viability [18,19,48]. Above all, direct human contact is the main driving factor for both viral and microbial infections. Museums and other institutions of cultural heritage can be considered as hot spots of increased biological air contamination due to their high human traffic. Therefore, their regular testing is of utmost importance.

Published studies investigating microbial air contamination usually focus on bacteria known to be etiological factors of respiratory diseases, such as *Mycobacterium tuberculosis*, *Legionella pneumophila*, *Pseudomonas aeruginosa*, or members of the *Staphylococcus* genus [49–51]. In this study, we detected a relatively high number of representatives of the *Pseudomonas* and *Staphylococcus* genera (Figure 3). While *Pseudomonas* spp. host many species that are able to colonize a wide range of niches, *Staphylococcus* spp. are instead associated with a multitude of mammalian infections [52,53]. Human pathogens can also be found in both genera, including: (i) an opportunistic pathogen *P. aeruginosa*, which is a common multidrug-resistant bacterium that causes severe infections, especially in patients with cystic fibrosis [54]; (ii) *Pseudomonas mendocina*, an emerging pathogen that causes mainly nosocomial endocarditis and meningitis [55]; (iii) *Staphylococcus aureus* and *Staphylococcus epidermidis*, which are known to cause various skin and soft-tissue infections [56]. Moreover, amongst the detected genera, some can host opportunistic human pathogens, as is the case for the *Chryseobacterium* genus. Mukerji et al. described several serious infections caused by *Chryseobacterium indologenes*, a bacterium that is usually found in soil and water [57].

In order to quantify the applicability of the used growth media for the detection of human pathogens in museums, we conducted a systematic literature review, as described in Section 2.5, to identify genera that potentially contain human pathogens. This analysis revealed that 64 out of 130 genera detected in this study contained bacterial species reported to be human pathogens (Table 3). Despite the very high contribution of genera containing human pathogens—49.23%, it should be noted that most of them are opportunistic pathogens, which are mostly dangerous, specifically for immunocompromised individuals. Table 3 shows that the FA and R2A media detected up to 46 genera containing human pathogens, while NA—42, LB—40, BHI and BA—37, and YEA only 19. These results were in line with general genus recovery rates presented in Table 2, showing that the media (FA and R2A) capable of recovering the highest number of different bacterial genera were also suitable for the detection of species that are potentially pathogenic to humans.

It is also important to highlight that one of the dominant genera found in the Museum of King John III's Palace at Wilanow, namely, *Acinetobacter*, contains two clinically important pathogenic species, namely, *A. baumannii* and *A. lwofii*, which are both often characterized by their ability to form complex biofilms that reduce their sensitivity to many antibacterial agents, such as antibiotics [58,59]. Although *A. baumannii* is rarely found outside hospitals, it is still possible that a museum microclimate, with its stable temperature and humidity, may provide suitable conditions for its development.

**Table 3.** Summary of the bacterial genera containing human pathogens that were recovered on various media.

| Medium | No. of Genera | Genera Containing Human Pathogens |
|---|---|---|
| FA | 46 | *Acinetobacter, Aerococcus, Agrococcus, Arthrobacter, Aureimonas, Bacillus, Bosea, Brachybacterium, Brevibacillus, Brevibacterium, Brevundimonas, Caulobacter, Chryseobacterium, Corynebacterium, Cupriavidus, Devosia, Enterococcus, Exiguobacterium, Flavobacterium, Gordonia, Kocuria, Kytococcus, Lactococcus, Macrococcus, Massilia, Methylobacterium, Microbacterium, Micrococcus, Mycobacterium, Novosphingobium, Paenibacillus, Pantoea, Paracoccus, Pseudomonas, Psychrobacter, Ralstonia, Rhizobium, Rhodococcus, Roseomonas, Rothia, Sphingobacterium, Sphingomonas, Staphylococcus, Stenotrophomonas, Streptococcus, Williamsia* |
| R2A | 46 | *Acinetobacter, Aerococcus, Agrococcus, Arthrobacter, Aureimonas, Bacillus, Brachybacterium, Brevibacillus, Brevibacterium, Brevundimonas, Caulobacter, Cellulomonas, Chryseobacterium, Corynebacterium, Curtobacterium, Dermacoccus, Dietzia, Enterococcus, Escherichia-Shigella, Exiguobacterium, Gordonia, Kocuria, Kytococcus, Lactococcus, Macrococcus, Massilia, Methylobacterium, Microbacterium, Micrococcus, Mycobacterium, Ornithinibacillus, Paenibacillus, Pantoea, Paracoccus, Pseudoclavibacter, Pseudomonas, Psychrobacter, Ralstonia, Rhizobium, Rhodococcus, Roseomonas, Rothia, Sphingomonas, Staphylococcus, Stenotrophomonas, Streptococcus* |
| NA | 42 | *Acinetobacter, Aerococcus, Arthrobacter, Aureimonas, Bacillus, Bergeyella, Brachybacterium, Brevibacillus, Brevibacterium, Brevundimonas, Cellulosimicrobium, Chryseobacterium, Clostridioides, Corynebacterium, Curtobacterium, Dermacoccus, Dietzia, Enterococcus, Exiguobacterium, Gordonia, Kocuria, Kytococcus, Macrococcus, Massilia, Microbacterium, Micrococcus, Novosphingobium, Paenibacillus, Paracoccus, Pseudomonas, Psychrobacter, Ralstonia, Rhizobium, Rhodococcus, Roseomonas, Rothia, Sphingobacterium, Sphingomonas, Staphylococcus, Stenotrophomonas, Streptococcus, Streptomyces* |
| LB | 40 | *Achromobacter, Acinetobacter, Aerococcus, Agrococcus, Aureimonas, Bacillus, Brachybacterium, Brevibacillus, Brevibacterium, Brevundimonas, Chryseobacterium, Clostridioides, Corynebacterium, Cupriavidus, Dietzia, Enterobacter, Enterococcus, Escherichia-Shigella, Exiguobacterium, Kocuria, Kytococcus, Macrococcus, Massilia, Microbacterium, Micrococcus, Paenibacillus, Pantoea, Paracoccus, Pseudomonas, Psychrobacter, Ralstonia, Rhizobium, Rhodococcus, Roseomonas, Rothia, Sphingomonas, Staphylococcus, Stenotrophomonas, Streptococcus, Vagococcus* |
| BHI | 37 | *Achromobacter, Acinetobacter, Aerococcus, Aureimonas, Bacillus, Brachybacterium, Brevibacterium, Brevundimonas, Cellulosimicrobium, Corynebacterium, Curtobacterium, Dermabacter, Dermacoccus, Dietzia, Enterobacter, Enterococcus, Exiguobacterium, Kocuria, Kytococcus, Macrococcus, Massilia, Microbacterium, Micrococcus, Novosphingobium, Paenibacillus, Pantoea, Paracoccus, Pseudomonas, Psychrobacter, Ralstonia, Roseomonas, Rothia, Sphingobacterium, Staphylococcus, Stenotrophomonas, Streptococcus, Streptomyces* |
| BA | 37 | *Acidovorax, Acinetobacter, Aerococcus, Arthrobacter, Aureimonas, Bacillus, Brachybacterium, Brevibacillus, Brevibacterium, Brevundimonas, Cellulomonas, Chryseobacterium, Corynebacterium, Curtobacterium, Dietzia, Enterococcus, Kocuria, Macrococcus, Massilia, Microbacterium, Micrococcus, Mycobacterium, Ochrobactrum, Paenibacillus, Pantoea, Paracoccus, Pseudomonas, Ralstonia, Rhizobium, Rhodococcus, Roseomonas, Rothia, Sphingobacterium, Sphingomonas, Staphylococcus, Stenotrophomonas, Streptococcus* |
| YEA | 19 | *Acinetobacter, Bacillus, Brachybacterium, Brevibacterium, Brevundimonas, Corynebacterium, Dietzia, Exiguobacterium, Kocuria, Leuconostoc, Massilia, Microbacterium, Micrococcus, Paenibacillus, Paracoccus, Pseudomonas, Ralstonia, Roseomonas, Staphylococcus* |

NA—nutrient agar, LB—lysogeny broth, BHI—brain heart infusion agar, BA—Blickfeldt agar, FA—Frazier agar, R2A—Reasoner's 2A agar, YEA—yeast extract agar.

## 4. Conclusions

The results obtained in this work emphasize the need to further develop standardized operational protocols for microbial air pollution assessment of objects of cultural heritage. The NGS analysis demonstrated that the selection of an appropriate medium played a crucial role in assessing the diversity of cultivable bacterial communities.

The results obtained in this work clearly show that rich media, which are currently the most commonly used type of media for the assessment of microbial air contamination, produced heavily biased results. The use of rich media promoted fast-growing bacteria, which in turn, led to the inhibition of the growth of other species. This may prevent the detection of the biodeteriorating and pathogenic bacteria present in a museum. To solve this problem, we combined currently used classical microbiology methods with a novel NGS-based approach. Our results showed that Frazier medium, which is a type of minimal medium, led to the recovery of the most diverse bacterial community. Therefore, it was suggested that rich media should be replaced by FA (preferentially in combination with R2A) medium in protocols applied in museums. This combination would allow for the detection of more bacterial species in general, as well as specifically human pathogens, and therefore lead to a better understanding of the biodeteriorating and pathogenic microflora that are present in museums.

**Supplementary Materials:** The following is available online at http://www.mdpi.com/2076-3417/10/20/7128/s1, Table S1: *p*-values of the pairwise comparisons between growth media CFU/m$^3$ counts.

**Author Contributions:** Conceptualization, M.D. (Mikolaj Dziurzynski), M.D. (Magdalena Dyda), and L.D.; data curation, M.D. (Mikolaj Dziurzynski); formal analysis, M.D. (Mikolaj Dziurzynski), M.D. (Magdalena Dyda), and L.D.; funding acquisition, M.D. (Magdalena Dyda) and A.L.; investigation, M.D. (Mikolaj Dziurzynski), M.D. (Magdalena Dyda), A.S., P.D., and L.D.; methodology, M.D. (Mikolaj Dziurzynski), M.D. (Magdalena Dyda), and L.D.; project administration, M.D. (Magdalena Dyda) and A.L.; resources, M.D. (Magdalena Dyda), A.L., and L.D.; software, M.D. (Mikolaj Dziurzynski); supervision, M.D. (Magdalena Dyda) and L.D.; validation, M.D. (Mikolaj Dziurzynski) and L.D.; visualization, M.D. (Mikolaj Dziurzynski) and L.D.; writing—original draft, M.D. (Mikolaj Dziurzynski), K.C., and L.D.; writing—review and editing, M.D. (Mikola Dziurzynski), K.C., M.D. (Magdalena Dyda), A.L., and L.D. All authors have read and agreed to the published version of the manuscript.

**Funding:** This work was supported by the "Intelligent system for measuring environmental parameters with built-in predictive models for the protection of cultural heritage" project, which is co-financed by the European Regional Development Fund as part of the European Regional Development Fund: Priority axis I "Use of research and development activities in the economy", measure 1.2 "R&D activities of enterprises", Regional Operational Programme of the Mazowieckie Voivodeship for the years 2014–2020, agreement no. RPMA.01.02.00-14-5780/16-00.

**Acknowledgments:** The authors are grateful to Antonina Ignatenko and Damian Kakietek for the technical support regarding the microbiological sampling. Special thanks to Michalina Grzegorczyk for the efficient organization of the measurements in the Museum.

**Conflicts of Interest:** The authors declare no conflict of interest.

## References

1. Wells, W.F. *Airborne Contagion and Air Hygiene. An Ecological Study of Droplet Infections*; Harvard University Press: Cambridge, UK, 1955.
2. Górny, R.L.; Cyprowski, M.; Ławniczek-Wałczyk, A.; Gołofit-Szymczak, M.; Zapór, L. Biohazards in the Indoor Environment—A Role for Threshold Limit Values in Exposure Assessment. In *The Management of Indoor Air Quality*; Dudzinska, M.R., Ed.; CRC Press: Boca Raton, FL, USA, 2011; ISBN 9780415672665.
3. Scaltriti, S.; Cencetti, S.; Rovesti, S.; Marchesi, I.; Bargellini, A.; Borella, P. Risk factors for particulate and microbial contamination of air in operating theatres. *J. Hosp. Infect.* **2007**, *66*, 320–326. [CrossRef] [PubMed]
4. Napoli, C.; Marcotrigiano, V.; Montagna, M.T. Air sampling procedures to evaluate microbial contamination: A comparison between active and passive methods in operating theatres. *BMC Public Health* **2012**, *12*, 594. [CrossRef] [PubMed]
5. Hayleeyesus, S.F.; Manaye, A.M. Microbiological quality of indoor air in university libraries. *Asian Pac. J. Trop. Biomed.* **2014**, *4*, S312–S317. [CrossRef] [PubMed]

6. Kovats, N.; Horvath, E.; Jancsek-Turoczi, B.; Hoffer, A.; Gelencser, A.; Urban, P.; Kiss, I.E.; Bihari, Z.; Fekete, C. Microbiological characterization of stable resuspended dust. *Int. J. Occup. Med. Environ. Health* **2016**, *29*, 375–380. [CrossRef]
7. Valeriani, F.; Cianfanelli, C.; Gianfranceschi, G.; Santucci, S.; Romano Spica, V.; Mucci, N. Monitoring biodiversity in libraries: A pilot study and perspectives for indoor air quality. *J. Prev. Med. Hyg.* **2017**, *58*, E238–E251.
8. Taberlet, P.; Coissac, E.; Pompanon, F.; Brochmann, C.; Willerslev, E. Towards next-generation biodiversity assessment using DNA metabarcoding. *Mol. Ecol.* **2012**, *21*, 2045–2050. [CrossRef]
9. Tarsitani, G.; Moroni, C.; Cappitelli, F.; Pasquariello, G.; Maggi, O. Microbiological analysis of surfaces of Leonardo da Vinci's Atlantic Codex: Biodeterioration risk. *Int. J. Microbiol.* **2014**. [CrossRef]
10. Pinar, G.; Krakova, L.; Pangallo, D.; Piombino-Mascali, D.; Maixner, F.; Zink, A.; Sterflinger, K. Halophilic bacteria are colonizing the exhibition areas of the Capuchin Catacombs in Palermo, Italy. *Extremophiles* **2014**, *18*, 677–691. [CrossRef]
11. Saiz-Jimenez, C.; Miller, A.Z.; Martin-Sanchez, P.M.; Hernandez-Marine, M. Uncovering the origin of the black stains in Lascaux Cave in France. *Environ. Microbiol.* **2012**, *14*, 3220–3231. [CrossRef]
12. Kusumi, A.; Li, X.; Osuga, Y.; Kawashima, A.; Gu, J.-D.; Nasu, M.; Katayama, Y. Bacterial communities in pigmented biofilms formed on the sandstone bas-relief walls of the Bayon Temple, Angkor Thom, Cambodia. *Microbes Environ.* **2013**, *28*, 422–431. [CrossRef]
13. Osimani, A.; Aquilanti, L.; Tavoletti, S.; Clementi, F. Microbiological monitoring of air quality in a university canteen: An 11-year report. *Environ. Monit. Assess.* **2013**, *185*, 4765–4774. [CrossRef] [PubMed]
14. Giovannoni, S.J.; Britschgi, T.B.; Moyer, C.L.; Field, K.G. Genetic diversity in Sargasso Sea bacterioplankton. *Nature* **1990**, *345*, 60–63. [CrossRef] [PubMed]
15. Ward, D.M.; Weller, R.; Bateson, M.M. 16S rRNA sequences reveal numerous uncultured microorganisms in a natural community. *Nature* **1990**, *345*, 63–65. [CrossRef] [PubMed]
16. Hugenholtz, P. Exploring prokaryotic diversity in the genomic era. *Genome Biol.* **2002**, *3*, reviews0003-1. [CrossRef]
17. Mirhoseini, S.H.; Nikaeen, M.; Khanahmad, H.; Hassanzadeh, A. Occurrence of airborne vancomycin- and gentamicin-resistant bacteria in various hospital wards in Isfahan, Iran. *Adv. Biomed. Res.* **2016**, *5*, 143. [CrossRef]
18. Herfst, S.; Bohringer, M.; Karo, B.; Lawrence, P.; Lewis, N.S.; Mina, M.J.; Russell, C.J.; Steel, J.; de Swart, R.L.; Menge, C. Drivers of airborne human-to-human pathogen transmission. *Curr. Opin. Virol.* **2017**, *22*, 22–29. [CrossRef]
19. Luongo, J.C.; Fennelly, K.P.; Keen, J.A.; Zhai, Z.J.; Jones, B.W.; Miller, S.L. Role of mechanical ventilation in the airborne transmission of infectious agents in buildings. *Indoor Air* **2016**, *26*, 666–678. [CrossRef]
20. Dyda, M.; Decewicz, P.; Romaniuk, K. International Biodeterioration & Biodegradation Application of metagenomic methods for selection of an optimal growth medium for bacterial diversity analysis of microbiocenoses on historical stone surfaces. *Int. Biodeterior. Biodegrad.* **2017**, 2–10. [CrossRef]
21. Chen, S.; Zhou, Y.; Chen, Y.; Gu, J. Fastp: An ultra-fast all-in-one FASTQ preprocessor. *Bioinformatics* **2018**, *34*, i884–i890. [CrossRef]
22. Callahan, B.J.; McMurdie, P.J.; Rosen, M.J.; Han, A.W.; Johnson, A.J.A.; Holmes, S.P. DADA2: High-resolution sample inference from Illumina amplicon data. *Nat. Methods* **2016**, *13*, 581–583. [CrossRef]
23. Bolyen, E.; Rideout, J.R.; Dillon, M.R.; Bokulich, N.A.; Abnet, C.C.; Al-Ghalith, G.A.; Alexander, H.; Alm, E.J.; Arumugam, M.; Asnicar, F.; et al. Reproducible, interactive, scalable and extensible microbiome data science using QIIME 2. *Nat. Biotechnol.* **2019**, *37*, 852–857. [CrossRef] [PubMed]
24. Callahan, B.J.; McMurdie, P.J.; Holmes, S.P. Exact sequence variants should replace operational taxonomic units in marker-gene data analysis. *ISME J.* **2017**, *11*, 2639–2643. [CrossRef] [PubMed]
25. Quast, C.; Pruesse, E.; Yilmaz, P.; Gerken, J.; Schweer, T.; Yarza, P.; Peplies, J.; Glockner, F.O. The SILVA ribosomal RNA gene database project: Improved data processing and web-based tools. *Nucleic Acids Res.* **2013**, *41*, 590–596. [CrossRef] [PubMed]
26. Hunter, J.D. Matplotlib: A 2D graphics environment. *Comput. Sci. Eng.* **2007**, *9*, 90–95. [CrossRef]
27. Waskom, M.; Botvinnik, O.; Ostblom, J.; Saulius, L.; Hobson, P. Mwaskom/Seaborn:v0.10.0 2020. Available online: https://github.com/mwaskom/seaborn/tree/v0.9.0 (accessed on 11 October 2020).

28. R Core Team. *R: A Language and Environment for Statistical Computing*; R Foundation for Statistical Computing: Vienna, Austria, 2017.
29. RStudio Team. *RStudio: Integrated Development Environment for R*; RStudio, PBC: Boston, MA, USA, 2015.
30. Grisoli, P.; Albertoni, M.; Rodolfi, M. Application of airborne microorganism indexes in offices, gyms, and libraries. *Appl. Sci.* **2019**, *9*, 1101. [CrossRef]
31. Godish, D.R.; Godish, T.J. Relationship between sampling duration and concentration of culturable airborne mould and bacteria on selected culture media. *J. Appl. Microbiol.* **2007**, *102*, 1479–1484. [CrossRef]
32. Dacarro, C.; Picco, A.M.; Grisoli, P.; Rodolfi, M. Determination of aerial microbiological contamination in scholastic sports environments. *J. Appl. Microbiol.* **2003**, *95*, 904–912. [CrossRef]
33. Hyvärinen, A.M.; Martikainen, P.J.; Nevalainen, A.I. Suitability of poor medium in counting total viable airborne bacteria. *Grana* **1991**, *30*, 414–417. [CrossRef]
34. Brągoszewska, E.; Bogacka, M.; Pikoń, K. Efficiency and eco-costs of air purifiers in terms of improving microbiological indoor air quality in dwellings—A case study. *Atmosphere* **2019**, *10*, 742. [CrossRef]
35. Brągoszewska, E.; Biedroń, I.; Kozielska, B.; Pastuszka, J.S. Microbiological indoor air quality in an office building in Gliwice, Poland: Analysis of the case study. *Air Qual. Atmos. Heal.* **2018**, *11*, 729–740. [CrossRef]
36. Kalwasińska, A.; Burkowska, A.; Wilk, I. Microbial air contamination in indoor environment of a university library. *Ann. Agric. Environ. Med.* **2012**, *19*, 25–29. [PubMed]
37. Castelino, M.; Eyre, S.; Moat, J.; Fox, G.; Martin, P.; Ho, P.; Upton, M.; Barton, A. Optimisation of methods for bacterial skin microbiome investigation: Primer selection and comparison of the 454 versus MiSeq platform. *BMC Microbiol.* **2017**, *17*, 23. [CrossRef] [PubMed]
38. Mazzoli, R.; Giuffrida, M.G.; Pessione, E. Back to the past: "Find the guilty bug-microorganisms involved in the biodeterioration of archeological and historical artifacts". *Appl. Microbiol. Biotechnol.* **2018**, *102*, 6393–6407. [CrossRef] [PubMed]
39. Szostak-Kotowa, J. Biodeterioration of textiles. *Int. Biodeterior. Biodegradation* **2004**, *53*, 165–170. [CrossRef]
40. Sterflinger, K.; Piñar, G. Microbial deterioration of cultural heritage and works of art–tilting at windmills? *Appl. Microbiol. Biotechnol.* **2013**, *97*, 9637–9646. [CrossRef]
41. Caselli, E.; Pancaldi, S.; Baldisserotto, C.; Petrucci, F.; Impallaria, A.; Volpe, L.; D'Accolti, M.; Soffritti, I.; Coccagna, M.; Sassu, G.; et al. Characterization of biodegradation in a 17th century easel painting and potential for a biological approach. *PLoS ONE* **2018**, *13*, e0207630. [CrossRef]
42. Saarela, M.; Alakomi, H.-L.; Suihko, M.-L.; Maunuksela, L.; Raaska, L.; Mattila-Sandholm, T. Heterotrophic microorganisms in air and biofilm samples from Roman catacombs, with special emphasis on actinobacteria and fungi. *Int. Biodeterior. Biodegrad.* **2004**, *54*, 27–37. [CrossRef]
43. Sorlini, C.; Zanardini, E.; Albo, S.; Praderio, G.; Cariati, F.; Bruni, S. Research on chromatic alterations of marbles from the fountain of Villa Litta (Lainate, Milan). *Int. Biodeterior. Biodegrad.* **1994**, *33*, 153–164. [CrossRef]
44. Nilsson, T.; Björdal, C.; Fällman, E. Culturing erosion bacteria: Procedures for obtaining purer cultures and pure strains. *Int. Biodeterior. Biodegrad.* **2008**, *61*, 17–23. [CrossRef]
45. Palla, F.; Mancuso, F.P.; Billeci, N. Multiple approaches to identify bacteria in archaeological waterlogged wood. *J. Cult. Herit.* **2013**, *14*, e61–e64. [CrossRef]
46. Pasquarella, C.; Balocco, C.; Pasquariello, G.; Petrone, G.; Saccani, E.; Manotti, P.; Ugolotti, M.; Palla, F.; Maggi, O.; Albertini, R. A multidisciplinary approach to the study of cultural heritage environments: Experience at the Palatina Library in Parma. *Sci. Total Environ.* **2015**, *536*, 557–567. [CrossRef] [PubMed]
47. Heyrman, J.; Swings, J. 16S rDNA sequence analysis of bacterial isolatesfrom biodeteriorated mural paintings in the Servilia tomb (Necropolis of Carmona, Seville, Spain). *Syst. Appl. Microbiol.* **2001**, *24*, 417–422. [CrossRef] [PubMed]
48. Tang, J.W. The effect of environmental parameters on the survival of airborne infectious agents. *J. R. Soc. Interface* **2009**, *6* (Suppl. 6), S737–S746. [CrossRef] [PubMed]
49. Nguyen, T.M.N.; Ilef, D.; Jarraud, S.; Rouil, L.; Campese, C.; Che, D.; Haeghebaert, S.; Ganiayre, F.; Marcel, F.; Etienne, J.; et al. A community-wide outbreak of legionnaires disease linked to industrial cooling towers–how far can contaminated aerosols spread? *J. Infect. Dis.* **2006**, *193*, 102–111. [CrossRef] [PubMed]
50. Fronczek, C.F.; Yoon, J.-Y. Biosensors for monitoring airborne pathogens. *J. Lab. Autom.* **2015**, *20*, 390–410. [CrossRef]

51. Nuermberger, E.; Bishai, W.R.; Grosset, J.H. Latent tuberculosis infection. *Semin. Respir. Crit. Care Med.* **2004**, *25*, 317–336. [CrossRef]
52. Jun, S.-R.; Wassenaar, T.M.; Nookaew, I.; Hauser, L.; Wanchai, V.; Land, M.; Timm, C.M.; Lu, T.-Y.S.; Schadt, C.W.; Doktycz, M.J.; et al. Diversity of *Pseudomonas* genomes, including populus-associated isolates, as revealed by comparative genome analysis. *Appl. Environ. Microbiol.* **2016**, *82*, 375–383. [CrossRef]
53. Haag, A.F.; Fitzgerald, J.R.; Penadés, J.R. *Staphylococcus aureus* in animals. *Microbiol. Spectr.* **2019**, *7*. [CrossRef]
54. Stefani, S.; Campana, S.; Cariani, L.; Carnovale, V.; Colombo, C.; Lleo, M.M.; Iula, V.D.; Minicucci, L.; Morelli, P.; Pizzamiglio, G.; et al. Relevance of multidrug-resistant *Pseudomonas aeruginosa* infections in cystic fibrosis. *Int. J. Med. Microbiol.* **2017**, *307*, 353–362. [CrossRef]
55. Gani, M.; Rao, S.; Miller, M.; Scoular, S. *Pseudomonas mendocina* bacteremia: A case study and review of literature. *Am. J. Case Rep.* **2019**, *20*, 453–458. [CrossRef]
56. Natsis, N.E.; Cohen, P.R. Coagulase-negative *Staphylococcus* skin and soft tissue infections. *Am. J. Clin. Dermatol.* **2018**, *19*, 671–677. [CrossRef] [PubMed]
57. Mukerji, R.; Kakarala, R.; Smith, S.J.; Kusz, H.G. *Chryseobacterium indologenes*: An emerging infection in the USA. *BMJ Case Rep.* **2016**. [CrossRef] [PubMed]
58. Antunes, L.C.S.; Visca, P.; Towner, K.J. *Acinetobacter baumannii*: Evolution of a global pathogen. *Pathog. Dis.* **2014**, *71*, 292–301. [CrossRef] [PubMed]
59. Pour, N.K.; Dusane, D.H.; Dhakephalkar, P.K.; Zamin, F.R.; Zinjarde, S.S.; Chopade, B.A. Biofilm formation by *Acinetobacter baumannii* strains isolated from urinary tract infection and urinary catheters. *FEMS Immunol. Med. Microbiol.* **2011**, *62*, 328–338. [CrossRef] [PubMed]

applied sciences

MDPI

Review

# A Review on Sampling Techniques and Analytical Methods for Microbiota of Cultural Properties and Historical Architecture

**Xinghua Ding [1], Wensheng Lan [2] and Ji-Dong Gu [1,*,†]**

[1] Laboratory of Environmental Microbiology and Toxicology, School of Biological Sciences, The University of Hong Kong, Pokfulam Road, Hong Kong, China; u3005522@connect.hku.hk

[2] Shenzhen R&D Key Laboratory of Alien Pest Detection Technology, The Shenzhen Academy of Inspection and Quarantine, Technology Center for Animal and Plant Inspection and Quarantine, Shenzhen Customs, 1011 Fuqiang Road, Shenzhen 518010, China; lanwshao@163.com

* Correspondence: jidong.gu@gtiit.edu.cn

† Current address: Environmental Engineering, Guangdong Technion Israel Institute of Technology, 241 Daxue Road, Shantou 515063, Guangdong, China.

Received: 28 September 2020; Accepted: 20 October 2020; Published: 16 November 2020

check for updates

**Abstract:** World cultural heritage suffers from deterioration caused by both natural and anthropogenic processes, among which microorganisms are significantly involved. Among the key issues of this topic, sampling techniques and analytical methods for revealing the microbiome are fundamental to obtaining useful results for understanding the key players and processes involved, and also for effective protection and management of the cultural heritage for humanity. A non-invasive and non-destructive sampling method is required for sampling of cultural properties prior to further analysis of the microbiome. One example is illustrated in this article. For many years, culture-dependent methods had been used before the invention of polymerase-chain reaction (PCR) methods and, more recently, specifically high-throughput next generation sequencing (NGS). NGS reveals the whole microbial community composition and the active microorganisms from genomic DNA and RNA, respectively. The recovered environmental DNA and RNA from samples provide the information on microbial community and composition, and the active members and biochemical processes of the microbial attributes. It should be emphasized that the metabolically-active members of functional microflora in the biofilm or microbiome on cultural heritage must be determined and identified from the RNA-based analysis to gain a substantially important insight of the active biodeterioration processes and also the effectiveness of the conservation strategies. The importance of the culture-independent technique, based on NGS, is that it can be used in combination with the conventional culturing methods to guide the isolation and enrichment of new microorganisms to gain further biochemical insights to advance the role of the specific microbial groups for biodeterioration of cultural heritage. At the same time, effective restoration and maintenance strategies can be formulated for the protection of world cultural heritage.

**Keywords:** cultural heritage; microbiome; biochemical processes; nitrogen cycle; sulfur cycle; acidic attack; stone and rock

## 1. Introduction

Protection of cultural heritage, a social 'resource', has been recognized by many countries as both an economic asset and also an important factor for promotion of social integration [1]. At the same time, world cultural heritage and archaeological sites are the bases for the global tourism industry and economy [2–4]. World cultural heritage is broadly defined by including physical or "tangible" assets, including archaeological sites, architecture, monuments, buildings, sculptures etc. and intangible attributes, such as performing arts, oral traditions, rituals etc. Cultural heritage can be physically movable, e.g., books, documents, movable artworks, machines, clothing, and other artifacts worthy of preservation for the future. Meanwhile, the physical non-movable heritage of buildings, architectural ornaments, other historic sites, and monuments is the legacy of history and humanity, allowing selective value to pass from generation to generation. The recognition of the necessity of the historical value of the civilization and of the objects that possess their value and significance promotes the necessity for preservation of culture and heritage for conservation and protection through scientific understanding and better management [5–16].

Conservation science studies the processes that lead to physical changes in historical objects or artifacts that affect their longevity and integrity as a result of various processes or mechanisms of deterioration. These deteriorative mechanisms inform the scientists or conservators about the potential risk factors and strategies to mitigate adverse environmental impacts through management policy to effectively assist long-term conservation [17–22]. Conservation science has gradually evolved into an interdisciplinary research subject by embracing physics, chemistry and biology, microbiology, engineering, materials science, and engineering as well as archaeology, anthropology, and art history [23–28]. In the current conservation research, significant efforts are being made by non-biological analyses, mostly chemistry and material characterization, while an emerging direction has been in analyzing microorganisms on these objects for a more comprehensive knowledge and preservation practice [29,30]. Fungi, bacteria, archaea, and lichens have been observed on a variety of historic artifacts such as mural paintings and bas-reliefs in caves, churches and catacombs, stone monuments and architectural surfaces in outdoor environments, and historical/archaeological relics and sites [31–39]. The biodeterioration caused by microorganisms not only influences the aesthetical appearance of the cultural heritage, but also sometimes results in destruction of the structural integrity of the valuable cultural heritage via (bio)chemical processes such as acidification, salt crystallization, calcium carbonate precipitation, water retention, accumulation of pollutants through biofilm formation, and progression [30,38,40–42]. As the world cultural heritage is both unique and irreplaceable, is hardly possible to estimate the economic value of the damage caused by microbial degradation.

Microbiological research on this subject is becoming recognized as an indispensable part of the overall conservation science and the application of current scientific analytical techniques enables microbiologists to determine the microbial community and composition as well as biodeterioration potential of the colonizing microorganisms on the cultural heritage under a given environmental condition. Such knowledge informs how deterioration is likely to occur on cultural heritage and reveals important clues for conservators to monitor or mitigate the adverse effects and to prevent the potential risk from deterioration of historical artworks and objects caused by microorganisms [43,44]. This article offers a synthesis of the important issues of microbiological research for cultural heritage conservation by focusing specifically on sampling and new analytical methods. Furthermore, an outlook is given on the future research needs and developments in this highly important and an emerging research field.

## 2. Sampling Techniques

With respect to cultural heritage, conservation science focuses mainly on protection and maintenance of physical or "tangible" cultural heritage such as artworks, architecture, monuments, and museum collections. The code of ethics in the conservation of cultural heritage outlines that the conservation–restoration of any given physical or "tangible" heritage is to only do the minimum required for preservation [45]. Thus, a detailed diagnostic analysis of the current status of a cultural heritage object by a non-invasive/non-destructive technique is a prerequisite to avoiding unnecessary or potential damage and to keeping the amount of handling to a minimum. Before the advent of the specialized instruments used for conservation studies, visual inspections were the only non-invasive inspection method available in the field of heritage conservation to examine for visual detectable signs of damage, decay, or deterioration. Because of the methodological limitations of visual inspection, the earliest conservation studies could not yield the detailed quantitative information on deterioration processes and mechanisms, and always led to inconclusive results [46,47]. Microorganisms were not previously considered in this field. Currently, a number of generic or specialized research techniques have begun to be applied to heritage conservation studies to provide the powerful means to analyze the current state of an object for preservation or to detect any products of deterioration and decay that ought to be removed or cleaned. On the collection of samples involved in heritage conservation studies, the dogma which outlines the best practice for conservators and scientists alike for scientific analysis is "non-invasive and non-destructive method for bio-molecular and chemical analysis that requires very small quantities of sample from inconspicuous locations on artworks and objects" [48]. As the requirement for sampling techniques has suggested, any analytical techniques to be used must be highly sensitive, requiring minimum amounts of sample or none at all. Because of this, conservation microbiologists need to maintain a close communication with conservators, curators, and heritage authorities and work closely on research projects to determine the most appropriate sampling schemes and the proposed measurements prior to carry out the sampling.

A significant proportion of the cultural heritage relics and physical objects that suffer varying levels of biodeterioration caused by microorganisms have been rescued by salvage archaeology effort. Under non-ideal preservation and storage conditions, artifacts that have always been in storage such as archives, books, documents and paintings, and their constituent materials, both organic and inorganic, can provide a suitable surface or substratum for colonization by different types of microorganisms to form biofilm [30,38,49,50]. Microorganisms grow on the surface of artifacts mainly in clusters or aggregates called microbial biofilms, that have the potential to cause adverse reactions chemically or biochemically with cultural heritage objects [30,31,36]. A microbial biofilm is a collection of microbial cells and extracellular polymer substances (EPSs) together adhering onto a solid substratum surface (Figure 1). They are ubiquitous and play an important role in ecosystems. Numerous studies have confirmed that biofilms are not simple microbial assemblages, but have well-defined spatial distribution characteristics and microbial community composition and structures to assist in their ecological functions [33,51]. Spatial microbial distribution has been reported in some environmental biofilms, especially in aquatic ecosystems [52–54]. Microorganisms communicate via chemical cues prior to and during the biofilm formation. A stratified bacterial structural organization exhibited in biofilms has been reported on the limestone of historical monuments at a Mayan site [55], and on sandstone of Angkor temples with both spatial and temporal dynamics [33,56]. These adverse (bio)chemical reactions range from simply stains to acidification, discoloration, and utilization of the materials as nutrients and weaken the structure causing further disintegration and delamination.

**Figure 1.** A SEM micrograph showing a matured biofilm consisting of both grown cells and extracellular polymeric materials on surface of a rock under moist conditions (scale bar, 2 μm).

Earlier sampling techniques required relatively large amount of materials to be taken from the object before chemical and especially microbiological analyses could be carried out due to the lower sensitivity of the analytical techniques involved. Because of this, the focus of this article is on the current sampling techniques with a minimum and non-invasive approach coupled with highly-sensitive microbiological analysis. Currently, a slightly different version of the same principle has been developed as a sampling technique using sterilized plastic adhesive sheet or tape to remove microbial biofilms and loosely-attached materials including degradation products from surfaces, regardless of the surface physical morphologies and material types [57] (Figure 2). This sampling device was initially developed by the Japan Space Agency for sampling surfaces for microbiological assessment on the International Space Station and then it was then adopted in the cultural heritage conservation work at Angkor for more than a couple of decades [31,33,58–62]. This sampling technique, with some small variations, has also been used in Europe for microbiological analysis of cultural heritage (Urzi and De Leo, 2001). The device, consisting of an adhesive on a plastic sheet and a supporting paperboard, is made under sterile conditions, and can be conveniently carried to sampling locations. At the sampling site, after peeling the paperboard from the adhesive and attaching the adhesive surface to the sampling location to obtain the materials from the surface onto adhesive (Figure 3), further microbiological analyses can be carried out on the materials picked up on the adhesive. This sampling technique shows a higher recovery rate for microbial populations than the conventional cotton swab method [63] or stamp agar [57]. By applying the adhesive technique in sampling, samples could be collected as many times as required at a selected location to obtain spatial distribution information and composition of the microbial community from the sandstone bas-relief wall of Bayon temple of Angkor Thom in Cambodia [31,33]. Specifically, samples were analyzed to establish the relationship between the specific colors on the sandstone wall and the dominant microorganisms [33]. By working closely with Apsara Authority and the Ministry of the Environment, Kingdom of Cambodia, this sampling technique was systematically applied in the conservation and protection research of Angkor monuments over the past 20 years or more to accumulate data on multiple dimensions of the research objectives [31,33,51].

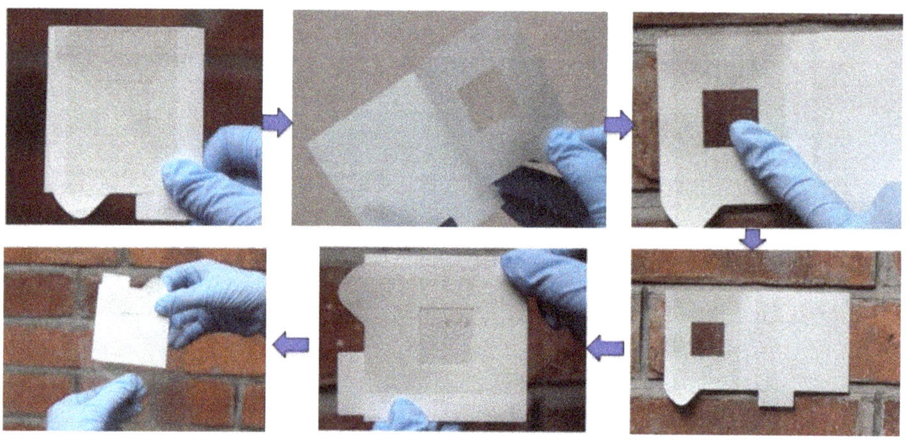

**Figure 2.** The adhesive sheet sampling device developed for non-invasive/non-destructive sampling from various solid surfaces. The adhesive sheet can collect biofilm (the microbial community and the decomposed materials on the solid surface) from the surface to the deepest layer by a series of continuous peeling-off sampling of the biofilm at the same sampling position to analyze the stratified microbiome structure.

**Figure 3.** The adhesive sampling device in a slightly different configuration from that shown in Figure 2 used for sampling the surface colonizers on sandstone wall of Angkor temple in Cambodia and then transferred into a sterile plastic bag ready for preservation and transportation.

The adhesive sampling technique allows analysis of samples for microbial community and composition on these sheets afterward [31,33]. For example, the characteristics of the stratified structure of bacteria in the biofilm were analyzed and characterized by using 16S rRNA-based PCR-denaturing gradient gel electrophoresis (DGGE) analyses among bacterial communities after non-destructive collection of biofilm samples onto adhesive sheets from the sandstone surface of Bayon temple at Angkor Thom, Cambodia [33]. The study disclosed a rich community of bacteria, especially cyanobacteria-related bacteria, on the outer surface layer of the biofilm, but fewer in the lower layer; while Chloroflexi-related bacteria appeared only in the lower layer of the biofilm. A similar distribution of bacterial structure has already been reported in microbial mats in a hot spring which was very rich in cyanobacteria near the surface and Chloroflexi in deeper layers [64–67]. Considering the differences between the optical absorption spectra of the Chloroflexi and Cyanobacteria together with the phylum Chloroflexi, which are an order of heterotrophic bacteria and an order of anoxygenic phototrophs [67–69], these results suggest that the bacteria residing within the biofilm are very specific spatially, depending on the availability of light and organic nutrients and the oxygen concentration. The biofilm formation and development on cultural heritage surfaces affect not only the aesthetic appearance of the artistic presentation and clarity but also the integrity of its underling materials by the colonization of microorganisms, and biofilm formation and development [30,31,40,49,70]. Information about the spatial distribution characteristics and microbial compositional structures of the microbial communities in the biofilm is therefore indispensable for understanding the microbial community involved in the biofilm formation and also controlling biofilm colonization of the cultural heritage.

## 3. Microbiome Analysis by Culture-Independent Methods

Knowledge of the microorganisms inhabiting cultural heritage and their involvement in the material deterioration is basic for conservators to formulate actions to preserve and protect cultural heritage. For decades, culture-dependent approaches have been applied to isolate and identity any culturable microorganisms from the microbiome inhabiting cultural heritage [60–62,71,72], but this approach has many shortcomings. Many naturally-occurring microorganisms that can be cultured in laboratory media are selected by the medium composition and also the culturing conditions [73], not the dominant members in the niche or in situ conditions, so the subsequent results of the physiology and biochemistry of these microorganisms have very limited relevance or value to the whole community composition and the biochemical processed involved. In addition, microorganisms cultured under laboratory conditions experience much higher concentrations of nutrients than the ambient environmental conditions on the cultural heritage, and this critical factor has been largely ignored in practice (Figure 4). The microbiome on sandstone is dominated by halophile and microorganisms are capable of adaptation to the local conditions and also enter a physiological state, called viable but non-culturable (VNC) state under adverse condition of nutrient limitation and low temperature [74,75]. Under natural condition, a complex microbial community is the normal state rather than the pure culture of a single species in artificial medium. The halophilic condition on sandstone selects relevant microorganisms and similarly antibiotic-resistant populations can be developed after such chemicals have been applied for a while to alter the indigenous microbiome.

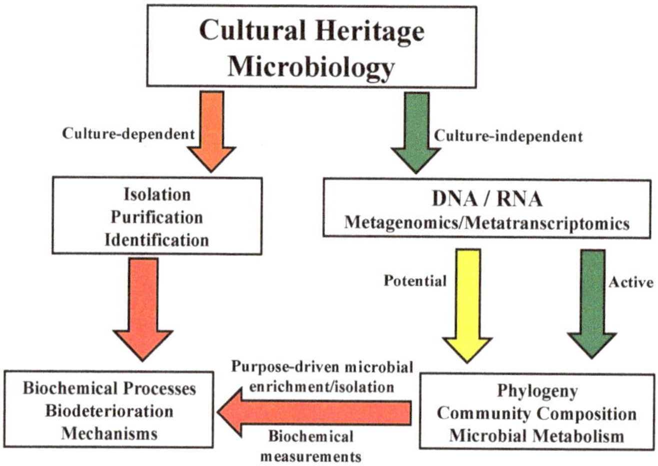


**Figure 4.** A schematic diagram showing the general framework in microbiological analyses of cultural properties and protection.

Culturing of microorganisms is a critically-important issue in environmental microbiology and it prevents researchers gaining further detailed insights into the biochemical function from members of a complex microbial community through isolation and pure culture techniques [73]. On the other hand, the development and application of DNA-based molecular methods in conservation sciences overcomes the culture-dependent limitations to reveal the microbial diversity and composition of the whole microbiome with greater depth on cultural heritage of any material type and at any time. Earlier molecular techniques used in conservation sciences include the genetic fingerprinting analyses such as DGGE/temperature gradient gel electrophoresis (TGGE), amplified ribosomal DNA restriction analysis (ARDRA), single-strand conformation polymorphism (SSCP), terminal restriction fragment length polymorphism (t-RFLR), and automated ribosomal intergenic spacer analysis (ARISA) after PCR [33]. For instance, a previous study investigated the abundance and diversity of microbial communities thriving on damaged mural paintings in Portugal, by combining culture-dependent methods and culture-independent methods, such as DGGE. However, it was found that the coupling of culture-dependent methods and DGGE still did not yield enough information to understand the abundance and the diversity of microorganisms present on the wall paintings [76]. Because of the methodological limitation and sensitivity of these genetic fingerprinting analyses, they failed to identify different members of a particular microbial group taxonomically to a satisfactory resolution with confidence.

### 3.1. High Through-Put Sequencing

The culture-independent methods based on PCR amplification of specific target genes, commonly the 16S rRNA gene, which serves as an evolutionary chronometer and allows phylogenetic analysis of obtained sequences from the microbial community [77], are becoming a routine technique used with a sharp increase of its acceptance today. Initially, the DNA-based techniques were applied to single colonies or libraries of PCR-amplified nucleotide sequences from environmental samples, after successful cloning of the single PCR products, but in recent years, the application of next generation sequencing (NGS) technology to microbial communities has revolutionized research on ecological microbiology, enabling, at an affordable cost, both deep information and also large-scale sequencing of DNA samples without culturing and cloning. This simplifies the analytical procedures involved and also makes this technology accessible based on paid-for-service. Metagenomics for the characterization of microbial community phylogeny and diversity is a new line of molecular microbiological methods based on genomic DNA extraction directly from environmental samples and then NGS technology to obtain the short reads of nucleotides for further assembly and analysis [78–80].

Metagenomic technologies can be either target-specific or non-specific. For the target-specific ones, the microbial community of bacteria and archaea is revealed based on the 16S rRNA gene, and that of eukaryotes based on the 18S rRNA or additional genes. Zhang et al. [31] investigated the bio-erosion of the sandstone of the Royal Palace of Angkor Thom and the temple Beng Mealea using the NGS targeting the 16S rRNA gene with samples of biofilm and exfoliated materials to obtain phylogeny and diversity information of the bacterial community in them. The results of this study using the adhesive sampling technique revealed that a higher bacterial diversity was detected in exfoliated sandstone materials than in the biofilms, but the diversity of bacteria was lowest in the lower layers of the biofilm after repetitively sampling at a selective location (discussed above). Firmicutes and Gemmatimonadetes were the dominant phyla and were detected only in exfoliated materials, while Cyanobacteria, Chloroflexi, and unassigned bacteria were more abundant in the biofilms. The results of this study suggested a microbial specificity on the sandstone monuments by the nature of the sample types and the different degree of deterioration. Biofilms grown on the surface of sandstone are in constant contact with the underlying inorganic substratum, the shortage of organic nutrients resulting in lower diversity of the community of heterotrophic bacteria adhering on the sandstone surface. Moreover, Cyanobacteria and Chloriflexi were identified as the key photoautotrophic microorganisms colonizing the surface of the sandstone. They appeared widely in the biofilms of Angkor monuments with a strong ability for utilizing direct or indirect sunlight, resistance to the high temperature, and the ability to survive the hypersaline environment of sandstone niche [33,38,56,62,81]. The active metabolism of these autotrophic microorganisms results in the fixation of the inorganic $CO_2$ as a carbon source and the ability to modify the surface layer for retention of water, providing opportunities for subsequent colonization by the heterotrophic bacteria and fungi [38,40,56,58,59,82]. Then, a diverse and stable microbial community can be formed on the exfoliation sandstone materials, and the microbial activity and abundant metabolites further accelerate the attack of the sandstone. Although the target-specific methods can amplify and sequence marker regions of the ribosomal DNA to obtain a genus-level picture of the microorganisms present in different environmental samples, the targeted metagenomic analysis alone could not yield enough information about the genes and the biochemical functions in a microbial community because of the lower depth of sequencing [83]. However, the target non-specific metagenomics approaches, with the rapid technological advancement in high-throughput NGS, as well as bioinformatics data processing and analysis, can extract strain-level information of the genes and the biochemical functions in a microbial community for a better understanding of the attacking mechanisms of microorganisms on historical materials. Microorganisms interact with underlying materials and with other microorganisms, e.g., biofilms, and such information can be very specific in evaluating the outcomes of the maintenance and conservation treatments on the function and community shifts of the microbiome on cultural heritage. A recent microbiological research studied microbial phylogenetic composition as well as biochemical capabilities of the microbiome inhabiting the sandstone surfaces of the Preah Vihear temple by using this non-targeted metagenomic approach. The results of this deep metagenomics analysis provided a detailed description of the microflora present on the surface of this cultural heritage [84]. The results indicated the presence of biochemical processes for carbon sequestration, and nitrogen and sulfur metabolism were identified as potentially the active microbial biochemical processes on the sandstone surface at this site. Specifically, the identified acid-producing biochemical processes of sulfur-oxidizing bacteria and ammonia-oxidizing bacteria and archaea suggest that the microbial flora on the sandstone surface participate in the deterioration of sandstone cultural heritage by producing acids [32,84,85]. High-throughput NGS is a powerful method and has the advantage of characterizing the microbial community phylogeny and diversity quickly, accurately, and economically with a deeper data recovery [31,86]. For investigations of microbiome of cultural heritage and the surrounding environment, where the biomass is relatively lower and the sample size is a serious concern due to the value of the object concerned, NGS is a good candidate method for such microbiological analysis to obtain the basic necessary data with high confidence as a first step.

### 3.2. Genomic DNA vs. RNA

The total microbial biomass is a useful indicator of microbial population and contamination on solid surfaces, but the viable microbes are important estimators of microbial activity and metabolic potential. The expression analysis of genes can yield detailed information about the metabolically-active members and microbial metabolic state of the microbiome and about the microbe-mediated biodeterioration process and potential, for example, following the microbial metabolic activities of ammonia oxidation by bacteria and archaea [32,85] and sulfur oxidation [60,61] on cultural relics with RNA. Assessment of the metabolically-active members or groups in the microbial community is a more selective and specific approach to identifying the specific culprits than the overall information recovered from the genomic DNA of samples.

Currently, it is necessary to determine the level of metabolic or physiological activity of specific functional microflora in biofilm or microbiome on cultural heritage because this information is the basis for conservators and scientists to gain a deeper understanding of the biodeterioration processes, to monitor the effectiveness and success of the conservation treatments, and to develop alternative and non-toxic conservation strategies to inhibit or slow down microbe-mediated biodeterioration (Figure 4). Although real-time quantitative PCR (RT qPCR) is currently a routine tool for solving many scientific problems, it is not useful for routine monitoring of biodeterioration and has been used only occasionally on samples of cultural heritage. In a recent study, the community of ammonia-oxidizing microorganisms as a biodeteriogen on the Angkor monuments, the ammonia-oxidizing archaea (AOA), were determined to be the more active members over ammonia-oxidizing bacteria (AOB) in the nitrifying reaction of the microbiome on Angkor monuments. They can play an important role in biodeterioration according to the results from the RT qPCR analyzing the expression of *amoA* gene of samples [32]. Further confirmation of the biochemical capabilities are still expected to establish the microbial transformation processes responsible for destruction of materials (see next section).

Gene-by-gene molecular analysis is too time-consuming and requires a high level of technical skill in conservation science. As a result, it is not cost-effective in application. Recent metatranscriptomics approaches that rely on NGS platforms offer more possibilities for understanding the metabolic pathways and the activity of the specific metabolic processes within microbial communities [30,83,87]. However, metatranscriptomics approaches applied to the study of biodeterioration of cultural heritage are still in their infancy and their application in conservation science is not yet widely available. In general, the costs for molecular analysis are going down rapidly and operational issues with RNA protection and extraction are also improving. Bioinformatics tools to process the metatranscriptome datasets are also becoming routine for processing to microbial communities and gene assembly associated with biodeterioration.

## 4. Biochemical Functional Approaches

Both growth on surface as biofilms and specific metabolic activity (i.e., acidification) of the active microorganisms are important information for the understanding and prediction of damage to the materials of cultural heritage. Earlier investigations suffered from the isolation of limited numbers of cultured bacteria in pure cultures from the general community of microorganisms due to the technical means available. There were several early attempts to quantify the activity of the microbiome based on chemical reactions [30,88]. Detection and quantifying ATP can be used as an estimation of biological activity on surfaces like paper, paintings, or other materials in swab samples using the commercially-available products of luminometers [89,90]. However, the active microbial groups and members on cultural heritage may not be the directly destructive ones responsible for the damage caused to the cultural heritage and the general purpose-driven biochemical analysis of the samples from cultural heritage only provides limited insights because a close relationship between the destruction and specific microbial group cannot be established.

The "tangible" or physical cultural heritage, no matter whether moveable artifacts or immovable historic sites, is in a constant state of chemical transformation and equilibrium with the surrounding environmental conditions [91] and selective biochemical processes accelerate the chemical transformation of the underlying materials given the colonization by microorganisms. Selective microbial groups carry out the important biochemical transformation reactions with high specificity regulated by relevant genes and the corresponding enzymes/proteins. It is known that autotrophic microorganisms are pioneering organisms that colonized the bare surfaces of sandstone when the cultural heritage was initially built, but the roles of these microorganisms are not straight-forwardly detrimental because their protective role for the sandstone, for example from lichen, was significant in the beginning [40]. After sequestration of atmospheric $CO_2$ onto the surface as biomass, the physical surface properties of the heritage are significantly altered, allowing for diversification of the microbial community with different physiological functions and also for accumulation of pollutants to promote the degradative microorganisms [63]. Information on these has been reviewed recently [40,92].

More recently, $NO_3^-$ production and accumulation of the nitrogen cycle have been widely observed and confirmed on Angkor sandstone monuments under the tropical climate in Cambodia [32,85]. The concentration of $NO_3^-$ is positively correlated with the acidity and pH value, among the different environmental variables of these sandstone monuments. The constant acidification on the cultural heritage initiates and accelerates the dissolution of mineral constituents in the sandstone and produces the phenomenon of salt attack by crystallization/efflorescence on/in the substratum material which leads to structural weakening and crack formation of the cultural heritage [16,29,30,35,70,93]. After successive sampling over a 3-year period and analyses of the ammonia-oxidizing archaea (AOA) and bacteria (AOB) by both genomic DNA and RNA from samples for real-time PCR, AOA was apparently higher than AOB on these sandstone monuments and detection of $NO_3^-$ was successful [32,85]. Furthermore, a stable isotope tracer experiment by addition of the $^{15}N$ in the chemical forms of $NH_4^+$ or $NO_3^-$ was initiated and incubated with samples collected from several Angkor monuments to quantify the nitrogen transformation rate from nitrate removal as nitrogen gas and determine the contribution of the microbial community to the removal of available nitrate. The results of this study revealed that the heterotrophic denitrification ($NO_3^- \rightarrow NO_2^- \rightarrow NO \rightarrow N_2$) activity in the microbiome of the sandstone monument was low, and the activity of anammox ($NH_4^+ + NO_2^- \rightarrow N_2$) in the sample from the monument was negligible [84]. According to data in this study, the nitrate removal reaction of denitrification was active, but not strong enough to remove the produced and accumulated $NO_3^-$ because ammonia oxidation is aerobic while denitrification is anoxic. The physical conditions of available oxygen most likely play the regulation role, resulting in the accumulation of nitrate on the Angkorian sandstone.

Sulfur is also involved in the sandstone attack through microbial transformation of S at different valent states. *Fusarium* sp. and *Mycobacterium* spp. were isolated from decayed materials of the sandstone monuments and elucidated their metabolic capability of sulfur oxidation to acidic sulphate via chemolithoautotrophic metabolism in laboratory experiments [61,62]. The metabolic products of these sulfur-oxidizing bacteria are capable of producing inorganic acid through oxidation of S and the acidity in the form of $H^+$ interacts with minerals causing dissolution and re-crystallization regulated through water retention and loss by natural force, resulting in structural weakening of the stone material integrity and disintegration [29,40,70]. These findings suggested that these sulfur-oxidizing *Fusarium* sp. and *Mycobacterium* spp. participate in the deterioration process (i.e., acidification) of the sandstone monuments under tropical conditions. The accumulation of acids, $NO_3^-$ and $SO_4^{2-}$, which are corrosive to the inorganic material substratum of the cultural heritage, can be observed by microbially-mediated biochemical transformations.

As the former section stated, NGS techniques can provide thorough information of the microbial community and composition inhabiting cultural heritage to allow for identifying of the microorganisms without culturing them, which cannot be achieved by a culture-dependent approach. However, the involvement of the new microorganisms in the biodeterioration needs further careful verification even if high percentages of sequence reads with a metabolic potential are detected. The results of NGS-based investigations can provide useful information for the purpose-driven biochemical functional analysis and identify the physiological characteristics of the abundant genes and associated groups for developing culturing strategies to enrich and isolate the implicated members for further verification of their biochemical and ecophysiological roles in deterioration (Figure 4). Therefore, NGS technology can be employed to obtain basic information to assist in further research directions and strategies to directly confirm the scientific questions prior to culture-based strategies or activity-based biochemical functional analysis. It is also important to use NGS to gain an overview or summary of the whole microbial community prior to a genuine scientific question being executed. In doing this way, culture-dependent and -independent methods can be integrated effectively to advance scientific research on cultural heritage and also provide useful understanding for protection and management.

## 5. Summary and Future Perspectives

Current research addressing the issue of biodeterioration of cultural heritage needs to focus on sampling methods and appropriate analytical techniques used afterward. Application of non-invasive/non-destructive sampling strategies is required in microbiological studies for conservation. The NGS-based metagenomics and metatransriptomics should be applied to the study of biodeterioration of cultural heritage for the microbial structure (especially spatial distribution structure) of the microbiome and the potential of specific microbial groups and microbe-mediated transformation processes contributing to biodeterioration. The NGS-based results can provide sound basis for the purpose-driven biochemical functional analysis or develop enrichment and culturing strategies to isolate the implicated groups of microorganisms with significant roles in biodeterioration. There are apparent challenges for investigations to develop the purpose-specific biochemical functional analysis for measuring the valid deterioration effects of microbiome and for enhancing knowledge of cultivation of microorganisms from a specific material.

Delineation of the processes and mechanisms involved for (bio)deterioration should be the basis for understanding the current status with a selective cultural heritage. When the destructive reactions are known and also the specific microorganisms responsible for the processes, remediation strategies and management plan can be proposed accordingly [94]. Sulfur-driving acid production can be dealt with from restriction on fuel quality and air quality mandate in policy and management while intervention on site can be carried out through mineralization to immobilize produced $SO_4^{2-}$. $NO_3^-$ is an issue that is more difficult to address than $SO_4^{2-}$, but an effective way to deal with it should be to focus on the source of $NH_4^+$ to eliminate its contribution from the beginning. In addition, new emerging science on materials can also take an important role in the effective protection and management of culture heritage to decrease the detrimental contribution from microorganisms and their biochemical capabilities. Overall, protection of cultural heritage is a complex subject and synergies of different subjects will contribute to the advancement of this field for both basic knowledge and also applications.

**Author Contributions:** Conceptualization, W.L. and J.-D.G.; writing—original draft preparation, X.D.; writing—review and editing, X.D., W.L. and J.-D.G.; supervision, J.-D.G.; project administration and funding acquisition, W.L. and J.-D.G. All authors have read and agreed to the published version of the manuscript.

**Funding:** This research was funded by Shenzhen Scientific Project Fund (JCYJ20170410103015603), and Hong Kong RGC GRF Grant (No. 17302119).

**Acknowledgments:** This project was financially supported by Shenzhen Scientific Project Fund (JCYJ20170410103015603), and Hong Kong RGC GRF Grant (No. 17302119). We would like to thank the Apsara Authority of the Kingdom of Cambodia for permission to conduct sampling and research for more than a decade, Ka Hong Cheung for the SEM in Figure 1, and Lin Gao for help with the photographs presented in Figure 2.

**Conflicts of Interest:** The authors declare no conflict of interest.

## References

1. Ertan, T.; Egercioglu, Y. The Impact of UNESCO World Heritage List on Historic Urban City Centers and Its Place in Urban Regeneration: The Case of Melaka, Malaysia and Tire, Turkey. *Procedia Soc. Behav. Sci.* **2016**, *216*, 591–602. [CrossRef]
2. Silverman, H.; Ruggles, D.F. *Cultural Heritage and Human Rights*; Springer: New York, NY, USA, 2007; p. ix. 205p.
3. Hilbert, G.S. Protection against Theft and Willful Damage + Museum Security. *Museum* **1985**, *37*, 115–118. [CrossRef]
4. Martin-Sanchez, P.M.; Novakova, A.; Bastian, F.; Alabouvette, C.; Saiz-Jimenez, C. Use of Biocides for the Control of Fungal Outbreaks in Subterranean Environments: The Case of the Lascaux Cave in France. *Environ. Sci. Technol.* **2012**, *46*, 3762–3770. [CrossRef]
5. Tanselle, G.T. *Literature and Artifacts*; Bibliographical Society of the University of Virginia: Charlottesville, VA, USA, 1998; p. xvii. 356p.
6. Mckenna, G.A. Security and the Spirit of Cooperation—Protection Systems in Museums Large and Small. *Mus. News* **1979**, *58*, 7–9.
7. Skaggs, D. An Ounce of Prevention—A Handbook on Disaster Contingency Planning for Archives, Libraries and Record Centers—Barton, Jp, Wellheiser, Jg. *Am. Arch.* **1986**, *49*, 471–472.
8. Castanier, S.; Le Metayer-Levrel, G.; Perthuisot, J.P. Ca-carbonates precipitation and limestone genesis—The microbiogeologist point of view. *Sediment. Geol.* **1999**, *126*, 9–23. [CrossRef]
9. Magaudda, G. The recovery of biodeteriorated books and archive documents through gamma radiation: Some considerations on the results achieved. *J. Cult. Herit.* **2004**, *5*, 113–118. [CrossRef]
10. Fernandes, P. Applied microbiology and biotechnology in the conservation of stone cultural heritage materials. *Appl. Microbiol. Biotechnol.* **2006**, *73*, 291–296. [CrossRef]
11. Rives, V.; Garcia-Talegon, J. Decay and conservation of building stones on cultural heritage monuments. *Mater. Sci. Forum* **2006**, *514–516*, 1689–1694. [CrossRef]
12. Mesquita, N.; Portugal, A.; Videira, S.; Rodriguez-Echeverria, S.; Bandeira, A.M.L.; Santos, M.J.A.; Freitas, H. Fungal diversity in ancient documents. A case study on the Archive of the University of Coimbra. *Int. Biodeterior. Biodegrad.* **2009**, *63*, 626–629. [CrossRef]
13. Pinar, G.; Jimenez-Lopez, C.; Sterflinger, K.; Ettenauer, J.; Jroundi, F.; Fernandez-Vivas, A.; Gonzalez-Munoz, M.T. Bacterial Community Dynamics During the Application of a Myxococcus xanthus-Inoculated Culture Medium Used for Consolidation of Ornamental Limestone. *Microb. Ecol.* **2010**, *60*, 15–28. [CrossRef] [PubMed]
14. Gu, J.-D.; Kigawa, R.; Sato, Y.; Katayama, Y. Addressing the microbiological problems of cultural property and archive documents after earthquake and tsunami. *Int. Biodeterior. Biodegrad.* **2013**, *85*, 345–346. [CrossRef]
15. Pinar, G.; Garcia-Valles, M.; Gimeno-Torrente, D.; Fernandez-Turiel, J.L.; Ettenauer, J.; Sterflinger, K. Microscopic, chemical, and molecular-biological investigation of the decayed medieval stained window glasses of two Catalonian churches. *Int. Biodeterior. Biodegrad.* **2013**, *84*, 388–400. [CrossRef] [PubMed]

16. Pinar, G.; Piombino-Mascali, D.; Maixner, F.; Zink, A.; Sterflinger, K. Microbial survey of the mummies from the Capuchin Catacombs of Palermo, Italy: Biodeterioration risk and contamination of the indoor air. *FEMS Microbiol. Ecol.* **2013**, *86*, 341–356. [CrossRef] [PubMed]
17. Nitterus, M. Ethanol as fungal sanitizer in paper conservation. *Restaurator* **2000**, *21*, 101–115. [CrossRef]
18. Nitterus, M. Fungi in archives and libraries—A literary survey. *Restaurator* **2000**, *21*, 25–40. [CrossRef]
19. Bohm, C.B.; Zehnder, K.; Domeisen, H.; Arnold, A. Climate control for the passive conservation of the Romanesque painted wooden ceiling in the church of Zillis (Switzerland). *Stud. Conserv.* **2001**, *46*, 251–268. [CrossRef]
20. Morgan, G.C. The Materials of Cultural Heritage in their Environment. *Mediev. Archaeol.* **2007**, *51*, 400–401.
21. Bastian, F.; Alabouvette, C. Lights and shadows on the conservation of a rock art cave: The case of Lascaux Cave. *Int. J. Speleol.* **2009**, *38*, 55–60. [CrossRef]
22. Cappitelli, F.; Fermo, P.; Vecchi, R.; Piazzalunga, A.; Valli, G.; Zanardini, E.; Sorlini, C. Chemical-physical and Microbiological Measurements for Indoor Air Quality Assessment at the Ca' Granda Historical Archive, Milan (Italy). *Water Air Soil Pollut.* **2009**, *201*, 109–120. [CrossRef]
23. Gheyle, W.; Dossche, R.; Bourgeois, J.; Stichelbaut, B.; Van Eetvelde, V. Integrating Archaeology and Landscape Analysis for the Cultural Heritage Management of a World War I Militarised Landscape: The German Field Defences in Antwerp. *Landsc. Res.* **2014**, *39*, 502–522. [CrossRef]
24. Adamo, A.M.; Giovannotti, M.; Magaudda, G.; Zappala, M.P.; Rocchetti, F.; Rossi, G. Effect of gamma rays on pure cellulose paper as a model for the study of a treatment of "biological recovery" of biodeteriorated books. *Restaurator* **1998**, *19*, 41–59. [CrossRef]
25. Baer, N.S. Microclimate for cultural heritage. *Stud. Conserv.* **2000**, *45*, 143. [CrossRef]
26. Allsopp, D.; Seal, K.J.; Gaylarde, C.C. *Introduction to Biodeterioration*, 2nd ed.; Cambridge University Press: Cambridge, UK; New York, NY, USA, 2004; p. xii. 237p.
27. Albertano, P.; Bruno, L.; Bellezza, S. New strategies for the monitoring and control of cyanobacterial films on valuable lithic faces. *Plant. Biosyst.* **2005**, *139*, 311–322. [CrossRef]
28. Saiz-Jimenez, C.; Miller, A.Z.; Martin-Sanchez, P.M.; Hernandez-Marine, M. Uncovering the origin of the black stains in Lascaux Cave in France. *Environ. Microbiol.* **2012**, *14*, 3220–3231. [CrossRef]
29. Scheerer, S.; Ortega-Morales, O.; Gaylarde, C. Microbial Deterioration of Stone Monuments-An Updated Overview. *Adv. Appl. Microbiol.* **2009**, *66*, 97–139. [CrossRef]
30. Sterflinger, K.; Pinar, G. Microbial deterioration of cultural heritage and works of art—Tilting at windmills? *Appl. Microbiol. Biotechnol.* **2013**, *97*, 9637–9646. [CrossRef]
31. Zhang, X.W.; Ge, Q.Y.; Zhu, Z.B.; Deng, Y.M.; Gu, J.-D. Microbiological community of the Royal Palace in Angkor Thom and Beng Mealea of Cambodia by Illumina sequencing based on 16S rRNA gene. *Int. Biodeterior. Biodegrad.* **2018**, *134*, 127–135. [CrossRef]
32. Meng, H.; Katayama, Y.; Gu, J.-D. More wide occurrence and dominance of ammonia-oxidizing archaea than bacteria at three Angkor sandstone temples of Bayon, Phnom Krom and Wat Athvea in Cambodia. *Int. Biodeterior. Biodegrad.* **2017**, *117*, 78–88. [CrossRef]
33. Kusumi, A.; Li, X.S.; Osuga, Y.; Kawashima, A.; Gu, J.-D.; Nasu, M.; Katayama, Y. Bacterial Communities in Pigmented Biofilms Formed on the Sandstone Bas-Relief Walls of the Bayon Temple, Angkor Thom, Cambodia. *Microbes Environ.* **2013**, *28*, 422–431. [CrossRef]
34. Ettenauer, J.; Sterflinger, K.; Pinar, G. Cultivation and molecular monitoring of halophilic microorganisms inhabiting an extreme environment presented by a salt-attacked monument. *Int. J. Astrobiol.* **2010**, *9*, 59–72. [CrossRef]
35. Pinar, G.; Ripka, K.; Weber, J.; Sterflinger, K. The micro-biota of a sub-surface monument the medieval chapel of St. Virgil (Vienna, Austria). *Int. Biodeterior. Biodegrad.* **2009**, *63*, 851–859. [CrossRef]
36. Saarela, M.; Alakomi, H.L.; Suihko, M.L.; Maunuksela, L.; Raaska, L.; Mattila-Sandholm, T. Heterotrophic microorganisms in air and biofilm samples from Roman catacombs, with special emphasis on actinobacteria and fungi. *Int. Biodeterior. Biodegrad.* **2004**, *54*, 27–37. [CrossRef]
37. Sterflinger, K. Fungi as geologic agents. *Geomicrobiol. J.* **2000**, *17*, 97–124. [CrossRef]
38. Liu, X.B.; Koestler, R.J.; Warscheid, T.; Katayama, Y.; Gu, J.-D. Microbial deterioration and sustainable conservation of stone monuments and buildings. *Nat. Sustain.* **2020**, 1–14. [CrossRef]

39. Ma, Y.T.; Zhang, H.; Du, Y.; Tian, T.; Xiang, T.; Liu, X.D.; Wu, F.S.; An, L.Z.; Wang, W.F.; Gu, J.-D.; et al. The community distribution of bacteria and fungi on ancient wall paintings of the Mogao Grottoes. *Sci. Rep.* **2015**, *5*, 7752. [CrossRef]
40. Liu, X.B.; Meng, H.; Wang, Y.L.; Katayama, Y.; Gu, J.-D. Water is a critical factor in evaluating and assessing microbial colonization and destruction of Angkor sandstone monuments. *Int. Biodeterior. Biodegrad.* **2018**, *133*, 9–16. [CrossRef]
41. Saijo, Y.; Nakagi, Y.; Ito, T.; Sugioka, Y.; Endo, H.; Yoshida, T. Relation of Dampness to Sick Building Syndrome in Japanese Public Apartment Houses. *Epidemiology* **2009**, *20*, S150. [CrossRef]
42. Polo, A.; Cappitelli, F.; Brusetti, L.; Principi, P.; Villa, F.; Giacomucci, L.; Ranalli, G.; Sorlini, C. Feasibility of Removing Surface Deposits on Stone Using Biological and Chemical Remediation Methods. *Microb. Ecol.* **2010**, *60*, 1–14. [CrossRef]
43. Dicus, D.H. One response to a collection-wide mold outbreak: How bad can it be, how good can it get? *J. Am. Inst. Conserv.* **2000**, *39*, 85–105. [CrossRef]
44. Diaz-Herraiz, M.; Jurado, V.; Cuezva, S.; Laiz, L.; Pallecchi, P.; Tiano, P.; Sanchez-Moral, S.; Saiz-Jimenez, C. The Actinobacterial Colonization of Etruscan Paintings. *Sci. Rep.* **2013**, *3*, 1440. [CrossRef]
45. American Institute for Conservation of Historic and Artistic Works. *Strategic Plan*; American Institute for Conservation of Historic and Artistic Works: Washington, DC, USA, 1990; 54p.
46. Arai, H. Foxing caused by Fungi: Twenty-five years of study. *Int. Biodeterior. Biodegrad.* **2000**, *46*, 181–188. [CrossRef]
47. Choi, S. Foxing on paper: A literature review. *J. Am. Inst. Conserv.* **2007**, *46*, 137–152. [CrossRef]
48. Gates, G.A. Discovering the material secrets of art: Tools of cultural heritage science. *Am. Ceram. Soc. Bull.* **2014**, *93*, 20–27.
49. Gu, J.-D. Microbiological deterioration and degradation of synthetic polymeric materials: Recent research advances. *Int. Biodeterior. Biodegrad.* **2003**, *52*, 69–91. [CrossRef]
50. Duan, Y.L. The microbial community characteristics of ancient painted sculptures in Maijishan Grottoes, China (vol 12, 2017). *PLoS ONE* **2018**, *13*, e0179718. [CrossRef]
51. Xu, H.B.; Tsukuda, M.; Takahara, Y.; Sato, T.; Gu, J.-D.; Katayama, Y. Lithoautotrophical oxidation of elemental sulfur by fungi including Fusarium solani isolated from sandstone Angkor temples. *Int. Biodeterior. Biodegrad.* **2018**, *126*, 95–102. [CrossRef]
52. Cole, A.C.; Semmens, M.J.; LaPara, T.M. Stratification of activity and bacterial community structure in biofilms grown on membranes transferring oxygen. *Appl. Environ. Microb.* **2004**, *70*, 1982–1989. [CrossRef]
53. Fdz-Polanco, F.; Mendez, E.; Uruena, M.A.; Villaverde, S.; Garcia, P.A. Spatial distribution of heterotrophs and nitrifiers in a submerged biofilter for nitrification. *Water Res.* **2000**, *34*, 4081–4089. [CrossRef]
54. Stewart, P.S.; Camper, A.K.; Handran, S.D.; Huang, C.T.; Warnecke, M. Spatial distribution and coexistence of Klebsiella pneumoniae and Pseudomonas aeruginosa in biofilms. *Microb. Ecol.* **1997**, *33*, 2–10. [CrossRef]
55. McNamara, C.J.; Perry, T.D.; Bearce, K.A.; Hernandez-Duque, G.; Mitchell, R. Epilithic and endolithic bacterial communities in limestone from a Maya archaeological site. *Microb. Ecol.* **2006**, *51*, 51–64. [CrossRef]
56. Lan, W.S.; Li, H.; Wang, W.D.; Katayama, Y.; Gu, J.-D. Microbial Community Analysis of Fresh and Old Microbial Biofilms on Bayon Temple Sandstone of Angkor Thom, Cambodia. *Microb. Ecol.* **2010**, *60*, 105–115. [CrossRef] [PubMed]
57. Yamaguchi, N.; Ishidoshiro, A.; Yoshida, Y.; Saika, T.; Senda, S.; Nasu, M. Development of an adhesive sheet for direct counting of bacteria on solid surfaces. *J. Microbiol. Methods* **2003**, *53*, 405–410. [CrossRef]
58. Hu, H.L.; van den Brink, J.; Gruben, B.S.; Wosten, H.A.B.; Gu, J.-D.; de Vries, R.P. Improved enzyme production by co-cultivation of Aspergillus niger and Aspergillus oryzae and with other fungi. *Int. Biodeterior. Biodegrad.* **2011**, *65*, 248–252. [CrossRef]
59. Hu, H.L.; Ding, S.P.; Katayama, Y.; Kusumi, A.; Li, S.X.; de Vries, R.P.; Wang, J.; Yu, X.Z.; Gu, J.-D. Occurrence of Aspergillus allahabadii on sandstone at Bayon temple, Angkor Thom, Cambodia. *Int. Biodeterior. Biodegrad.* **2013**, *76*, 112–117. [CrossRef]
60. Kusumi, A.; Li, X.S.; Katayama, Y. Mycobacteria isolated from Angkor monument sandstones grow chemolithoautotrophically by oxidizing elemental sulfur. *Front. Microbiol* **2011**, *2*, 104. [CrossRef] [PubMed]
61. Li, X.S.; Arai, H.; Shimoda, I.; Kuraishi, H.; Katayama, Y. Enumeration of Sulfur-Oxidizing Microorganisms on Deteriorating Stone of the Angkor Monuments, Cambodia. *Microbes Environ.* **2008**, *23*, 293–298. [CrossRef]

62. Li, X.S.; Sato, T.; Ooiwa, Y.; Kusumi, A.; Gu, J.-D.; Katayama, Y. Oxidation of Elemental Sulfur by Fusarium solani Strain THIF01 Harboring Endobacterium *Bradyrhizobium* sp. *Microb. Ecol.* **2010**, *60*, 96–104. [CrossRef]
63. Mitchell, R.; Gu, J.-D. Changes in the biofilm microflora of limestone caused by atmospheric pollutants. *Int. Biodeterior. Biodegrad.* **2000**, *46*, 299–303. [CrossRef]
64. Hanada, S.; Hiraishi, A.; Shimada, K.; Matsuura, K. *Chloroflexus Aggregans* Sp-Nov, a Filamentous Phototrophic Bacterium Which Forms Dense Cell Aggregates by Active Gliding Movement. *Int. J. Syst. Bacteriol.* **1995**, *45*, 676–681. [CrossRef]
65. Hanada, S.; Takaichi, S.; Matsuura, K.; Nakamura, K. *Roseiflexus castenholzii* gen. nov., sp nov., a thermophilic, filamentous, photosynthetic bacterium that lacks chlorosomes. *Int. J. Syst. Evolut. Microbiol.* **2002**, *52*, 187–193. [CrossRef]
66. Pierson, B.K.; Giovannoni, S.J.; Stahl, D.A.; Castenholz, R.W. Heliothrix-Oregonensis, Gen-Nov, Sp-Nov, a Phototrophic Filamentous Gliding Bacterium Containing Bacteriochlorophyll-A. *Arch. Microbiol.* **1985**, *142*, 164–167. [CrossRef]
67. Nubel, U.; Bateson, M.M.; Madigan, M.T.; Kuhl, M.; Ward, D.M. Diversity and distribution in hypersaline microbial mats of bacteria related to *Chloroflexus* spp. *Appl. Environ. Microb.* **2001**, *67*, 4365–4371. [CrossRef]
68. Sekiguchi, Y.; Yamada, T.; Hanada, S.; Ohashi, A.; Harada, H.; Kamagata, Y. Anaerolinea thermophila gen. nov., sp nov and Caldilinea aerophila gen. nov., sp nov., novel filamentous thermophiles that represent a previously uncultured lineage of the domain Bacteria at the subphylum level. *Int. J. Syst. Evolut. Microbiol.* **2003**, *53*, 1843–1851. [CrossRef] [PubMed]
69. Keppen, O.I.; Tourova, T.P.; Kuznetsov, B.B.; Ivanovsky, R.N.; Gorlenko, V.M. Proposal of Oscillochloridaceae fam. nov, on the basis of a phylogenetic analysis of the filamentous anoxygenic phototrophic bacteria, and emended description of Oscillochloris and Oscillochloris trichoides in comparison with further new isolates. *Int. J. Syst. Evolut. Microbiol.* **2000**, *50*, 1529–1537. [CrossRef] [PubMed]
70. Warscheid, T.; Braams, J. Biodeterioration of stone: A review. *Int. Biodeterior. Biodegrad.* **2000**, *46*, 343–368. [CrossRef]
71. Laiz, L.; Pinar, G.; Lubitz, W.; Saiz-Jimenez, C. Monitoring the colonization of monuments by bacteria: Cultivation versus molecular methods. *Environ. Microbiol.* **2003**, *5*, 72–74. [CrossRef] [PubMed]
72. Kaarakainen, P.; Rintala, H.; Vepsalainen, A.; Hyvarinen, A.; Nevalainen, A.; Meklin, T. Microbial content of house dust samples determined with qPCR. *Sci. Total Environ.* **2009**, *407*, 4673–4680. [CrossRef] [PubMed]
73. Amann, R.I.; Ludwig, W.; Schleifer, K.H. Phylogenetic Identification and in-Situ Detection of Individual Microbial-Cells without Cultivation. *Microbiol. Rev.* **1995**, *59*, 143–169. [CrossRef] [PubMed]
74. Colwell, R.R.; Brayton, P.R.; Grimes, D.J.; Roszak, D.B.; Huq, S.A.; Palmer, L.M. Viable but Non-Culturable Vibrio-Cholerae and Related Pathogens in the Environment—Implications for Release of Genetically Engineered Microorganisms. *Nat. Biotechnol.* **1985**, *3*, 817–820. [CrossRef]
75. Rahman, I.; Shahamat, M.; Chowdhury, M.A.R.; Colwell, R.R. Potential virulence of viable but nonculturable Shigella dysenteriae type 1. *Appl. Environ. Microb.* **1996**, *62*, 115–120. [CrossRef]
76. Rosado, T.; Mirao, J.; Candeias, A.; Caldeira, A.T. Microbial communities analysis assessed by pyrosequencing-a new approach applied to conservation state studies of mural paintings. *Anal. Bioanal. Chem.* **2014**, *406*, 887–895. [CrossRef]
77. Riley, M.; Anilionis, A. Evolution of Bacterial Genome. *Annu. Rev. Microbiol.* **1978**, *32*, 519–560. [CrossRef]
78. Otlewska, A.; Adamiak, J.; Gutarowska, B. Application of molecular techniques for the assessment of microorganism diversity on cultural heritage objects. *Acta Biochim. Pol.* **2014**, *61*, 217–225. [CrossRef]
79. Adamiak, J.; Otlewska, A.; Tafer, H.; Lopandic, K.; Gutarowska, B.; Sterflinger, K.; Pinar, G. First evaluation of the microbiome of built cultural heritage by using the Ion Torrent next generation sequencing platform. *Int. Biodeterior. Biodegrad.* **2018**, *131*, 11–18. [CrossRef]
80. Gutarowska, B.; Celikkol-Aydin, S.; Bonifay, V.; Otlewska, A.; Aydin, E.; Oldham, A.L.; Brauer, J.I.; Duncan, K.E.; Adamiak, J.; Sunner, J.A.; et al. Metabolomic and high-throughput sequencing analysis-modern approach for the assessment of biodeterioration of materials from historic buildings. *Front. Microbiol.* **2015**, *6*, 979. [CrossRef] [PubMed]
81. Bartoli, F.; Municchia, A.C.; Futagami, Y.; Kashiwadani, H.; Moon, K.H.; Caneva, G. Biological colonization patterns on the ruins of Angkor temples (Cambodia) in the biodeterioration vs bioprotection debate. *Int. Biodeterior. Biodegrad.* **2014**, *96*, 157–165. [CrossRef]

82. Selbmann, L.; Onofri, S.; Zucconi, L.; Isola, D.; Rottigni, M.; Ghiglione, C.; Piazza, P.; Alvaro, M.C.; Schiaparelli, S. Distributional records of Antarctic fungi based on strains preserved in the Culture Collection of Fungi from Extreme Environments (CCFEE) Mycological Section associated with the Italian National Antarctic Museum (MNA). *Mycokeys* **2015**, *10*, 57–71. [CrossRef]
83. Perito, B.; Cavalieri, D. Innovative metagenomic approaches for detection of microbial communities involved in biodeteriorattion of cultural heritage. *IOP Conf. Ser. Mater. Sci. Eng.* **2018**, *364*. [CrossRef]
84. Ding, X.H.; Lan, W.S.; Wu, J.P.; Hong, Y.G.; Li, Y.L.; Ge, Q.; Urzi, C.; Katayama, Y.; Gu, J.-D. Microbiome and nitrate removal processes by microorganisms on the ancient Preah Vihear temple of Cambodia revealed by metagenomics and N-15 isotope analyses. *Appl. Microbiol. Biotechnol.* **2020**, 1–15. [CrossRef]
85. Meng, H.; Luo, L.; Chan, H.W.; Katayama, Y.; Gu, J.-D. Higher diversity and abundance of ammonia-oxidizing archaea than bacteria detected at the Bayon Temple of Angkor Thom in Cambodia. *Int. Biodeterior. Biodegrad.* **2016**, *115*, 234–243. [CrossRef]
86. de Felice, B.; Pasquale, V.; Tancredi, N.; Scherillo, S.; Guida, M. Genetic fingerprint of microorganisms associated with the deterioration of an historical tuff monument in Italy. *J. Genet.* **2010**, *89*, 253–257. [CrossRef] [PubMed]
87. Ettenauer, J.; Pinar, G.; Sterflinger, K.; Gonzalez-Munoz, M.T.; Jroundi, F. Molecular monitoring of the microbial dynamics occurring on historical limestone buildings during and after the in situ application of different bio-consolidation treatments. *Sci. Total Environ.* **2011**, *409*, 5337–5352. [CrossRef] [PubMed]
88. Rodriguez, G.G.; Phipps, D.; Ishiguro, K.; Ridgway, H.F. Use of a Fluorescent Redox Probe for Direct Visualization of Actively Respiring Bacteria. *Appl. Environ. Microb.* **1992**, *58*, 1801–1808. [CrossRef]
89. Berthold, F.; Tarkkanen, V. Luminometer development in the last four decades: Recollections of two entrepreneurs. *Luminescence* **2013**, *28*, 1–6. [CrossRef]
90. Rakotonirainy, M.S.; Arnold, S. Development of a new procedure based on the energy charge measurement using ATP bioluminescence assay for the detection of living mould from graphic documents. *Luminescence* **2008**, *23*, 182–186. [CrossRef]
91. Lowenthal, D. *The Past is a Foreign Country*; Cambridge University Press: Cambridge Cambridgeshire, UK; New York, NY, USA, 1985; p. xxvii. 489p.
92. Zhang, G.X.; Gong, C.J.; Gu, J.G.; Katayama, Y.; Someya, T.; Gu, J.-D. Biochemical reactions and mechanisms involved in the biodeterioration of stone world cultural heritage under the tropical climate conditions. *Int. Biodeterior. Biodegrad.* **2019**, *143*, 104723. [CrossRef]
93. Saiz-Jimenez, C.; Laiz, L. Occurrence of halotolerant/halophilic bacterial communities in deteriorated monuments. *Int. Biodeterior. Biodegrad.* **2000**, *46*, 319–326. [CrossRef]
94. Jimenez-Lopez, C.; Rodriguez-Navarro, C.; Pinar, G.; Carrillo-Rosua, F.J.; Rodriguez-Gallego, M.; Gonzalez-Munoz, M.T. Consolidation of degraded ornamental porous limestone stone by calcium carbonate precipitation induced by the microbiota inhabiting the stone. *Chemosphere* **2007**, *68*, 1929–1936. [CrossRef]


**Publisher's Note:** MDPI stays neutral with regard to jurisdictional claims in published maps and institutional affiliations.

MDPI
St. Alban-Anlage 66
4052 Basel
Switzerland
Tel. +41 61 683 77 34
Fax +41 61 302 89 18
www.mdpi.com

*Applied Sciences* Editorial Office
E-mail: applsci@mdpi.com
www.mdpi.com/journal/applsci

www.ingramcontent.com/pod-product-compliance
Lightning Source LLC
Chambersburg PA
CBHW040743250926
51850CB00007BB/1395
* 9 7 8 3 0 3 6 5 3 0 5 0 5 *